FROM
5G TO 6G
AND BEYOND

The 7 Cs of Future Communications

Other World Scientific Titles by the Author

CMOS Millimeter-Wave Integrated Circuits for Next Generation Wireless Communication Systems
ISBN: 978-981-120-260-5

Design of CMOS RF Integrated Circuits and Systems
ISBN: 978-981-4271-55-4

FROM
5G TO 6G
AND BEYOND

The 7 Cs of Future Communications

Editor

Kiat Seng YEO

Singapore University of Technology and Design, Singapore
& Tianjin University, China

World Scientific

NEW JERSEY · LONDON · SINGAPORE · BEIJING · SHANGHAI · HONG KONG · TAIPEI · CHENNAI · TOKYO

Published by

World Scientific Publishing Co. Pte. Ltd.

5 Toh Tuck Link, Singapore 596224

USA office: 27 Warren Street, Suite 401-402, Hackensack, NJ 07601

UK office: 57 Shelton Street, Covent Garden, London WC2H 9HE

Library of Congress Cataloging-in-Publication Data
Names: Yeo, Kiat Seng, 1964– editor.
Title: From 5G to 6G and beyond : the 7 Cs of future communications /
    editor, Kiat Seng Yeo, Singapore University of Technology and Design, Singapore.
Other titles: From Five G to Six G and beyond
Description: First. | New Jersey : World Scientific, [2023] |
    Includes bibliographical references and index.
Identifiers: LCCN 2023000941 | ISBN 9789811270840 (hardcover) |
    ISBN 9789811270857 (ebook for institutions) | ISBN 9789811270864 (ebook for individuals)
Subjects: LCSH: 6G mobile communication systems.
Classification: LCC TK5103.252 .F76 2023 | DDC 621.382--dc23/eng/20230202
LC record available at https://lccn.loc.gov/2023000941

British Library Cataloguing-in-Publication Data
A catalogue record for this book is available from the British Library.

For any available supplementary material, please visit
https://www.worldscientific.com/worldscibooks/10.1142/13265#t=suppl

Desk Editors: Gregory Lee/Amanda Yun

Typeset by Stallion Press
Email: enquiries@stallionpress.com

Small success depends on intelligence, big success depends on wisdom;

Short-term success depends on luck, long-term success depends on determination;

Past or current success depends on insight, future success depends on foresight;

Individual success depends on ability, team success depends on trust.

Kiat Seng Yeo

Preface

Future communication will be a complex and interconnected network consisting of a wide variety of products, services, and systems. As the future network will become extremely heterogeneous, it will need a multidisciplinary team to address challenges to wireless connection, spectrum share, as well as data fusion. With speeds of up to 20 gigabits per second and the ability to support up to one million devices per square kilometer, 5G — the latest generation of mobile communications technology — may already seem impressive, but 6G is set to take these capabilities even further. For instance, with 5G, it would take 4 seconds to download a two-hour movie. However, with 6G, it would only take 1 second to download a 150-hour Netflix show. With wireless connectivity becoming increasingly common, there will be more applications like virtual and augmented reality, artificial intelligence (AI), machine learning, and even autonomous vehicles requiring high data rates combined with extremely low latencies, or delays, due to processing time. Accordingly, 6G, with its peak data rate of 1 terabit per second and latency of 100 microseconds, should be able to smoothly support these use cases. At the same time, individuals and organizations will expect a trustworthy network that can seamlessly and securely deliver data — meaning that further research must focus on building a holistic 6G network security architecture.

Since 2020, the standardization of 5G wireless communication networks and the simultaneous deployment across many countries worldwide is underway. Until the start of 5G communications, network utilization was predominantly for entertainment video and audio file streaming and buffering applications. One of the core distinctive benefits of 5G communication is the low deterministic latency to assure the end-to-end reliability and accuracy that future applications mandate. However, 5G networks consume more energy than 4G networks. This is because off-the-shelf servers are used against the application-specific, integrated circuits in a radio access network. In disparity, we must target to deliver the next generation networks, which at their introduction, do not surpass the previous generation's energy needs. Therefore, the requirements for future networks should not be fulfilled just by merely using the 5G network. Hence, the focus has started on the research for low-power 6G wireless communication networks.

Recently, mathematical computing has developed variations such as cloud computing, fog computing, and edge computing. As these new variations are being evolved, new application scenarios utilizing them are hopeful of resolving some inherent network concerns, such as network flexibility, distributed computing, low latency, and precise time synchronization. These are also some of the limitations of 5G caused by short packets. This restriction from 5G causes degradation of some of the requirements, such as high-reliability and low-latency. Hence, 6G is bound to be addressing more challenging applications. 6G will likely be an efficient and self-contained ecosystem based on AI, which is progressively evolving from being human-centric to being both human- and machine-centric. 6G will bring low latency and unrestricted complete wireless connectivity. Hence, enormous efforts are underway to consider beyond state of-the-art protocols and mechanisms for wireless communication. Thus, the next generation of wireless communication is expected to meet the demands of various challenging use cases that go far beyond distribution of voice, video, and data.

This book covers all aspects of future communications, from key technologies, design challenges, network requirements and user experiences to standardization, chip design, and industry applications from 5G to 6G. Specifically, the Convergence (Chapter 1), Communication (Chapter 2), Connectivity (Chapter 3), Component (Chapter 4), Conformity (Chapter 5), Circuit (Chapter 6), and

Chip (Chapter 7) design and the requirements for future communication are discussed. The use cases of 6G, RF transceivers roadmap for 2030 and beyond, as well as modeling of RF devices for 5G/6G are presented. Additionally, a modified Shannon's capacity formula that is critical for future advanced wireless communication, such as 6G and beyond, is examined for the first time. The standardization of 6G wireless communication systems with emphasis on Standard Development Organizations (SDOs), regulatory bodies and administrations, ITU, industry forums, and 6G standard timeline are reported. Finally, RF/mm-wave integrated circuit design for future communication provides readers with easy understanding of voltage-controlled oscillator, power amplifier, low-noise amplifier, frequency synthesizer, high-frequency divider, and chip-to-chip communications isolation technology is described.

About the Editor

Professor Kiat Seng Yeo (M'00–SM'09–F'16) received his B.Eng. (EE) in 1993 and Ph.D. (EE) in 1996, both from Nanyang Technological University (NTU). Currently, he is Advisor (Global Partnerships) at Singapore University of Technology and Design (SUTD) and Distinguished Professor at Tianjin University, China. He was Chairman of the University Research Board and Associate Provost for Research and International Relations at SUTD. Professor Yeo is a widely known authority in low-power RF/mm-wave IC design and a recognized expert in CMOS technology. He was a Member of the Board of Advisors of the Singapore Semiconductor Industry Association. Before his appointment at SUTD, he was Associate Chair (Research), Head of Division of Circuits and Systems, and Founding Director of VIRTUS (IC Design Centre of Excellence) of the School of Electrical and Electronic Engineering at NTU. Professor Yeo has secured over S$70 million of research funding as Principal Investigator from various funding agencies and the industry since 2000. He has published 10 books, 7 book chapters, over 600 international top-tier refereed journal and conference papers, and holds 38 patents. Professor Yeo holds/held key positions in many international conferences as Advisor, General Chair, Co-General Chair, and Technical Chair. He was awarded the Public Administration Medal (Bronze) on National Day 2009 by the President of the Republic of Singapore and the Nanyang Alumni Achievement

Award in 2009 for his outstanding contributions to the university and society. In 2020, he was conferred the Long Service Medal on National Day by the President of the Republic of Singapore. Professor Yeo is a Fellow of the Singapore Academy of Engineering (SAEng), a Fellow of the Singapore National Academy of Science (SNAS), a Fellow of the Asia-Pacific Artificial Intelligence Association (AAIA), and a Fellow of IEEE for his contributions to low-power integrated circuit design. He is the principal author of *World University Research Rankings (WURR) 2020*. Professor Yeo was recognized among the top 2% of scientists worldwide by Stanford University in 2020, 2021 and 2022.

About the Contributors

Thangarasu Bharatha Kumar (M'17–SM'19) received his B.E. (E&C) degree from Ratreeya Vidyalaya College of Engineering (RVCE), Bangalore, India, in 2002 and his M.Sc. degree from the German Institute of Science and Technology (GIST) in 2010. He completed his Ph.D. (EE) in 2015 at Nanyang Technological University (NTU), Singapore.

From Jan. 2010 to Aug. 2015, he was with VIRTUS, IC Design Centre for Excellence, NTU, as a Research Associate, where he worked on SiGe HBT and CMOS-based reconfigurable amplifiers for microwave and millimeter wave RF integrated circuit design. He was also a Research Fellow II at the Singapore University of Technology and Design (SUTD), Singapore. In 2023, he joined Tianjin University, China, as an Associate Professor. His research interests include RF and millimeter-wave reconfigurable integrated circuit design. He has authored/co-authored over 40 journals and conference papers. He was a recipient of the DAAD scholarship during his M.Sc. study.

Nagarajan Mahalingam (M'16–SM'22) received his B.E. degree in electronics and communication engineering from Bharathidasan University, India, in 2001, his M.S. degree in electrical engineering from the University of Texas at Arlington, Arlington, US, in 2005, and his Ph.D. from Nanyang Technological University (NTU), Singapore, in 2016.

In 2006, he joined Advanced RFIC (S) Private Limited as an IC Design Engineer and worked on frequency synthesizers for portable wireless and data converter applications. He joined the Circuits & Systems Division, School of Electrical and Electronic Engineering, NTU, in 2008 as a Research Associate and focused on low-power designs for wireless and biomedical applications. He was an integral member of the 60 GHz team at NTU, which developed the VIRTUS chipset. He was also a Research Fellow II at the Singapore University of Technology and Design (SUTD). In 2023, he joined Tianjin University, China, as an Associate Professor. His current research interests include radio and millimeter-wave integrated circuit design with a focus on oscillators and frequency synthesizers.

Dipu Bhaskar received his Bachelor's degree from Anna University, India, in 2006, and his Master's degree from Nanyang Technological University and Technology University of Munich in 2010. His research interests include the design and testing of low-power analog RF circuits specialized in energy harvesting-based design. His research activities include analog design automation and design optimization using artificial intelligence and machine learning.

Jianguo Ma (F'16) received his Ph.D. in engineering in 1996 from Duisburg University, Duisburg, Germany. From Sep. 1997 to Dec. 2005, he was a faculty member of Nanyang Technological University (NTU), Singapore, after his postdoctoral fellowship with Dalhousie University, Canada from Apr. 1996 to Sep. 1997. He was with the University of Electronic Science and Technology of China from Jan. 2006 to Oct. 2009, and he served as the Dean for the School of Electronic Information Engineering and the founding director of the Qingdao Institute of Oceanic Engineering of Tianjin University from Oct. 2009 to Aug. 2016. He was a distinguished professor at the Guangdong University of Technology from Sep. 2016 to Aug. 2021. Since Sep. 2021, Dr. Ma serves as the Vice Dean for the School of Micro-Nano Electronics, Zhejiang University.

His research interests are microwave electronics, RFIC applications to wireless infrastructures, and microwave and THz microelectronic systems. He is also a Fellow of IEEE for the Leadership in Microwave Electronics and RFICs Applications. Dr. Ma was a member of the Editorial Board for Proceedings of IEEE from 2013 to 2018. Currently, he is the Editor-in-Chief of IEEE Transactions on Microwave Theory and Techniques.

Yayu Gao received her B.S. degree in electronic engineering from the Nanjing University of Posts and Telecommunications, China, in 2009 and her Ph.D. in electronic engineering from the City University of Hong Kong in 2014. Since Mar. 2014, she has been an Associate Professor with the School of Electronic Information and Communication, Huazhong University of Science and Technology, Wuhan, China. Her current research interests include random access networks, next-generation Wi-Fi networks, heterogeneous network coexistence, and network intelligence.

Wen Zhan received his B.S. and M.S. degrees from the University of Electronic Science and Technology of China in 2012 and 2015, respectively. He received his Ph.D. from the City University of Hong Kong (CityU) in 2019. He was also a Research Assistant and postdoc at CityU. Since 2020, he has been with the School of Electronics and Communication Engineering, Sun Yat-sen University, China, where he is currently an Assistant Professor. His research interests include Internet of Things, modeling, and performance optimization of next-generation mobile communication systems.

Lin Dai received her B.S. degree in electronic engineering from the Huazhong University of Science and Technology, China, in 1998 and her Ph.D. in electronic engineering from Tsinghua University, China, in 2003. Since 2007, she has been with the City University of Hong Kong, where she is currently a full professor. She has broad interests in communications and networking theory, with a special interest in wireless communications. She was a co-recipient of the Best Paper Award at the IEEE Wireless Communications and Networking Conference (WCNC) 2007 and the IEEE Guglielmo Marconi Prize Paper Award (the annual Best Paper Award of IEEE Transactions on Wireless Communications) in 2009. She served as a co-chair of the PHY Track of IEEE WCNC 2013, the leading co-chair of Wireless Communications Symposium of IEEE International Conference on Communications (ICC) 2015, and TPC vice chair of ICC 2019. She received the President's Award of City University of Hong Kong in 2017.

Tejinder Singh (Senior Member, IEEE) received his Ph.D. (with the highest academic honor) in electrical and computer engineering from the University of Waterloo (UW), ON, Canada, in 2020. He is currently a Principal Member of Technical Staff within the Office of the CTO at Dell Technologies, ON, Canada, and an Adjunct Assistant Professor at UW. He was a Postdoctoral Fellow at NASA/JPL-Caltech, CA, US, from 2020 to 2021 and a Research Assistant Professor at UW from 2020 to 2022. He held a Microelectronics R&D Engineer position in industry from 2010 to 2012 and served as a University Instructor from 2014 to 2015. He served as a Chair of the ECE-GSA at UW from 2017 to 2018. He has authored or co-authored several research publications in the field of developing phase change materials (PCMs) and microelectromechanical systems (MEMS)-based devices for microwave and mmWave applications.

Dr. Singh is a recipient of the Governor General's Gold Medal, one of the highest Canadian honors in academia and research, for his academic excellence and outstanding doctoral research. His research contributions have received numerous accolades at the international level, including Govt. of Canada Vanier Graduate Scholarship, NSERC Postdoctoral Fellowship, EuMA Young Engineer Award, 9 Best Paper Awards at prestigious conferences, and CMC's Brian L. Barge Microsystems Integration Award.

Currently, he is the Editor-in-Chief of IEEE MTT-S Student Newsletters, Managing Editor for the IEEE MTT-S e-Newsletters, and an associate editor of *Microsystems Technologies Journal* published by Springer-Nature.

Raafat R. Mansour (Life Fellow, IEEE) is a Professor of Electrical and Computer Engineering at the University of Waterloo (UW) and holds Tier 1 — Canada Research Chair (CRC) in Micro-Nano Integrated RF Systems. He held an NSERC Industrial Research Chair (IRC) for two terms (2001–2005 and 2006–2010). Prior to joining UW in Jan. 2000, Prof. Mansour was with COM DEV Ltd, a satellite subsystem company in Ontario, Canada, from 1986 to 1999, where he held various technical and management positions in COM DEV's Corporate R&D Department. He holds 43 US and Canadian patents and has more than 400 refereed publications to his credit. He founded the Centre for Integrated RF Engineering (CIRFE) at UW.

Professor Mansour is a Fellow of IEEE, the Canadian Academy of Engineering (CAE), and the Engineering Institute of Canada (EIC). He served as the Chair of the Technical Program Committee of the IEEE MTT-S International Microwave Symposium in 2012. He was the recipient of Several Best Paper Awards, the 2014 Professional Engineers Ontario (PEO) Engineering Medal for Research and Development, and the 2019 IEEE Canada A.G.L. McNaughton Gold Medal Award.

Kaixue Ma (M'05–SM'09) received his B.E. and M.E. degrees from Northwestern Polytechnical University (NWPU), Xi'an, China, and his Ph.D. from Nanyang Technological University (NTU), Singapore. From 1997 to 2002, he worked at the Chinese Academy of Space Technology (Xi'an) as a group leader. From 2005 to 2007, he was with MEDs Technologies as an R&D Manager. From 2007 to 2010, he was with the Singapore-based public-listed company ST Electronics as R&D manager, Project Leader, and technique management committee. From 2010 to 2013, he was with NTU as a Senior Research Fellow and Millimetre-wave RFIC team leader for the 60-GHz Flagship Chipset project. From 2013 to 2018, he was a full professor at the University of Electronic Science and Technology of China (UESTC), Chengdu, China. Since Feb. 2018, he has been the Dean and Distinguished Professor of the School of Microelectronics, Tianjin University, the Director of Tianjin Key Laboratory of Imaging and Sensing Microelectronics Technology, and the Chairperson of Tianjin IC Association and Principal Investigator for National Xinhuo Platform of Tianjin for Innovation and Entrepreneurship.

Dr. Ma is also a Fellow of the Chinese Institute of Electronics and an awardee of the Chinese National Science Fund for Distinguished Young Scholars. He received 10 technique awards, including the best paper award. He has filed over 40 patents and published two books, 160 SCI international Journal papers (over 130 IEEE journal articles), and 180 international conference papers. He is currently working on GaAs and silicon-based RF mm-wave and THz integrated circuits and systems for 5G/6G wireless communication and sensing applications. He is an associate editor of *IEEE Transactions on Microwave Theory and Techniques*.

Chenyang Meng received his B.S. degree from Xidian University, China, in 2021. He is currently pursuing his Ph.D. with the School of Microelectronics and the Tianjin Key Laboratory of Imaging and Sensing Microelectronic Technology, Tianjin University, China.

His current research interests include integrated sensing and communication and wake-up radio solutions.

Xinbo Zhang received his B.S. degree in electronic science and technology from Hebei University of Technology, China, in 2021. Currently, he is studying for his Ph.D. at Tianjin University, China. His current research interests include frequency synthesizer techniques and integrated circuits for RF and millimeter-wave transceivers.

Zhen Yang received his B.S. degree from the University of Electronic Science and Tech of China, Chengdu, China, in 2017. He is currently pursuing his Ph.D. in radio physics and is a member of the Tianjin Key Laboratory of Imaging and Sensing Microelectronic Technology, Tianjin University. His current research interests include millimeter wave and terahertz IC design.

Richard Lum (M'88–SM'10) received his B.Eng. in electrical engineering from the National University of Singapore and his MBA from the UK Open University in 1988 and 1998, respectively.

Mr. Lum is the CEO, co-founder and Microchip Architect of MPics Innovations Pte Ltd, which specializes in custom analog circuit design and manufacturing. He has more than 20 patents in the field of analog circuits and isolation technology. His current interests involve the design and development of new and novel methods of signal communications across galvanically isolated barriers in multi-chip modules, and he has filed a patent for a new isolation method suitable for broadband power devices such as GaN and SiC. He is also an industrial entrepreneur and has successfully started-up integrated design companies in Singapore for Aztech Systems and for BlueChips Technology, which subsequently IPO-ed in Malaysia's MESDAQ Board. Prior to MPics, he was the Senior Director of R&D for Broadcom's Isolation Products Division, responsible for product definition, technology development, and product design and development of advanced optical and non-optical isolation devices for Industrial and EV Markets. He is a Senior Member of IEEE.

If something is too successful, it creates problems.
If something is too powerful, it creates fear.

Kiat Seng Yeo

Contents

COMMUNICATION

CONNECTIVITY

COMPONENT

CONFORMITY

Chapter 5. Standardization of 6G Wireless Communication
Systems 187

Sumei Sun, Yonghong Zeng and Amnart Boonkajay

CIRCUIT

Chapter 6. Generalized Multiple Tanks Based
RF/mm-wave IC for Future Communications 245

*Kaixue Ma, Zhen Yang, Chenyang Meng, Xinbo Zhang
and Kiat Seng Yeo*

CHIP

CONVERGENCE

Resear**ch** starts with a **r**eview and ends with **h**ope.
Educatio**n** starts with **e**xploration and ends with **n**urture.
Arts starts with an **a**spiration and ends with **s**urprises.
Desi**gn** starts with a **d**ream and ends as a **n**ecessity.

Scien**ce** starts with a **s**earch and ends with an **e**xplanation.
Technolog**y** starts with a **t**hought and ends with **y**ou.
Engineerin**g** starts with **e**nthusiasm and ends with **g**reatness.
Mathematic**s** starts with a **m**odel and ends with a **s**olution.

When you rearrange the letters **READ STEM**, they form the word **MASTERED**.

It is not surprising that people who have **mastered** all disciplines **read stem**.

Kiat Seng Yeo

Chapter 1

Transceivers for Future Communications

Bharatha Kumar, Mahalingam Nagarajan, Dipu Bhaskar
and Kiat Seng Yeo

1.1. Fifth-Generation Wireless Network Communication

Since 2020, the standardization of fifth-generation (5G) wireless communication networks and the simultaneous deployment across many countries worldwide is underway. Until the start of 5G communications, the network utilization was predominantly for entertainment video and audio file streaming and buffering applications. One of the core distinctive benefits of 5G communications is the low deterministic latency to assure the end-to-end reliability and accuracy that future applications mandate.

The main abilities for 5G includes a peak data rate of more than 10 Gbps, a low latency of 1 ms, supports a mobility of about 500 km/h, a network connection density of 1 million devices/km^2, a minimum spectrum efficiency of 3, and an energy efficiency that is 100 times better when compared to the communication network of fourth-generation (4G) systems.

The trend for wireless communication is drastically progressing over the recent few years. The main contributions for this spontaneous improvement are from the drive of the various standards' regulatory bodies, research labs from both commercial companies and academic universities, and the end-user's product scope. Together, they are leading to visible spontaneous recent 5G developments, with the introduction of groundbreaking technologies such as massive multiple-input multiple-output (MIMO) systems and the inclusion of new frequency bands such as mmWave bands. In addition to these RF design techniques, new application-related techniques, such as ultra-densification networks and network softwarization with virtualization, etc., are also being significantly explored. As an outcome of such revolutionary ideas and techniques, new applications are evolving, such as virtual and augmented reality (VAR), contactless mobile payment, enhanced broadband, etc. This has clearly set the path for 5G as potential to advance and adjust itself to the wide diverse application scenarios [1].

1.2. Limitations of 5G

Commercial off-the-shelf servers are used against domain-specific integrated circuits in a radio access network (RAN), which is virtualized and proposed with a major increase in energy consumption, contradicting measures for enhancing energy efficiency. This result is concluded from the fact that 5G networks consume more energy than 4G networks, nevertheless delivering a higher bandwidth. In disparity, we should target to deliver next generation networks, which at their introduction time, do not exceed the previous generation's energy needs, though the requirements of networks beyond 2030 will not be fulfilled by merely using 5G. Hence, the focus has started on the research for sixth-generation (6G) wireless communication networks.

On one side, end-user application needs are continuously evolving, which are evidently visible. With the same trend, the application requirements to be acceptably fulfilled may surpass the capability of 5G communication networks. To bring applications such as VAR with more reality requires a data rate of least Tbps (10^{12} bps) and a fast response with low latency in the microseconds (10^{-6} s). Such extreme requirements are beyond the maximum capability of 5G, even by utilizing its allocated millimeter wave (mmWave) frequency band. As the number of users of such futuristic applications increases, the communication network density will exceed the 5G network's utilization capability. As a consequence of such a demanding future application market, ongoing research has been initiated towards addressing these critical concerns, and it is targeted towards the high bandwidth teraherz (THz) frequency band communication as well as in planning for better-controlled network automation. Across nations worldwide, the importance of such high-speed communication systems is recognized through 5G networks, and also the overall economic progress and emphasis on security have set the path towards the next generation of communication. Recently, many countries have publicized strategies to enable communication networks beyond 5G and expressed their desire to launch research initiatives towards 6G [1].

1.3. Requirements of 6G and Application Scenarios of 6G Wireless Networks

Mathematical computing recently has developed variations such as cloud computing, fog computing, edge computing, etc. As these new variations are evolving, new application scenarios utilizing them are hopeful of resolving some inherent network concerns, such as the network flexibility, distributed computing, low latency, and precise time synchronization. These are also some of the limitations of 5G caused by the short packets. This restriction from 5G causes degradation of some of the requirements like high-reliability and low-latency. Eventually, disrupting delivery of some of the services, such as high data rates, global coverage, and Internet of Everything (IoE), which are some of the demands of futuristic mobile communications of 2030 and beyond [1].

As shown in Fig. 1.1, a major scientific advancement is anticipated to realize the connectivity requirements of 6G networks such as: (a) a worldwide network operating at a high frequency, such as a THz band with much wider spectrum bandwidth, (b) smart communication environments using the wireless propagation, (c) persistent artificial intelligence (AI), (d) an all-spectrum reconfigurable wireless transceiver front-end for dynamic spectrum access and usage, (e) intelligent communication modes for energy savings, (f) unmanned aerial vehicles (UAVs), and (g) massive MIMO communication networks.

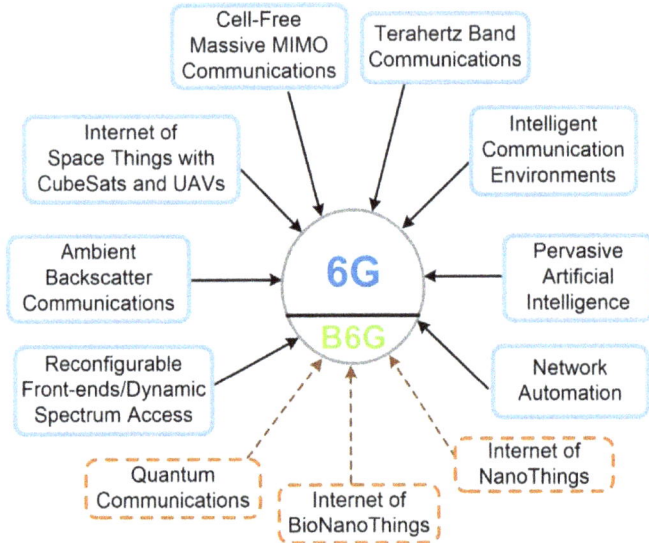

Fig. 1.1. The future-envisioned key-enabling technologies for 6G and beyond wireless communications systems (© IEEE [1]).

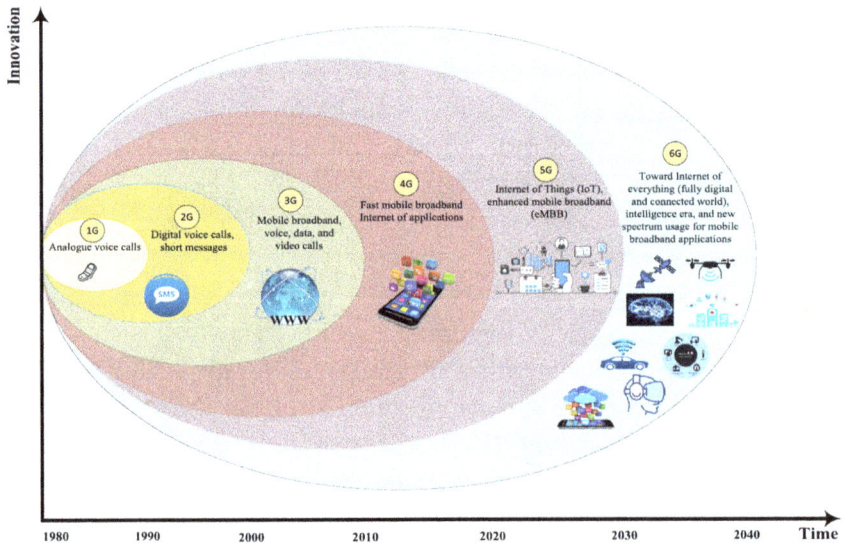

Fig. 1.2. The mobile network generation evolution from 1G to 6G (© IEEE [5]).

The evolution of each of the wireless network generation from 1G to 6G (as shown in Fig. 1.2) has consistently taken a considerable time — around one decade — to mature and move on to the next mobile generation.

In summary, a consolidation of end-user needs and technical breakthroughs that enable those requirements are the key motivation for a generational jump from 5G to beyond existing wireless systems, i.e., 6G. In the forthcoming decade, 6G is expected to bring together billions of things, humans, smart vehicles, robots, etc., which will generate large amount of digital data. Hence, 6G is bound to address more challenging applications. 6G will likely be an efficient self-contained ecosystem based on AI, which will progressively evolve from being human-centric to both human- and machine-centric. 6G will bring a low latency and unrestricted complete wireless connectivity.

In conclusion, a consolidation of the end users' needs and technical research breakthroughs serves to enable the critical needs for a wireless communication leap over prevalent 5G communication wireless systems. Collectively, this enables strong justification to proceed with the transition towards the next wireless communications, which are 6G systems [1].

One can foresee that 6G will not only allow a persistently smart, robust, multi-fold and, more importantly, secure wireless network system terrestrially, but also integrate outer space wireless communications that will result in a ubiquitous wireless communication network in setting the stage for an exact wireless omnipresence. The anticipated key performance indicators (KPIs) directing the evolution of 6G systems are discussed in this section [1].

While the ITU's Standardization Sector for Telecommunications (ITU-T) is already working on the 6G systems standard along with a set of authorized recommendations for 6G KPI metrics, initial draft values have recently appeared in the public domain [1]. Table 1.1 presents these KPI parameters with specified values as comparison against the 5G metrics. These KPIs serve as the vital target specifications to assess the capability and performance of 6G. From the table, we will find the classification of 6G KPIs.

Table 1.1.   The evolution from 5G to 6G wireless communication systems [1].

Key Performance Indicator	5G	6G
System Capacity		
Peak Data Rate (Gbps)	20	1,000
Experienced Data Rate (Gbps)	0.1	1
Peak Spectral Efficiency (b/s/Hz)	30	60
Experienced Spectral Efficiency (b/s/Hz)	0.3	3
Maximum Channel Bandwidth (GHz)	1	100
Area Traffic Capacity (Mbps/m^2)	10	1,000
Connection Density (devices/km^2)	10^6	10^7
System Latency		
End-to-end Latency (ms)	1	0.1
Delay Jitter (ms)	NA	10^{-3}
System Management		
Energy Efficiency (Tb/J)	NA	1
Reliability (Packet Error Rate)	10^{-5}	10^{-9}
Mobility (km/h)	500	1,000

System capacity: This group of KPI metrics mainly deals with system throughput. Within this framework, a distinguishing system consideration is that the user experienced data rate and the spectral efficiency metrics refers to the values that should be ensured to at least 95% of all user locations [1].

System latency: This KPI includes the metric for end-to-end latency along with system delay jitter. We note that this jitter is a new KPI metric for 6G that computes the variations of latency in the system and is not available for 5G.

System management: This class of KPIs chiefly manages with metrics related to the organization and arrangement of networks, such as overall energy efficiency, robustness, and mobility. For this metric as well, 5G does not include a target KPI. Hence it will require a revolutionary breakthrough for achieving the KPIs as specified in Fig. 1.3.

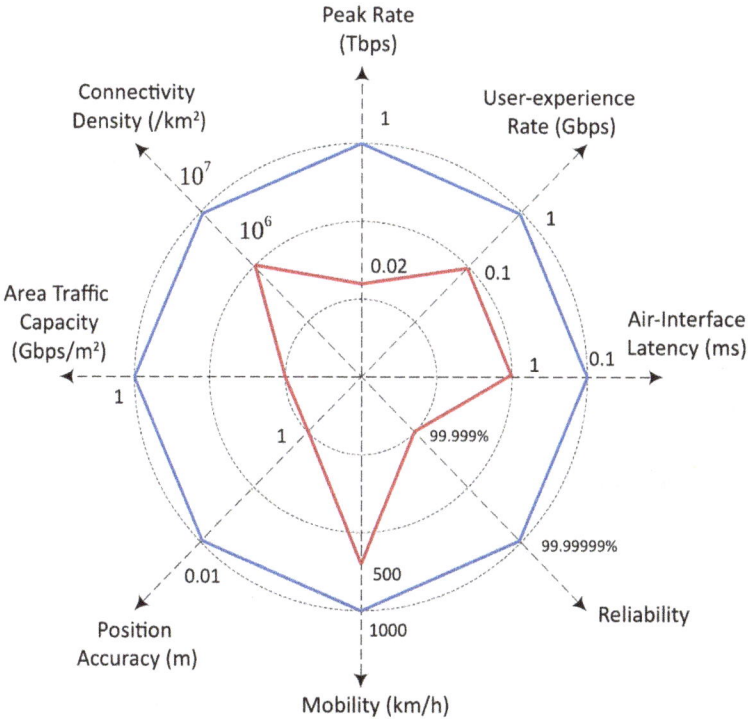

Fig. 1.3. A quantitative comparison of the requirements between 5G and 6G based on the KPIs (© IEEE [3]).

Newly allocated spectrum usage and radio design models: 5G adapted the mmWave spectrum to achieve higher data rates and consequently possess a larger channel bandwidth. However, the integration of terahertz (THz) and sub-THz spectrum bands will be demanded for 6G. Meantime, the inclusion of these new spectrum bands will also necessitate novel radio designs that can concurrently sense and communicate over the entire available spectrum.

Novel network architectures: The conventional cell-based architecture utilized in 5G wireless networks is unable to scale up to meet the larger area traffic capacity and larger connection density specifications as dictated by 6G.

Intelligence and automation: The stringent spectral usage efficiency, robustness, and latency specifications related to 6G infer that manual configuration of the network will no longer be applicable and feasible. Alternatively, network intelligence and network automation will become prevalent in building increasingly demanding autonomous networks.

Enhancing network coverage beyond the terrestrial links: A true wireless ubiquity using 6G can be achieved only by expanding the network beyond the terrestrial networks and by integrating both the near-Earth connectivity medium as well as deep-space connectivity. Towards the contentment of this outstanding visualization, several enabling solutions are perceived and actively studied for practical realization.

Figures 1.4 and 1.5 suggest the proposed technologies for enabling 6G, including (i) THz band operation with wider bandwidths, (ii) intelligent communication settings based on the environmental restrictions such as actively controlled signal transmission and reception, (iii) persistent artificial intelligence, (iv) large-scale engineering network automation, (v) an all-supported spectrum design with

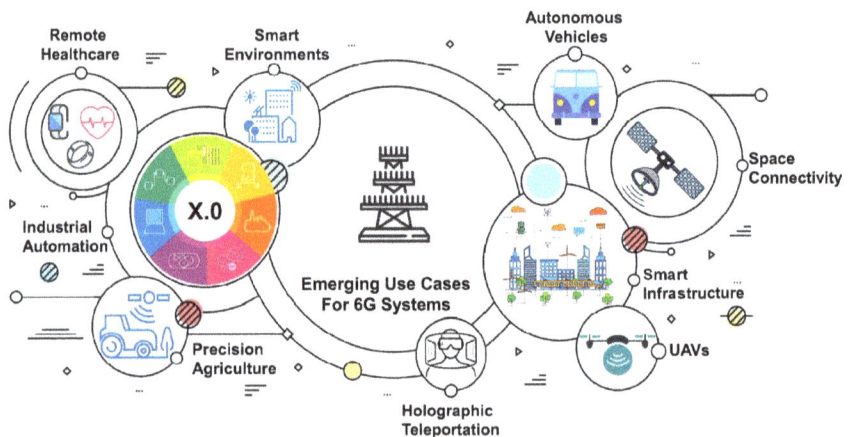

Fig. 1.4.   Application use cases provided by 6G systems (© IEEE [1]).

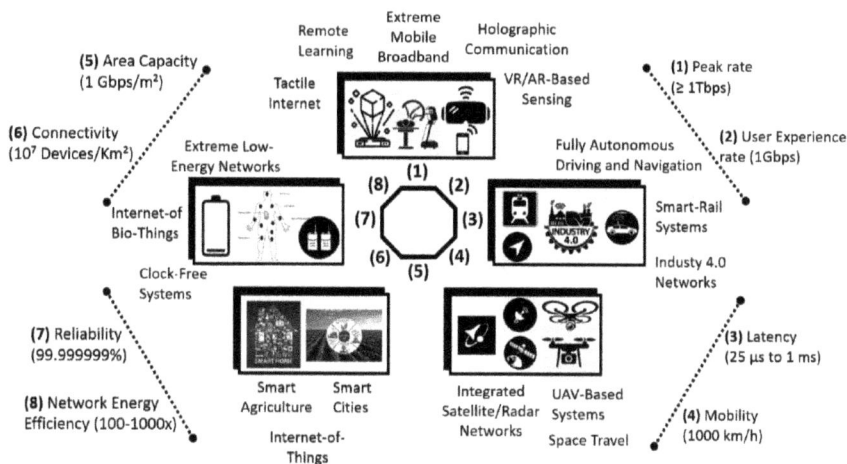

Fig. 1.5. A future vision for 6G systems and its underlying application use cases (© IEEE [4]).

the reconfigurable front-end for dynamic spectrum access without changing the communication link settings, (vi) the concept of enabling Internet of Space Things (IoST), and (viii) massive MIMO communication networks. We also ensure three additional promising technologies that are expected to shape the future of communications but yet may not be sufficiently mature for 6G. These include: (a) Internet of Nano Things, (b) Internet of BioNano Things, and (c) quantum communications.

While the prior art regarding 6G wireless systems is increasingly gaining importance over the past years [2–7], a majority of the technical publications in this field concentrate on a few selective topics. This limits an all-inclusive outline based on key characteristics that are likely to form the future of wireless communications in 6G. Furthermore, the above-mentioned novel topics for the research of 6G and beyond 6G systems, as shown in subsequent sections, are expected to be equally crucial but are still unavailable in existing publications. Research and development on the 6G wireless networks continue to advance and achieve new breakthroughs.

1.4. Futuristic Application Use Cases for 6G

The experience gained from the constantly evolving 5G systems serves as the basis for forming the application use cases for 6G communication networks. 5G first introduced some of fundamental application use cases, such as improved mobile broadband and highly reliable low latency data transmissions envisioned to aid in a wide variety of revolutionary applications. However, as noted, the 5G KPIs are not stringent enough to meet the futuristic needs. The reason for such relaxed performance trade-offs with 5G systems are based on the limitations from the output capacity, end-to-end latency, network area-coverage, robustness, energy efficiency, etc. Hence, a better hypothesis of the application scenarios that would benefit most end users from 6G becomes crucial, and the diverse critical use cases that will eventually be permitted in 6G are presented in Figs. 1.2 and 1.3 and illustrated in the next section [1].

1.4.1. *Multi-sensory holographic teleportation*

Virtual reality (VR), together with augmented reality (AR), has hugely advanced from the application use cases as introduced in 5G. There are plenty of practical application scenarios such as advanced healthcare systems, including virtual ailment diagnosis, performing remote controlled surgery, high-precision sensing for advanced remote excavation, and virtual real-time video conferencing, which cannot be adequately supported even by a combination of AR and VR.

Finally, the holographic teleportation technique has been realized as the accepted successor to AR and VR-based system solutions. The holographic teleportation contrasting the existing solutions functions in a true three-dimensional (3D) space constellation and influences all five human physical senses, i.e., sight, hearing, touch, smell, and taste. This enables a real-world appealing user experience. Meantime, the holographic teleportation demands for very high data rates close to 5 Tbps and a smallest end-to-end latency of less than 1 ms [1], both of which are not achievable with the 5G systems. Hence, 6G, with its anticipated throughput in the range of Tbps and sub-ms latency, will play a crucial part in the development of such an application scenario.

1.4.2. *Remote real-time healthcare system*

The idea of remote healthcare solutions is largely possible based on two predominant factors, namely, the quality of the established communication link as well as the technology availability for high-speed connectivity [1]. Regarding the improvement of both these factors, the need for THz band communication systems and network automation solutions becomes critical. With these factors, the 6G network will pioneer in the most conceivable wireless communications network, concentrating on very large throughput along with the extremely low latency. Nevertheless, the availability of the required technology, such as the IoST will serve as a predominant part in providing universal seamless connectivity. With these considerations, there is a reasonable possibility of achieving remote healthcare solutions.

1.4.3. *Autonomous cyber-physical systems*

A few of the most emerging cyber-physical systems are self-governing automobiles and UAVs, which are already in existence today with several limitations [1]. A reduced risk drivability of these autonomous systems is possible with the transfer of large volumes of data between the fundamental nodes. Such enormous amount of data mainly consists of high-resolution location information, precise real-time mapping of the surrounding physical terrain, traveling route optimization for both the shortest and fastest paths, the traffic situation along the route, and safety information. While the resulting huge quantity of data must be transitioned within constrained limits in a best possible error-free method, it is also evident that these fundamental nodes are mobile, typically traveling at speeds in excess of 100 kmph. Consequently, in addition to providing the support for such massive data transfers, latency in sub-ms and very high reliable communication become imperative. The connectivity, as well as quality of the communication using an autonomous connectivity solution, enables an improved cyber-physical system. Since the passenger's safety is of utmost priority, a robust system operation is possible at very high system speeds, which presently is not achievable with 5G systems [2].

1.4.4. *Space connectivity*

A combination of terrestrial, over the air, and outer-space communication connectivity are still emerging technologies even within 5G. With such a seamless connectivity established, there is a wide variety of application use cases, as shown in Fig. 1.6, that emerges. These application scenarios range from radio astronomy study, remote sensing to enable easy navigation, and a backhauling network. More precisely, such applications include merchandise tracking, terrestrial cellular loading, conservational monitoring, distant UAV control and coordination, etc. These use cases significantly benefit from universal connectivity, which is planned to be provided in 6G. Conclusively, 6G will provide the IoST to serve as the main aiding technology for the non-terrestrial connectivity portion.

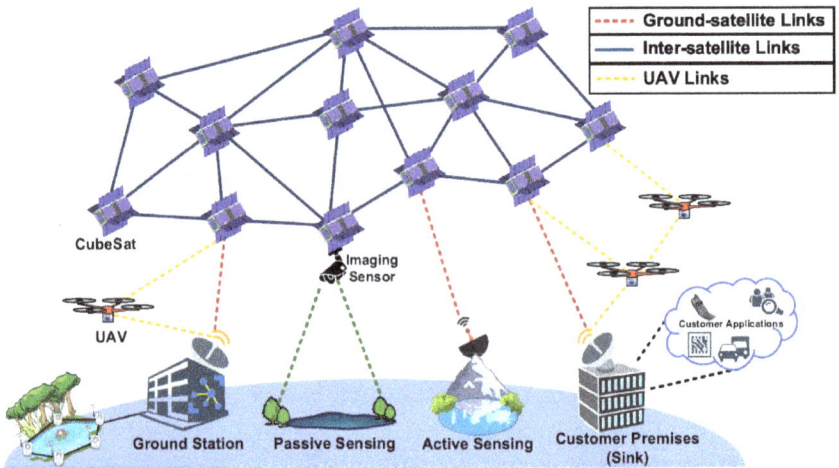

Fig. 1.6. The illustration of the IoST-based wireless communication (© IEEE [1]).

1.5. Pervasive Artificial Intelligence

In the field of communications and digital signal processing, AI is readily applicable to most mathematical intensive cognitive radios, remote sensing and control applications, computer vision, and overall communication network management, as shown in Fig. 1.7. More precisely, in the wireless communications domain, AI and its associated algorithms are gradually gaining importance by demonstrating their usefulness in numerous evolving technologies, such as the massive MIMO communications, which requires efficient and accurate channel estimation and precise symbol detection. Recently, the emerging field of AI has perceived massive growth, progressing to its practical implementation in a wide variety of domains supported by worldwide research across both academia and industry [1].

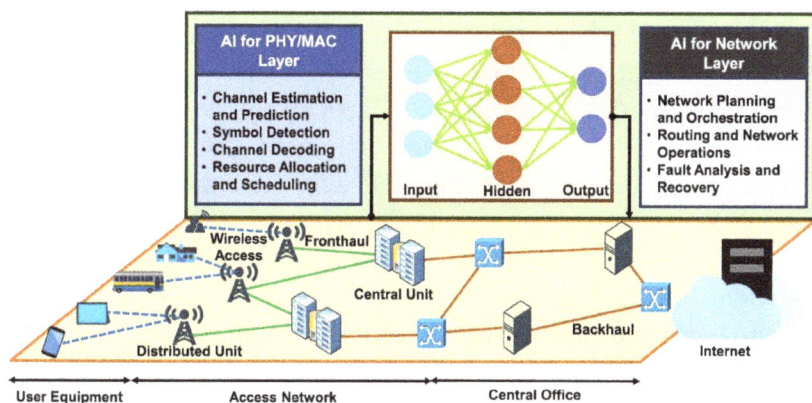

Fig. 1.7. Applications of AI at different protocol layers of wireless systems (© IEEE [1]).

1.5.1. *Possible problems for the pervasive AI*

During the initial standardization years of 5G, pioneer research investigators have illustrated the potential possibilities of including AI in 5G for achieving low latency and high spectrum efficiency. Precisely, AI algorithms can help simplify the subsequent tasks of network standardization, which would otherwise yield low efficiency with the conventional methods. So far, these AI-based solutions as well as inclusive machine learning (ML) have not been officially incorporated in 5G networks worldwide due to the following limitations. Recently in 2020, the ITU team instigated AI/ML incorporation in a 5G challenge to encourage researchers to recognize and resolve practical problems using AI/ML-based solutions in certain relevant 5G directions. Hence, it is intended that such efforts would be considered in later 5G developments but will only be materialized in a more concrete and universal manner in 6G.

There are still some issues prevalent to be resolved with AI in wireless communication networks, which are expected subsequently to bring a model shift towards data-oriented approaches. Firstly, there is still no predetermined method on which AI algorithms work the best to resolve the generalized problems in wireless networks, such as suitable modulation and coding scheme design, complex channel estimation, efficient resource allocation, etc. There has been a limitation in the approach as almost all the works in literature so far only claim substantial precision or reduced complexity with either mathematical theories or practical collected datasets [7]. Furthermore, we note that there is no effective method to set a reasonable comparison among all the proposed AI solutions. This is due to the variations in the selected datasets, considered assumptions, chosen evaluation criteria, etc. Nevertheless, a thorough consolidation of all the key conditions should be accomplished in order to recognize suitable AI algorithms without any loss of general relevance when it comes to practical deployments of communication networks. Secondly, the limited availability of good quality datasets is critical for the measurements and validation of the proposed sorting or iterative regression AI algorithms.

1.5.2. *Security concerns in 6G*

Though 6G networks will encounter an expanding number of mobile users and increasingly ample application scenarios, human-oriented mobile communication is still the core of 6G to achieve customized services. With the exploration of new 6G supporting communication techniques, the potential risk of security and privacy concerns are mainly present in 6G networks, which are listed as the following.

Due to the inclusion of AI in 6G communication networks, the main privacy-related problems include the risk of data security, security for the AI model and algorithm, susceptibilities in AI software systems and frameworks, and the possible harmful usage of AI methodologies.

To train AI models and algorithms, 6G network service providers need to collect a massive amount of data, which may contain user private information such as identification numbers and real-time location. During data transmission and processing, there is a high probability of sensitive information being leaked. In this condition, the dynamic protection of the AI model, such as the monitoring of the data flow and the output of the AI model, becomes critical for data security.

The user access and supervision of large-scale, as well as plenty of low-power Internet of Things (IoT) devices can effortlessly cause a sudden increase in data signaling, causing security risks for the IoT network. Hence, efficient validation techniques are required to support authorized massive IoT device deployments. Traditional security measures such as the encryption and decryption method mainly rely on unique secure key management systems and available computing resources. In the case of IoT devices with low power consumption, their security capabilities are limited by inadequate computing capability, small memory storage, and minimal battery energy utilization. Hence, it becomes necessary that a lightweight simplified security mechanism be designed to support energy-efficient access of low-power IoT devices. Alternatively, enormously complex device interconnection has expanded the chance for attacking the IoT network's security. Hence, the security protection for 6G should be strengthened for the network infrastructure as well as the distributed secure defense mechanism towards the deployment of an enormous number of IoT devices [1].

1.6. Available Resources — Frequency Spectrum

To support 5G, the sub-6 GHz band and mmWave band have already been extensively studied and utilized. Furthermore, to enable 6G network requirements, the THz and the optical frequency bands,

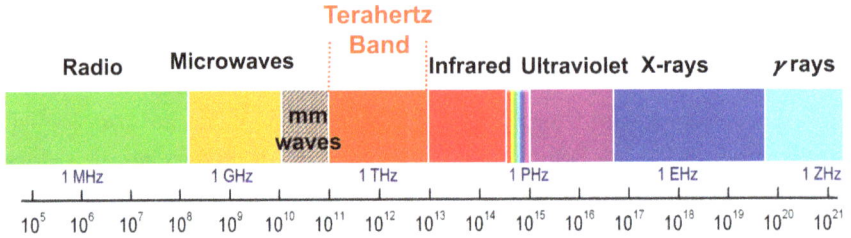

Fig. 1.8.   The THz band with plenty of usable spectrum resources for future wireless communications links (© IEEE [1]).

as shown in Fig. 1.8, need further investigation. Due to their contrasting frequency and bandwidth as well as their channel signal propagation characteristics with large differences, they should be carefully considered as feasible options for 6G networks [2].

1.6.1.   *Sub-6 GHz bands*

The success of 3G, 4G and 5G networks is mainly due to their utilization of sub-6 GHz bands with their fundamental wide coverage and low-cost deployments. The sub-6 GHz band inherits an exclusive wireless signal propagation property and channel characteristics, which were extensively examined during the 5G-network establishment. Many models have been formulated to accurately include the propagation effects. Two such propagation effects studied are the large-scale signal fading that depends on the path-loss and object material shadowing and the small-scale fading caused by the multipath propagation based on echoing or reflections. These factors are also crucial for 6G networks, and efforts need to be put forth towards including this band in the spectrum for 6G [2].

1.6.2.   *mmWave bands*

As the frequency spectrum at the traditional microwave bands are fully occupied, the next band, being the mmWave communication band operating between the 30 GHz and 300 GHz, is a promising option to aid ultra-high speed and high-capacity communication in the 6G wireless systems. Unlike the orthodox mobile communication systems using the sub-6 GHz bands, the mmWave band has severe propagation characteristics. This significant variation can be mitigated by recent technologies with reduced hardware limitations.

The targeted data rates between the sub-6 GHz and that of the mmWave 5G mobile communications are in the range of few Gbps and about 10 Gbps, respectively. Generally, to increase the wireless transmission capacity, there are two main methods: one is by using a large bandwidth fixed spectrum resource component, which directly provides a proportional capacity, and the other is by improving the spectral efficiency of a smaller bandwidth resource using complex modulation and multiplexing schemes.

The limited finite spectrum resource in the traditional microwave frequency band, as well as in the sub-6 GHz band, has been applied to various wireless systems till date and are almost fully utilized. Therefore, for the 5G network to achieve the target data rate of about 10 Gbps, the mmWave spectrum band was explored. The mmWave band from the spectrum plot of Fig. 1.8 suggests an abundant spectrum resource as compared relatively in addition to the microwave bands as well the sub-6 GHz band. Furthermore, the spectral efficiency improvement is still required to achieve the high data rates as demanded by 5G and to also support a larger communication density.

In addition to the available spectrum resources in the mmWave band, the 5G capacity requirement is satisfied by improving the spectral efficiency further by including techniques such as the massive MIMO. However, this improvement is at the expense of using plenty of RF transceiver chains and antennas, which result in high fabrication cost, power consumption, and form factor. Furthermore, to enable maximum utilization of such wide frequency bandwidth and to enable a real-time signal processing of massive input data, the hardware needs to be faster; mainly, the analog-to-digital (A/D) and digital-to-analog (D/A) converters need to be high-speed designs.

Based on the practical channel measurements as shown in Fig. 1.9, the path loss rapidly increases with carrier frequency, and the mmWave band signals propagation is affected by severe path loss. Therefore, highly directional transmission using antenna arrays at both the end user as well as base station becomes inevitable to

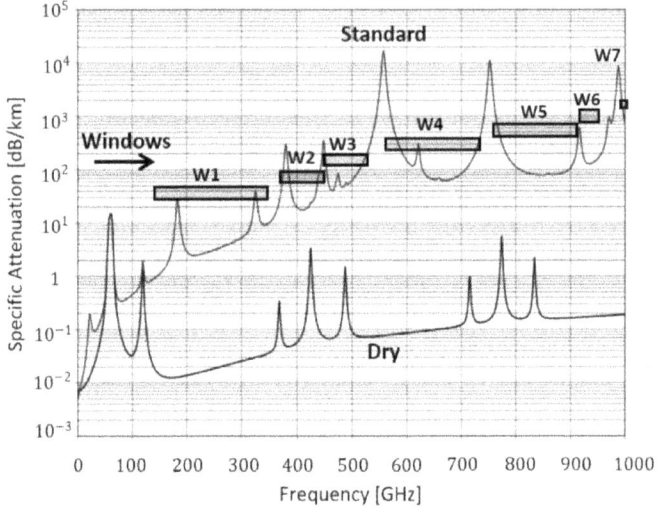

Fig. 1.9.   Atmospheric dry air and sea-level water absorption attenuation of radio waves to 1 THz (© [4]).

compensate for the long distance-dependency on the path loss by the high antenna gain. By this method, the multi-user interference is reduced, and the corresponding signal-to-interference-and-noise ratio (SINR) is increased.

Amid such challenges in using the mmWave bands for 5G applications, there exists a fundamental advantage of mmWave signals, which is its shorter wavelength; hence, this necessitates smaller form factor antennas. With this inherent benefit, the large antenna arrays can accommodate a greater number of mmWave antennas as compared to sub-6 GHz band antennas. In consolidation of such denser mmWave antenna arrays and the 3D beamforming technique, the mmWave communication systems can improve the beam directive gain and better beam steering to enhance the signal coverage range. However, the highly directional transmission and reception of mmWave signals in 5G is a promising option to increase the propagation distance — the ultra-dense small cell networks with multiple transmission paths become necessary to achieve the required network capacity. An added advantage to this technique is that the mmWave systems are less vulnerable to interferences as the signal penetration losses are considerably higher than the sub-6 GHz bands. Hence, the data exchange can be controlled within a limited location or facility [1, 2].

1.6.3. *Terahertz band communications*

As numerous wireless technologies are emerging, they are gradually getting incorporated in communication networks and gaining end-user access. Hence, a significant rise in wireless data traffic is witnessed. A recent demand for high-speed data transfer and wide coverage by the end users has resulted in the substantial growth of the data traffic [5]. The inclusion of the terahertz band (0.1 to 10 THz) into modern wireless communications is among the promising technology intentions of ongoing research. Beyond the mmWave band, the THz frequency band is the next candidate with even wider spectrum resources that can support massive data rates in the range of tera bits per second (Tbps), as shown in Fig. 1.10. This will enable numerous applications, with the requirement of ultra-high speed and simultaneous massive data transfer communications between nearby devices forming Terabit Wireless Personal Area Networks. With such an infrastructure support, many futuristic

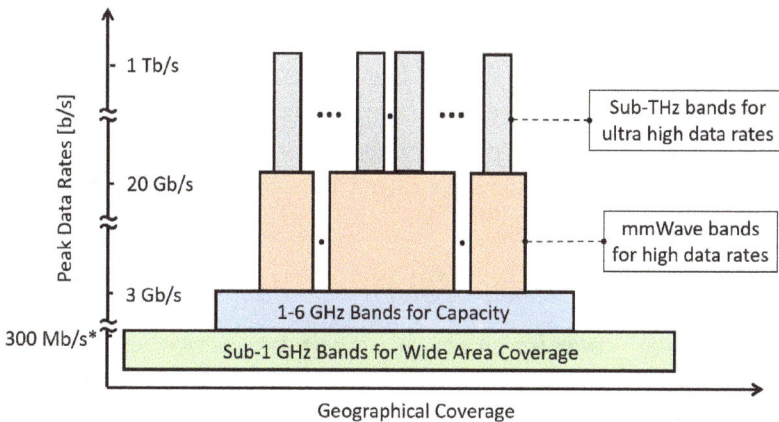

Fig. 1.10. The planned 6G operation across a wide frequency range starting from sub-1 GHz to sub-THz bands (© IEEE [7]).

applications are expected to be launched, such as the high-definition quality videoconferencing between the mobile devices.

Lately, some of the frequency bands above 0.1 THz are made available by the Federal Communications Commission (FCC) for promoting research activities in the wireless communications [6]. Although a few mobile operators have utilized low mmWave frequencies for providing their 5G network services with the purpose of achieving a large maximum data rate of 0.1 Tbps, the practical measurement results are far from the desired data rate, presenting a peak data rate of only around 1 Gbps. This discrepancy between the intended and actually attainable data rates is caused by multiple factors, such as the high complexity in the practical communication channels, limitations in integrated circuit design at such high frequencies, and real-world interferences from other systems or sources operating in the adjacent frequency bands, etc. Even though the THz band have already been implemented in past applications such as for imaging and object detection, as well as THz radiation spectrometry in astrophysical research, their application scenarios in wireless communications are very nascent and are still under investigation. The THz band, which is positioned in the frequency spectrum between the mmWave spectrum and the infrared light spectrum, as shown in Fig. 1.8, has abundant spectrum resources as wide bandwidth, previously considered inaccessible. Nevertheless, key progress in the research domains of the RF transceiver and antenna design has realized the THz transmission links develop into a promising possibility for implementing the indoor communications networks. Very recently, there has been substantial advancement in realizing the wireless network-on-chip using THz bands [7].

1.6.3.1. *Application scenarios of THz band communications*

Unlike present lower-frequency band wireless networks, THz-band wireless communications have several unconventional application scenarios, mainly due to the discrete practical electromagnetic (EM) wave and photonics characteristics of this extremely high-frequency band.

In addition to the assured high Tbps-level links for mobile systems, the THz-band spectrum can also be extended to the following application scenarios:

Local Area Networks: Some high-frequency spectrum bands are appropriate for short-range wireless links within 10 m, including the 625–725 GHz and 780–910 GHz bands [8]. The THz band communications are predicted to principally form the THz optical channel or bridge to aid unified conversion between fiber optics and THz-band signal links with minimum latency.

Personal Area Networks: Communications using the THz band can effortlessly deliver "optical fiber-like" data rates without the need for any physical wires or cables between multiple communicating devices at a distance of a few meters. Such communication application scenarios can be predominantly found in indoor facilities and multimedia streaming kiosks.

Data Center Networks: Traditional data centers maintain guaranteed connectivity in wired networks using robust, durable cables, resulting in high installation and reconfiguration costs. The multiple wires or cables can easily be tangled, causing chaos, and they also require more network deployment resources. Alternatively, the THz band data links provide an unquestionable prospect for unified connectivity at such ultra-high data speeds in fixed communication networks, and also enable adaptability by hardware reconfiguration.

Wireless Network-on-Chip: As the trend in the transceiver hardware design persuades for a higher level of integration and system compactness, the THz band optical bridges are foreseen as an optimistic candidate to launch the wireless connections on chip. With this technology, the problematic wired cable connections shall be easily substituted, and the conception of a high-speed THz band communication module is reinforced.

Nano-networks: The THz band has a nanometer (10^{-9} m) range wavelength and consequently becomes a viable option for deploying the nano-network infrastructure as compared to other frequency

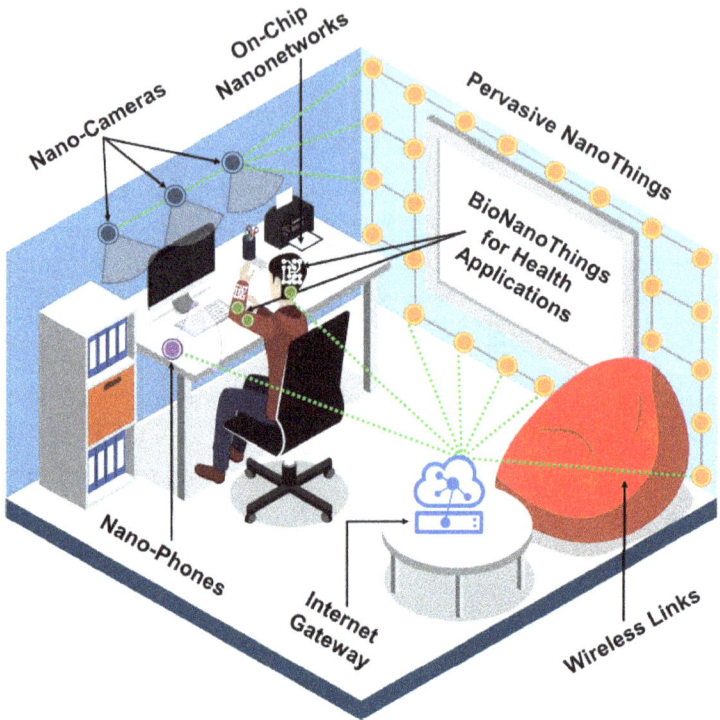

Fig. 1.11. Application scenarios of molecular communications in IoNT and IoBNT wireless systems (© IEEE [1]).

bands. With this context, a nano-network can be built using interlinked nano-devices that enable seamless data exchange, large storage, and fast computation, as shown in Fig. 1.11.

As compared to terrestrial communication links, the THz band is a more favorable candidate for outer space communication, such as the inter-satellite links, which are not constrained by the atmospheric attenuation, as shown in Fig. 1.9. With the larger bandwidth, the THz band can accommodate a greater number of satellites to form the network and hence improve the link performance over the existing spectrum bands used in the inter-satellite links. The THz band does not enforce strict requirements on beam alignment unlike the widely used optical links. Hence, a high level of outer space link stability is achieved.

1.6.3.2. *Possible problems in the THz band communications*

Presently, the semiconductor technology fabrication and measurements at such high THz frequencies remain a pertinent challenge. Some fabrication techniques, such as photolithography, are still able to produce transceiver front-end integrated circuits along with on-chip antenna arrays, as illustrated in the next section on THz band transceiver designs. Based on the fact that the THz band needs small dimension antenna array elements, results in the full utilization with a denser antenna array. This physical miniaturization and, eventually, the highly dense antenna can enhance the signal coverage by forming a narrow array radiation pattern. The main lobes of such radiation patterns have high directivity, thus focusing most of the RF energy towards a particular desired direction. Nevertheless, such highly directional beams limit the coverage to a narrow angle, causing low transmission energy efficiency. For this, a solution named "THz Prism" was recently proposed [5] that generates multiple beams with minute frequency shifts oriented in different directions to cover a larger region while maintaining good communication distance. This design technique utilizes the true propagation delays for RF frontend paths before the phase-shifters to form a prism-like formation, hence the name. This distributes and spreads the original single narrow beam into several beams, each with a slight frequency shift with respect to the center frequency.

Separately, a pursuit for more innovations in the antenna design, and resolving the remaining challenges that exist in the control and digital signal processing schemes associated with the frontend transceiver and baseband designs in the THz band, is ongoing. Adding to this, the real-time digital control algorithms with small latencies as well as the communication protocols for better coordination between the transmitter, receiver, and reflect arrays becomes needed. Among several recent literary works, the researchers reported a smart reflect array-assisted mmWave system compatible with the IEEE 802.11ad standard in [5]. A similar technique can also be adapted to the THz band and mitigate some of these known issues.

1.7. Projected Roadmap for Wireless Communication Systems for 6G and Beyond

The transition from 5G to 6G is governed by the needs of society accompanied by the overabundance of applications that meet the requirement of 6G is envisioned. The various KPIs linked with 6G, along with the key aiding technologies, are summarized in this section [1].

The technical preparedness and worldwide implementation of 5G network-based systems are setting the platform for a meaningful evolution of future wireless communications. The third-generation partnership project (3GPP) standards in the forthcoming years to 2024 are expected to cater for the 5G development, and the ITU initiated focus groups on technologies for Network 2030 to mainly study the possibilities of communication network for 2030 and beyond. From the previous sub-section, we can visualize the significant efforts towards research and development, setting the foundation for the next generation wireless communication.

In parallel to the academia, there is a significant increase in the contribution from the industry for the development of technologies to support both hardware and software demonstrations, followed by industrial testbeds for 6G targeted in 2026 and beyond. We expect these testbeds will enable a better platform to showcase the potential of 6G and its appropriateness for futuristic applications targeted from 6G, such as multiple-sensory holographical teleportation, remote healthcare monitoring and consultations with minimum update delay, remote controlled industrial automation, and smart organizations and environments.

Even though there are some enduring discussions within the wireless communication network community as to whether there is any need for a next generation after 5G, i.e., 6G, or whether the network generations ought to be stopped at 5G, some pioneering research works have already been started on the next-generation wireless networks. A tentative roadmap of possible definitions, specifications, standardizations, and regulations for 6G are projected, as shown in Fig. 1.12. A focus group formed under ITU-T for the technologies of networks beyond 2030 was established in July 2018. This focus group

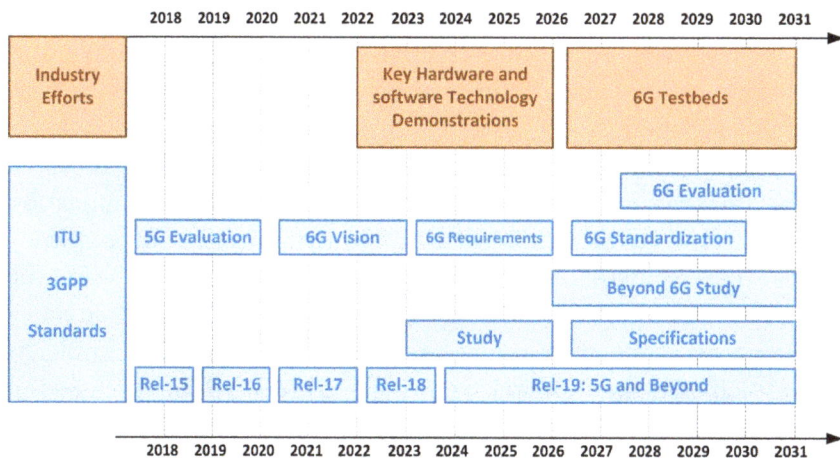

Fig. 1.12. The future envisioned key enabling technologies for 6G and beyond wireless communications systems (© IEEE [1]).

aims to understand the existing technology platforms and review the prevalent standards for recognizing the gaps for improvement and also the challenges towards implementing the capabilities for future networks of 2030 and beyond [1].

As the 6G conceptualization and standardization is evolving, there is an emergence of innovative progressive application scenarios as discussed previously. Although the main focus is presently on 6G fixed communication networks, this focus group is also exploring the possibility for definition of the 6G mobile system based on identified necessary visions, requirements, network architectures, and novel application scenarios. According to the practical timeline, as shown in Fig. 1.12, once the study of 6G vision is initiated, and by the middle of the 2020s, the ITU-R section will start publishing the requirements in IMT-2030. This will then set the standardization for 6G into the initial assessment stage. The ITU in the past has successfully launched the IMT-2000, IMT-Advanced and IMT-2020. Based on its past accomplishments, which is quite evident, by introducing IMT-2030 with similar activities, the ITU can foresee the evolution of IMT towards 2030 and beyond with fully operational deployment. The ITU-R working group 5D has planned to complete this projected study for IMT towards 2030 at the conference to be held in June 2022. For this conference, a preliminary IMT report will be drafted and associated information from various sources such as external institutions, academic organizations, and even from several countries/regional research programs shall be considered. It is the responsibility of ITU-R to organize the World Radiocommunication Conference (WRC) that mainly oversees the allocation of frequency bands. This conference is held every three to four years. In the past WRC, which was held in 2019 as WRC-19, the frequency spectrum assignment for the 5G network was considered. Similarly, it is anticipated that the spectrum issues for 6G will be discussed in the WRC scheduled in 2023 (WRC-23), and the spectrum assignment for 6G communication networks may be formally determined in 2027 (WRC-27).

In early 2019, the 3GPP team froze the specifications of Release 15, which is also called Rel. 15 or R15, and was considered as the first stage of 5G standardization. Rel. 15 mainly focused on extended mobile broadband and provided the basis for ultra-low latency, particularly to support low latency applications. The following release, i.e., Rel. 16, which was in July 2020, completed and is considered as the second stage of 5G standardization [5]. Subsequently, 3GPP was working on a more advanced release version (Rel. 17), which further standardized 5G and was completed by early 2021. From this roadmap it is evident that the standards organizations such as 3GPP and ITU are working on the standardizing of the 5G until mid-2025. Subsequently, research for 6G will be initiated with the vision and understanding of the technology trends towards 6G by 2030 and beyond.

In the 5G era, a close worldwide effort and cooperation was established that enabled the achievement of a globally accepted standard. For 6G, the international cooperation is expected to be enhanced further to not only cover the current cooperation from different countries and organizations, but also in contributing opportunities, even at the initial research stage of 6G. In addition, new materials for integrated semiconductor circuit devices and advanced test equipment will eventually be a challenge as we proceed towards their theoretical physical limitations. This will result in much closer cooperation and coordination among the industries [1].

1.8. RF Transceivers Roadmap for 2030 and Beyond

In the recent years, due to the availability of large bandwidth, wireless applications have been demonstrated in the millimeter-wave frequencies such as 60 GHz, 77 GHz automotive radar, 94 GHz imaging systems, and terahertz (THz) short-range wireless links in the 240/300 GHz frequency range. The definition of THz, which sometimes either refers to frequencies between 0.1 THz and 10 THz [8] or frequencies between 100 GHz and 300 GHz, defined both as millimeter-wave and THz [9]. Nevertheless, this frequency range offers greater potential for future ultra-high speed communication systems and is of greater interest. As shown in Fig. 1.13, the data rate of wireless communication has been steadily increasing, comparable to the wired networks achieving 1Tb/s by 2030.

For the next generation wireless communication, a 300-GHz frequency band has been attracting attention and the frequency allocation around 300 GHz is shown in Fig. 1.14. The frequencies from 252.72 to 321.84 GHz is proposed for IEEE 802.15.3d [10]. The

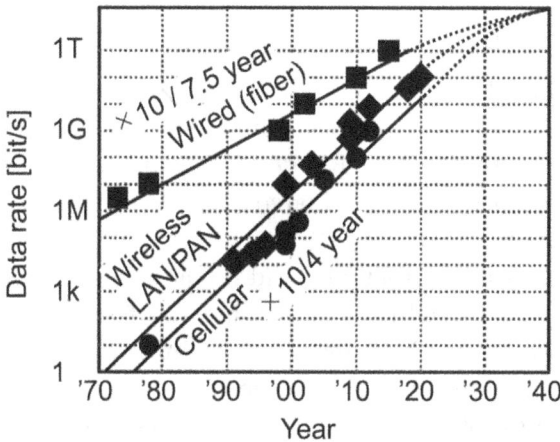

Fig. 1.13. Data-rate trends for wired and wireless communications [11].

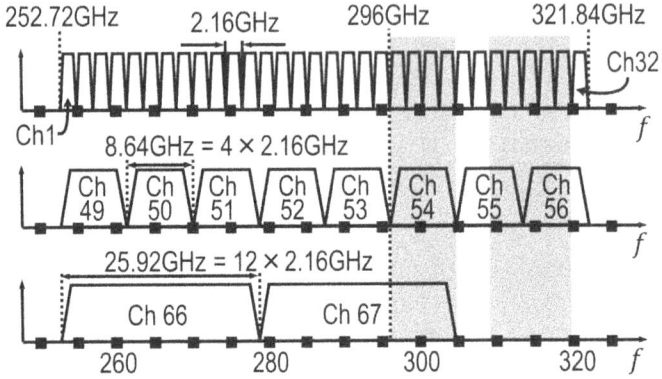

Fig. 1.14.   Proposed channel allocation by IEEE802.15.3d [10].

frequencies above 275 GHz have not been allocated but frequencies reaching until 296 GHz has been identified for wireless communication usage in WRC 2019 [12]. In this section, we will discuss the radio transceiver achieving Tb/s wireless communication.

1.8.1. *Technology for a future integrated transceiver*

The performance of the THz communication system is severely limited by the choice of integrated circuit technology, owing to cost effectiveness, integration capabilities, and recent advancements in the silicon.

Technologies, including the silicon germanium (SiGe) and complementary metal oxide semiconductor (CMOS), can support applications operating in the Teraherz range. For high performance Tb/s transceivers, the amplifying RF devices (transistor) speed is of the utmost importance and is evaluated based on the transistor figure-of-merit operating frequency (f_T) and unity gain frequency (fmax), where the transistor current gain and power gain is equal to 1 (0 dB), respectively. The f_T and fmax determine how fast the transistor will operate. Below fmax, the transistors exhibit power gain, and above fmax, the transistors exhibit power loss.

The f_T/fmax of state-of-the-art semiconductor technologies is shown in Fig. 1.15. The fmax of SiGe-based technology is greater than 500 GHz and can be used in the design of transceiver components

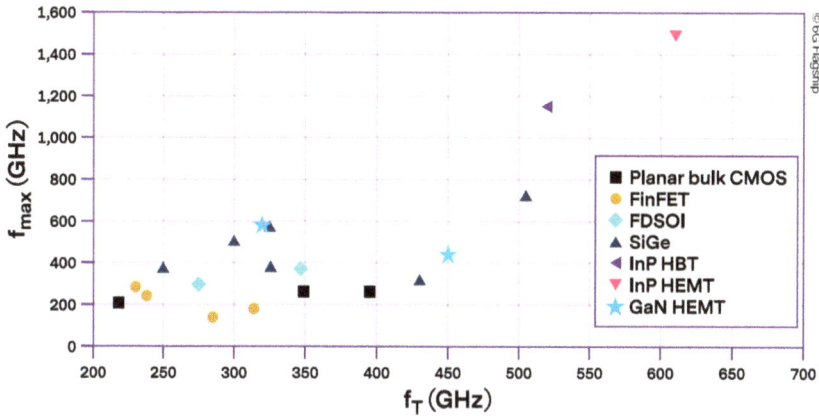

Fig. 1.15.   Overview of f_T and fmax of state-of-the-art semiconductor technologies (© 6G Flagship [13]).

for the next-generation communication at the 300 GHz band. The survey of output power and noise figure of the SiGe-based circuit in THz frequencies is presented in [14]. In the frequency of 325 GHz, the maximum output power is reported in −3 dBm in [15], and in 500 GHz, an output power of −8 dBm is reported in [16], and a noise figure of 12 dB is reported in [17] for the receiver operating at 220 GHz. In comparison, the continuous scaling of the CMOS technology has resulted in a higher speed for the digital circuits due to the increase in the transistor's intrinsic gain. Though the technology scaling has not benefitted the design of RF circuits, the compatibility of SiGe technology with CMOS technology can promise a greater level of performance, although the available CMOS devices in BiCMOS technology would not be scaled in comparison to CMOS-only technology.

1.8.2. *Integrated transceiver for future communication*

The system architecture block diagram of the THz system is shown in Fig. 1.16. The transceiver system typically consists of a digital baseband portion, the main radio frequency (RF) transceiver comprising the modulator to up-convert and down-convert the RF frequency to baseband, and the RF power amplifier (PA) and low

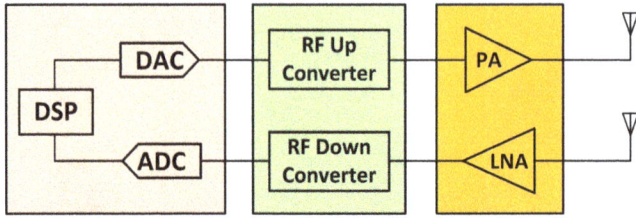

Fig. 1.16.   System architecture of the THz system.

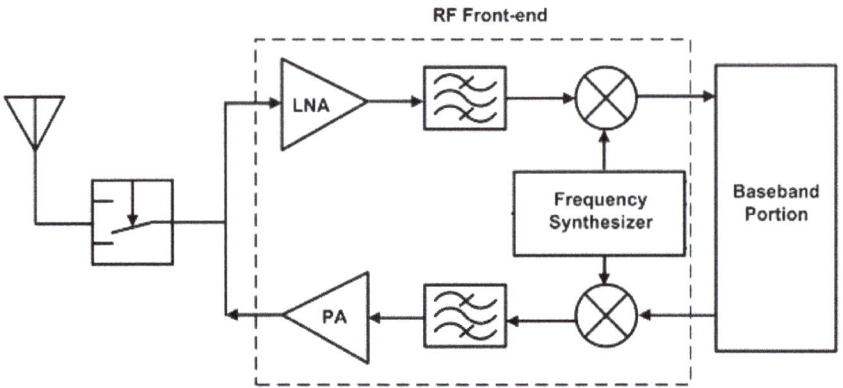

Fig. 1.17.   Radio frequency transceiver block diagram.

noise amplifier (LNA), which enable communication to the outside world. The block diagram of the RF transceiver is shown in Fig. 1.17. As shown in Fig. 1.17, the indispensable blocks in the RF transceiver are the phase-locked loop (PLL) synthesizers and the transmitter (Tx) and receiver (Rx) subsystem. The following sections provide an overview of the recent advances and challenges in the design of state-of-the-art THz transceiver subsystems for future applications, with a brief discussion of examples from the literature.

Due to the inherent challenges in the design of high-performance THz transceivers, such as low f_T/fmax, device scaling, low quality factor of capacitors and varactors, and the dominance of parasitic capacitance together with high interconnect losses, there has been an increasing interest in the research and development of high-performance THz transceivers.

1.8.2.1. *Transmitter*

The transmitter in the transceiver generates the modulated RF carrier for signal transmission via the antenna. The PA is the most important and critical component of the transmitter, and its performance directly impacts the performance of the wireless link. Though the design of RF PAs has evolved over the years, the design of THz PAs to deliver good performance is quite challenging. The design hexagon of the PA and the trade-off is shown in Fig. 1.18. The important consideration for the PA design is its reliability to deliver large output power with sufficient power density across the frequency band with a compact area. Typically, wireless links employ complex modulation schemes to maximize the channel throughput and often require large dynamic ranges. Therefore, the PA needs to have a high power-added efficiency (PAE). Also, a high PA linearity is a requirement to preserve the signal fidelity and link quality when complex modulation schemes are used. Therefore, the PA design would trade off efficiency, linearity, output power, carrier bandwidth, and reliability.

For the THz power amplifiers, SiGe-based PA has replaced the III–V semiconductor-based PA and is preferred ahead of CMOS PAs due to the advancement in silicon technology. The PA operating at a carrier frequency of 230–245 GHz, delivering a maximum speed of 90 Gbps, is presented in [18] and is shown in Fig. 1.19. The transmitter employs a 4-stage PA in the up-converter and consumes

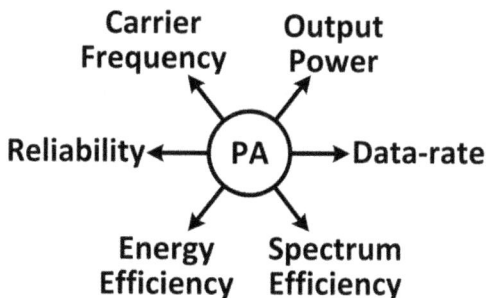

Fig. 1.18.   Power amplifier design hexagon: performance and trade-offs.

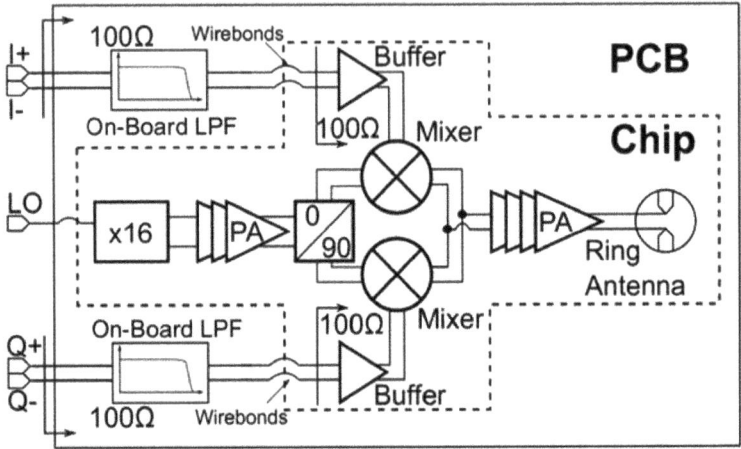

(a)

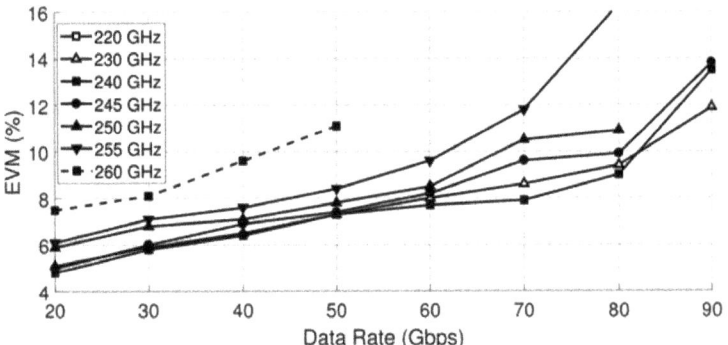

(b)

Fig. 1.19.   (a) Block diagram of the transmitter, and (b) EVM vs. data rate for different carrier frequencies (© IEEE [18]).

960 mW for a 3-dB RF bandwidth of 28 GHz and Psat of 8.3 dBm at the carrier frequency of 230 GHz.

The 240-GHz Tx-based on SiGe with a f_T/fmax of 300/450 GHz is shown in Fig. 1.20 [19]. The Tx has a multiplier-based THz LO, double-balanced gilbert-cell mixer, a 4-stage PA and a ring antenna delivering a power of −4.4 dBm at the carrier frequency of 236 GHz with a power consumption of 1.033 W. An improvement in the f_T/fmax to 350/550 GHz can be achieved by reducing the emitter

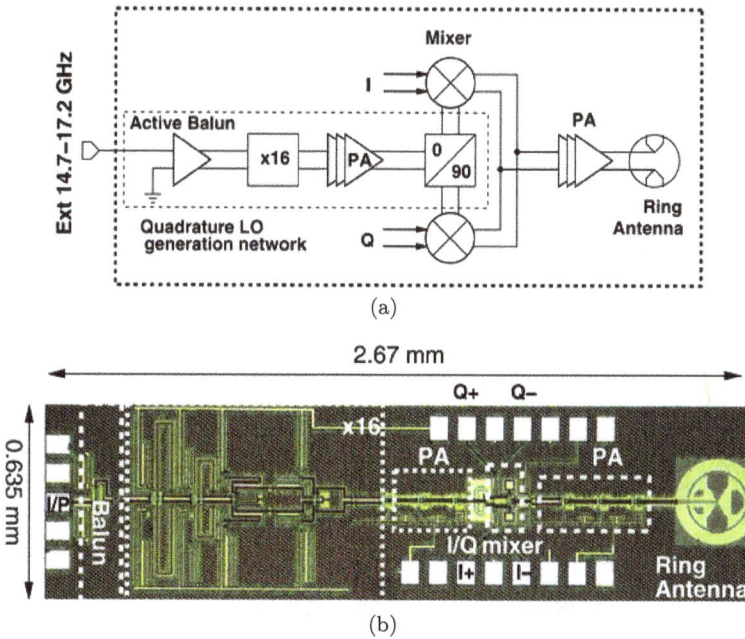

Fig. 1.20. (a) Block diagram of the 240 GHz transmitter, and (b) TX chip photo (© IEEE [19]).

area from 120 nm to 90 nm in [19] and the transmitter delivers a peak output power of 6 dBm at 240 GHz as shown in Fig. 1.21 [20].

More recently, due to the technology scaling, the fmax of the NMOS transistor has increased steadily [21], and a CMOS-based THz transmitter design has been reported [10, 22]. In comparison to the SiGe-based design, the CMOS-based Tx suffers from either lower output power [22] or the PA in the THz transmitter is based on other technologies such as SiGe or III–V technologies or omitted from the Tx chain [10] and added as an external component.

A CMOS-based transmitter delivering a data-rate of 105 Gbps at 300 GHz is shown in Fig. 1.22 [22]. The proposed Tx operates above the fmax and relies on multi-stage power combining to achieve the data-rate at a 300-GHz carrier frequency. However, the downside of the power combining is the presence of unwanted spurious due to the mixing products and image signals, which needs additional external

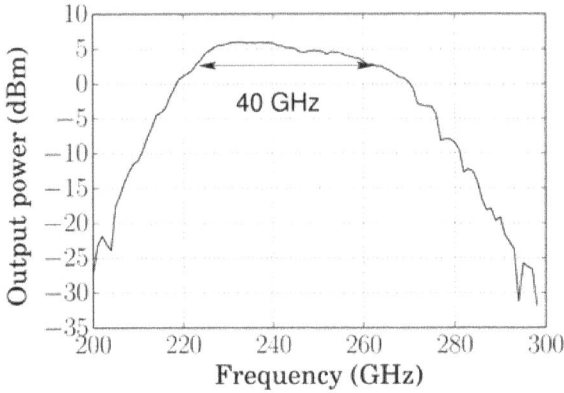

Fig. 1.21. TX output power with increased f_T/f_{max} to 350/550 in [19] (© IEEE [20]).

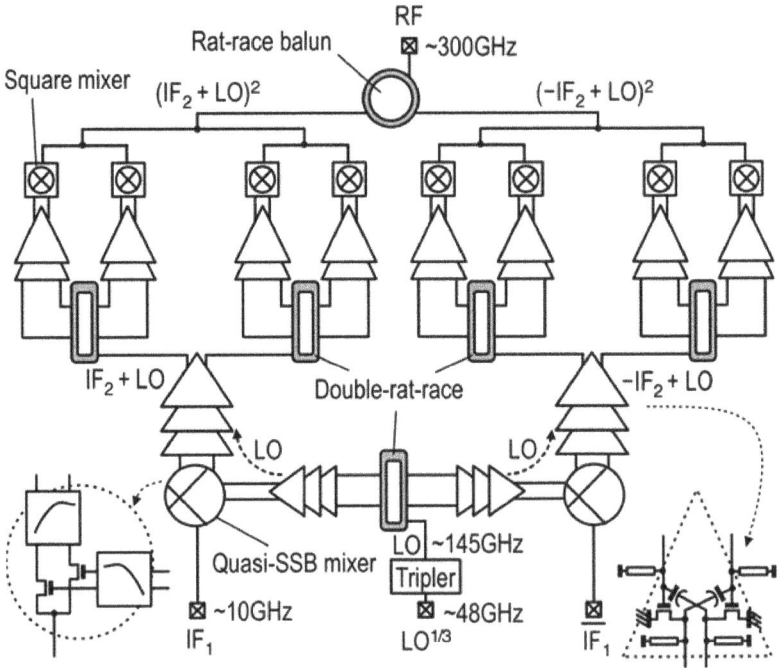

Fig. 1.22. Schematic of a 300-GHz transmitter with 105 Gbps data-rate (© IEEE [22]).

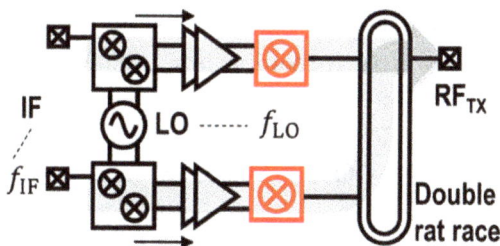

Fig. 1.23.   Schematic of a 300-GHz transmitter without PA [10].

filtering and high power dissipation of 1.4 W for the maximum Tx power of −5.5 dBm.

The 300 GHz Tx with PA omitted is shown in Fig. 1.23 [10]. In the proposed transceiver, a doubler is chosen as a mixer in the transmitter path and the final stage mixers are combined in parallel to generate the required output power. In the transceiver, the rat race circuit integrates the Tx and Rx into a single chip and the output of Tx is from the differential mode of the rat race circuit as shown in Fig. 1.23. The transceiver achieves a data rate of 80 Gbps, satisfying the IEEE 802.15.3d wireless standard with a maximum output power of −1.6 dBm, and the total power consumption of the transceiver is 1.8 W.

1.8.2.2. *THz LO generation*

One of the major building blocks in the RF transceiver is the PLL-based frequency synthesizer. The PLL in the RF transceiver generates a stable local oscillator (LO) signal for the up-conversion and down-conversion of the RF signal in the Tx and Rx subsystems, respectively. One of the main requirements of the PLL is to generate a LO signal covering the frequency range of the application, low phase noise or timing jitter, and low spurious emissions. In the LO generation, the VCO and divider following the VCO are the most important and critical blocks in the PLL. The LO generation in the THz band can be realized either based on a voltage controlled oscillator (VCO) at THz frequencies or using the frequency multiplier based approach. The multiplier-based THz LO generation offers advantages such as a high tuning range, better phase noise, and higher usable bandwidth. In comparison, the VCO-based THz LO suffers from a low tuning range due to the parasitic capacitance and higher power consumption due to the requirement of injection-locked dividers following the VCO when used in the PLL-based frequency generation.

Terahertz frequency synthesizer based on SiGe is quite popular due to the high f_T/fmax and lower noise of bipolar junction transistor devices in comparison to CMOS. Frequency synthesizers and harmonic VCOs operating at frequencies >200 GHz are reported in [23] to [29]. The most common VCO topology for THz signal generation is either the push-push oscillator topology or the frequency multiplier or doubler-based approach. A 320-GHz VCO with a push-push VCO and doubler-based approach is shown in Fig. 1.24 [23]. In the push-push oscillator, the second harmonic signal is extracted from the collector common node of the 160-GHz fundamental frequency VCO. In the doubler-based approach, the fundamental VCO is followed by a frequency doubler, though a push-push VCO offers the best performance in terms of power consumption and occupies a smaller area in comparison to the frequency doubler approach, though the VCO output power is much lower in comparison to the frequency doubler.

The THz PLL at 300 GHz with a harmonic VCO based on a triple push VCO is presented in Fig. 1.25 [24]. As the VCO's frequency is close to fmax, the VCO's negative resistance transistors could not provide any negative resistance to compensate the losses in the tank circuit. With the triple-push VCO, the VCO's third harmonic would be the output signal, and the first stage divider would operate at a 100-GHz frequency with lower power consumption compared to a divider operating at 300 GHz. In THz, the VCO frequency tuning is a concern. The continuous tuning of the VCO varactor in parallel to the tank would load the tank and is in the order of the parasitic capacitance. Also, the severe loss of the varactor at higher frequencies is a concern. Therefore, a Colpitts-based active varactor (CAV) and triple-push VCO (shown in Fig. 1.26) is proposed. For the divider following the VCO, the injection-locking divider is a popular choice

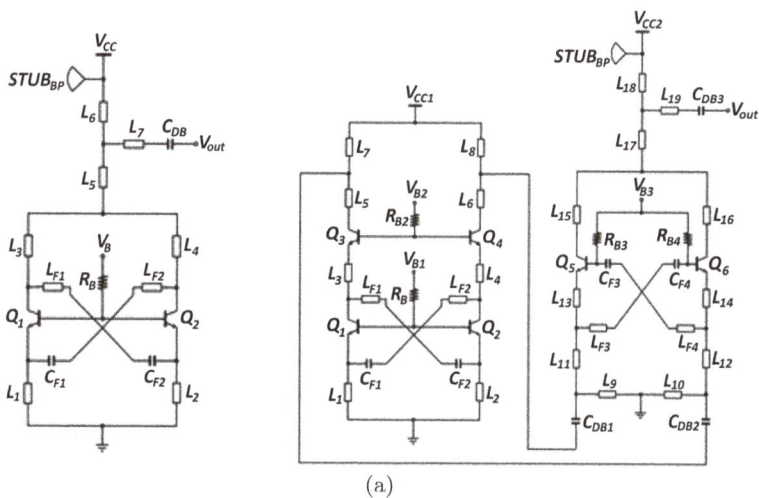

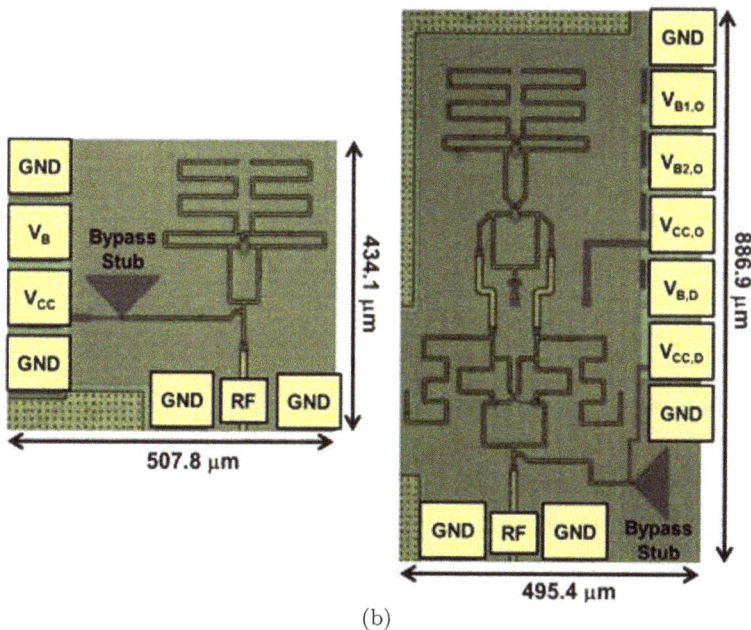

Fig. 1.24.   (a) Schematic of a 320-GHz VCO. (b) Chip microphoto of a 320-GHz VCO (© IEEE [23]).

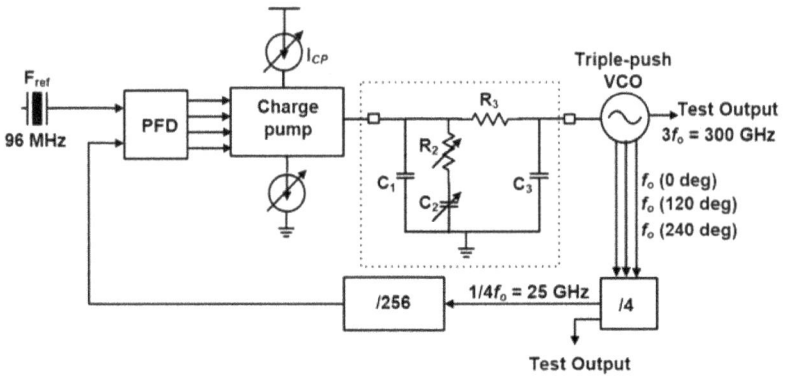

Fig. 1.25.   Block diagram of a 300-GHz frequency synthesizer.

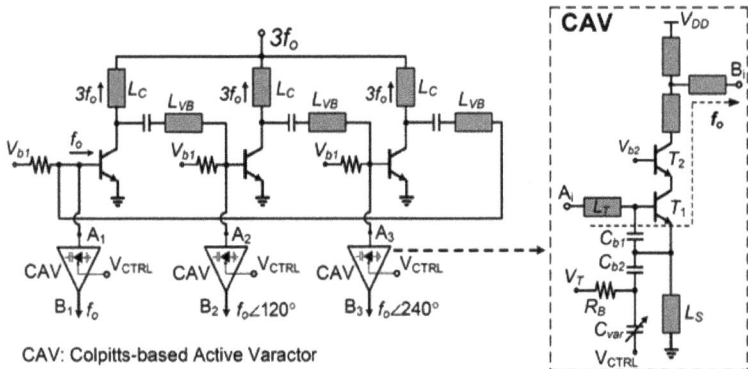

Fig. 1.26.   Schematics of a Colpitts active varactor (© IEEE [24]).

due to the higher operating frequency and lower power consumption. However, injection-locked dividers have a narrow tuning range, and a multi-phase injection frequency divider is proposed for frequency division, as shown in Fig. 1.27. Compared with a single injection, multi-phase injection achieves wider locking range and requires lower input power [24].

A low-power 300-GHz SiGe LO generator based on the second harmonic and frequency doubler is implemented in [25]. The schematic of the VCO and buffer is shown in Fig. 1.28. Operating with a fundamental frequency of 79 GHz, the VCO second harmonic,

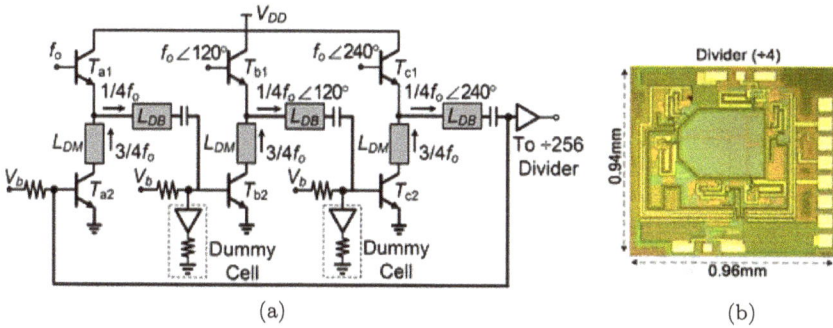

Fig. 1.27. Divide-by-4 injection-locked divider (a) schematics, and (b) die-photo (© IEEE [24]).

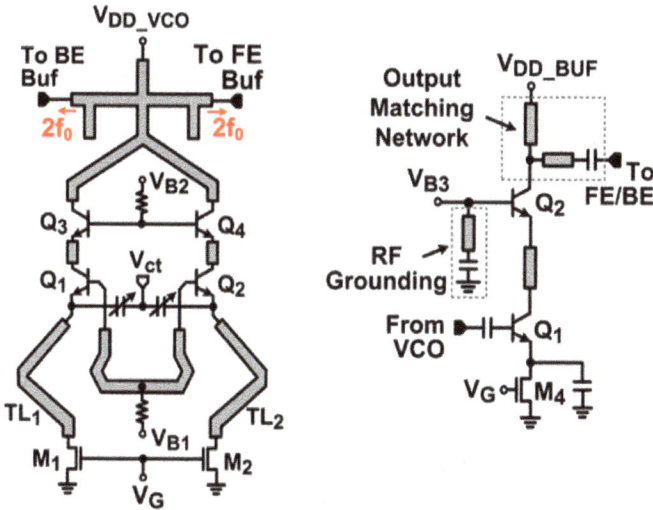

Fig. 1.28. Schematic of a VCO and buffers (© IEEE [25]).

in combination with front-end and back-end buffers, pushes the output frequency to the THz range. The feedback loop is at the lower frequency, as the power required for the feedback at THz frequency consumes higher DC power and the quality of passives increases quickly at high frequencies due to the skin effect. Also, at lower frequencies, the passives have larger geometries, which help to reduce the impact of process variations. In the THz frequency

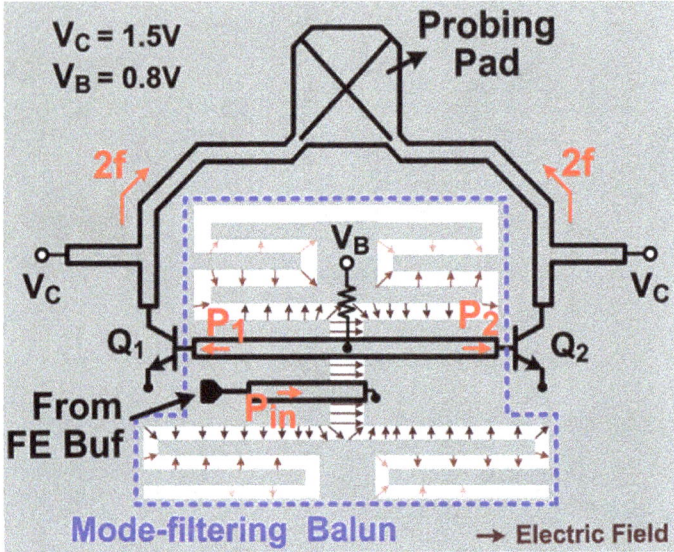

Fig. 1.29.   Schematic of a frequency doubler (© IEEE [25]).

generation stage, a frequency doubler, as shown in Fig. 1.29, is employed. Since the buffer preceding the doubler is single-ended, a mode-filtering balun is used to convert it to balanced waves for second harmonic generation. To double the frequency, the fundamental power is injected into the base of the amplifying transistor, and the second harmonic output is extracted from the collector. The matching network is built-in to reduce the signal loss. The die microphoto of the proposed THz source is shown in Fig. 1.30.

Though SiGe VCOs are popular for THz implementations, CMOS VCOs operating at frequencies above 200 GHz for future wireless communications are reported in [26] to [29], largely due to the increase in f_T/fmax due to the scaling of CMOS technology. A 200-GHz fundamental cross-coupled VCO with on-chip inductor and varactor with an overall tuning range of 8 GHz, as shown in Fig. 1.31, is implemented in 32 nm CMOS SOI process technology [26]. In order to increase the tuning range and reduce the effect of parasitic capacitance, an additional source inductor is used in the NMOS cross-coupled pair. The VCO measures a output power of −13.5 dBm

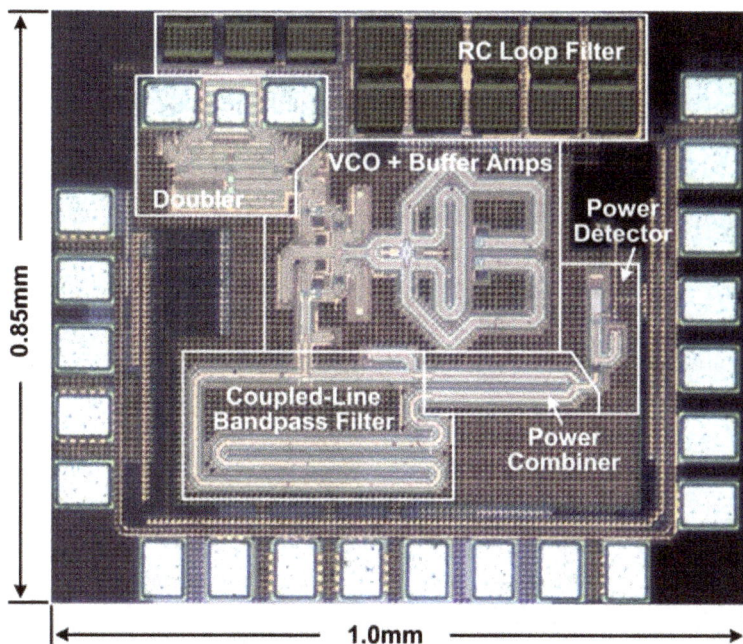

Fig. 1.30. Die microphoto of the proposed THz source (© IEEE [25]).

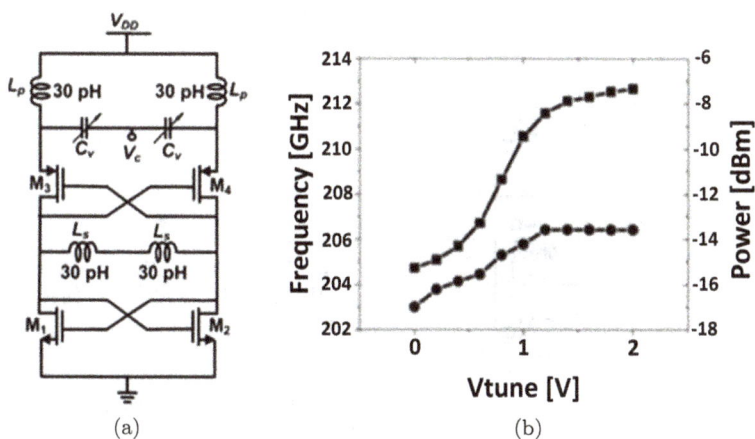

(a) (b)

Fig. 1.31. A 210-GHz CMOS fundamental VCO core and tuning range performance (© IEEE [26]).

and a phase noise of $-80.9\,\mathrm{dBc/Hz}$ at $1\,\mathrm{MHz}$ offset from the carrier of $209.2\,\mathrm{GHz}$ with a power consumption of $42\,\mathrm{mW}$.

Frequency multipliers are critical in THz signal generation, and a CMOS 270-GHz LO generation is shown in Figs. 1.32 to 1.34 [27]. The synthesizer employs a low-frequency VCO and injection-locked multipliers to generate a 270-GHz LO frequency. To improve the locking range of the injection-locked frequency multiplier (ILFM), a third harmonic and fourth harmonic extraction enhancement technique is proposed. The third harmonic of the VCO is fed to the buffer and is split using a triple coil transformer. One path is directly

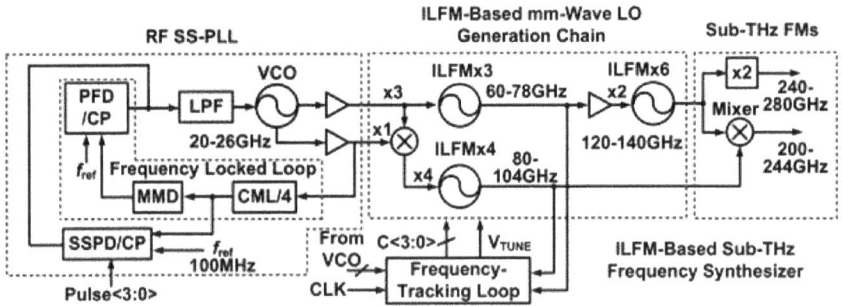

Fig. 1.32.   Block diagram of a 270-GHz CMOS synthesizer (© IEEE [27]).

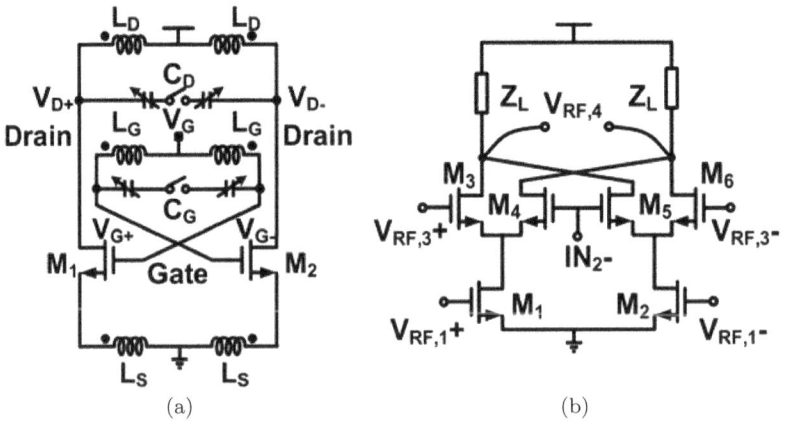

Fig. 1.33.   Schematic of a (a) VCO, and (b) frequency quadrupler (© IEEE [27]).

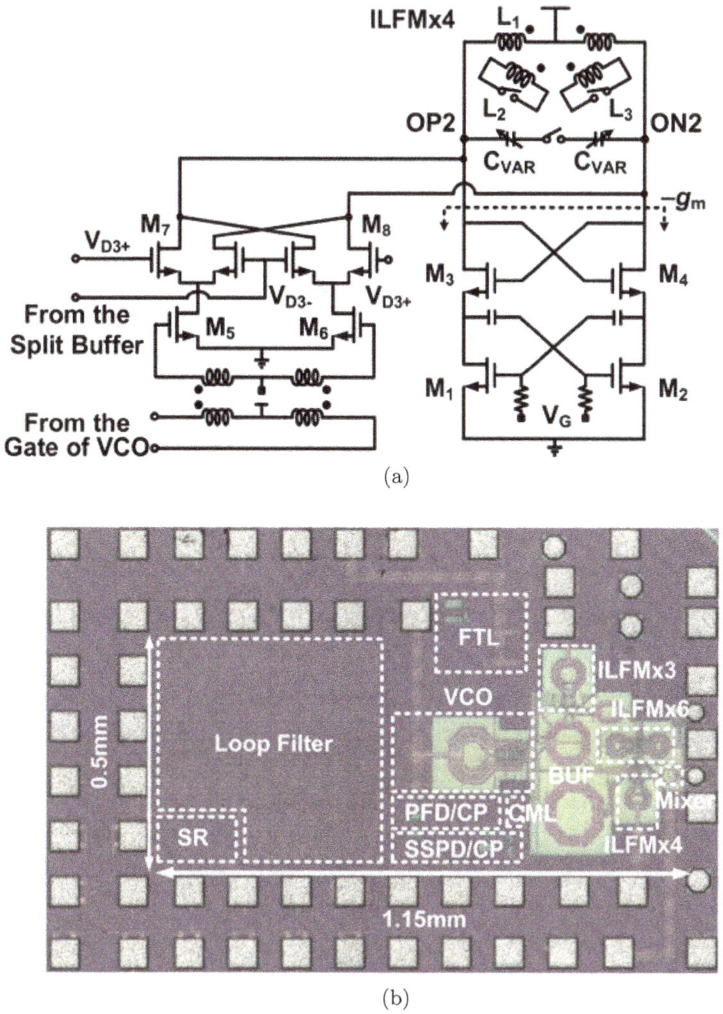

(a)

(b)

Fig. 1.34. (a) Schematic of ILFDM×4 mixing based injector, and (b) Synthesizer die microphoto (© IEEE [27]).

fed to the ILFM×3, and in the other path, the third harmonic is mixed with the fundamental signal to attain the fourth harmonic for ILFM×4. The third harmonic is further passed down through a push-push VCO to generate a sub-THz frequency in the ILFD×6. In the last stage, the output of the ILFD is doubled and mixed with the fourth harmonic to achieve a wide frequency tuning range. The die microphoto of the frequency synthesizer is shown in Fig. 1.34(b).

1.8.2.3. *THz receiver*

The receiver in the transceiver performs the down-conversion and demodulation of the RF signal in the presence of unwanted signals (blockers) and noise. In the early years, the receiver architecture was mostly based on super-hetrodyne with multiple intermediate frequency (IF) stages and a large number of external analog filtering components. Due to the advancement in process technology, state-of-the art receivers directly down-covert the RF to baseband. While a higher output power is the requirement for the transmitter, the receiver sensitivity to detect small signals in the presence of noise is an important specification for the receiver. In the receiver, the gain and noise figure of the receiver first stage low noise amplifier (LNA) is very important. The LNA noise figure is typically limited by the noise performance of the active transistor and is proportional to the ratio of operating frequency and f_T of the process technology [30]. However, in practice, the LNA noise figure is much higher than the minimum transistor noise due to the presence of other noise sources.

The typical implementation schemes for the THz receiver in SiGe and CMOS process technology are shown in Fig. 1.35. In the SiGe implementation, the LNA is employed as the first block [19, 31], and in CMOS implementation, the design of the LNA is typically

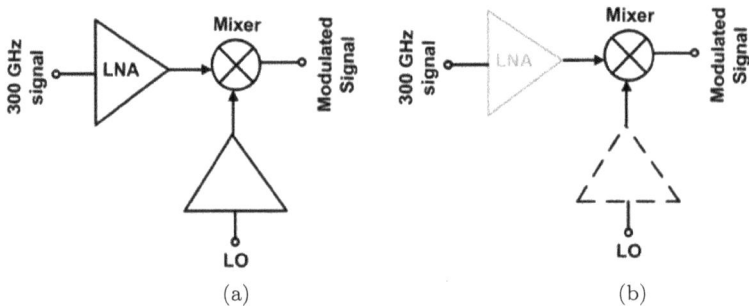

Fig. 1.35.   Typical receiver implementation: (a) SiGe implementation, and (b) CMOS implementation.

challenging due to the low f_T/fmax in comparison to SiGe. Furthermore, more recently, in SiGe THz, driver amplifiers operating in 300 GHz are added to the LO signal to improve the conversion gain of the fundamental mixer and omit the LNA. In CMOS THz receivers, when LNA is omitted, the down-conversion mixer is the first stage. Typically, in such a scenario, harmonic mixers are used as down-conversion mixers to lower the requirement on the LO frequency. However, harmonic mixer suffers from high conversion loss and poor noise figure, which degrades the overall receiver performance. Therefore, it is necessary to use the fundamental mixer as the first stage, where the LO frequency will also be in the THz frequency range. One of the major downsides is the requirement of high LO output power to reduce the mixer conversion loss and high power consumption required in the LO. Several high performance THz receivers are reported in the literature [19, 26, 31–34], as presented in this section.

A fully integrated 300–350-GHz receiver implemented in SiGe with an f_T/fmax of 250 GHz/370 GHz is reported in [32]. The receiver block diagram is shown in Fig. 1.36. The LNA-less receiver chain consists of a 3-stage PA and frequency doubler, which upconverts the LO to the fundamental mixer. The 3-stage PA in the LO path is necessary to boost the LO signal amplitude to drive the fundamental mixer active transistor into a non-linear region to generate the harmonics and increase the conversion gain of the mixer. In addition, to increase the common-mode LO signal rejection, an additional common-base stage is added to isolate the sensitive mixer core transistor.

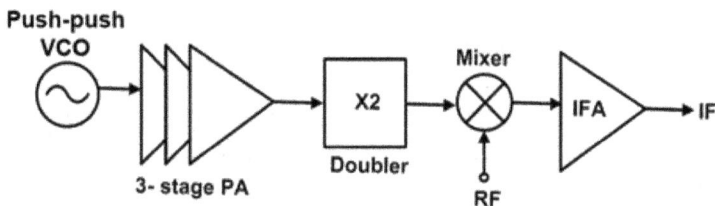

Fig. 1.36. Block diagram of a THz receiver.

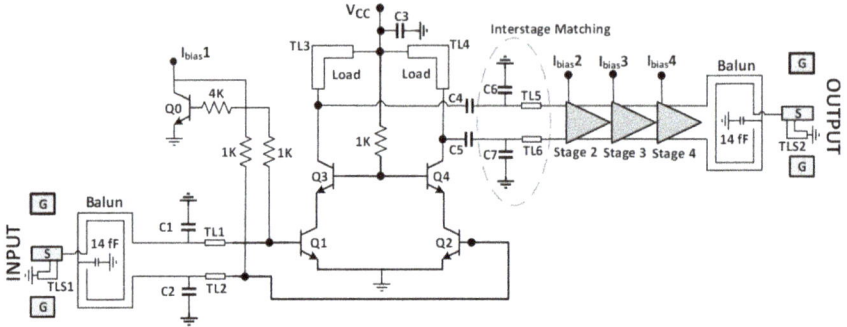

Fig. 1.37.   Schematic of a 300-GHz LNA (© IEEE [33]).

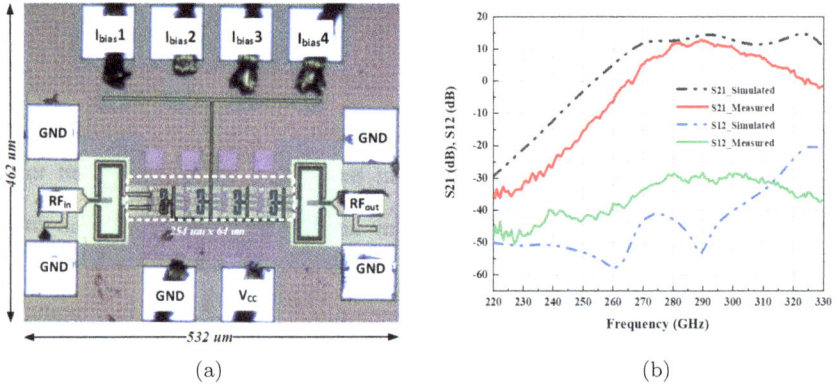

(a)   (b)

Fig. 1.38.   300-GHz LNA: (a) die microphoto, and (b) S21 plot (© IEEE [33]).

The design of an LNA operating at 2/3 of fmax without gain boosting at 300 GHz is reported in [33], as shown in Fig. 1.37. Based on the SiGe process technology and f_T/fmax of 300/450 GHz, the LNA achieves a gain >11 dB from 290 to 320 GHz, as shown in Fig. 1.38. A wideband direct conversion 240 GHz receiver in SiGe with f_T/fmax of 300/450 GHz is shown in Fig. 1.39 [19]. The receiver employs a 3-stage LNA and cascode topology for LNA individual stages, which maximizes the gain with better isolation. Transmission line inductors and capacitors are used in the output-matching network. The mixer for the direct conversion quadrature

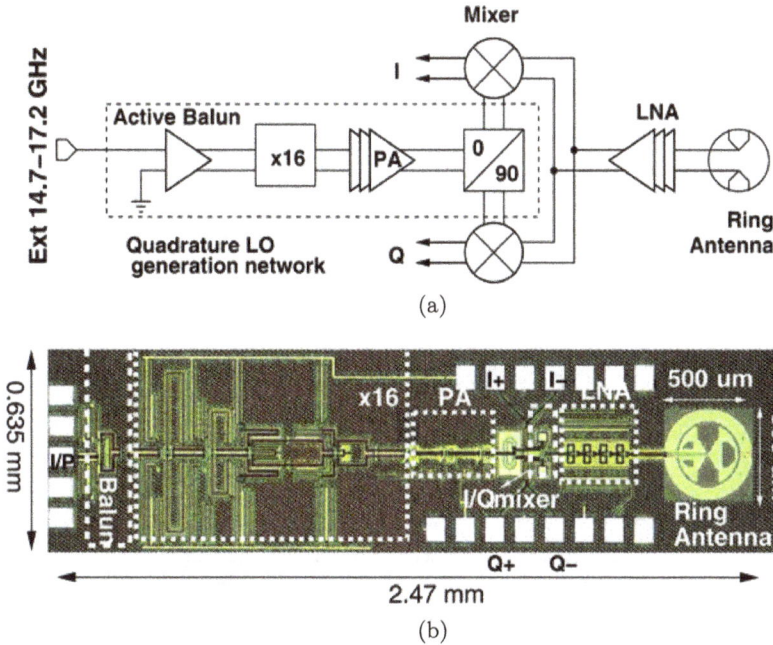

Fig. 1.39. 240-GHz Quadrature receiver (a) block diagram, and (b) die micro-photo (© IEEE [19]).

mixer is based on Gilbert cell topology, and a resistive load is used in the mixer. The receiver achieves a peak conversion gain of 10.5 dB at 236 GHz and a SNR >10 from 210 to 275 GHz. The receiver noise figure is 15 dB.

A 40-nm CMOS-based receiver operating at a frequency of 265 GHz and achieving a 76-Gbps data rate is reported in [34]. As shown in Fig. 1.40, the receiver chain consists of a down-conversion mixer, IF amplifier chain, and LO chain. Since the LNA is omitted from the proposed receiver, the LO generator must deliver a high output power. Therefore, an external 25-GHz LO source with 0 dBm power is chosen in the work. The external 25 GHz is multiplied nine times for the 225-GHz LO, and the power level is 4 dBm. To suppress the LO leakage, an open-stud notch filter is placed at the output of the mixer. The power consumption of the receiver is 467 mW and the noise figure is 17.4 dB.

(a)

(b)

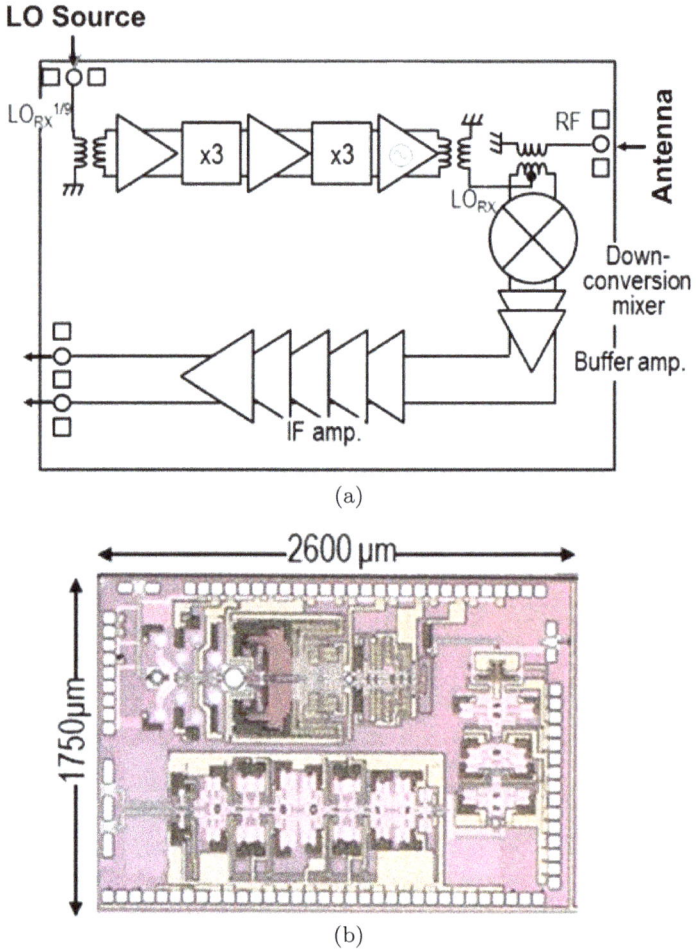

Fig. 1.40.   A 265-GHz CMOS receiver: (a) block diagram, and (b) die microphoto (© IEEE [34]).

The high-performance receiver with only one CMOS SOI LNA operating at 210 GHz [26] is shown in Fig. 1.41. At 210 GHz, to maintain the low noise figure in the receiver, multiple stages maximizing gain and minimizing noise along with inductive degeneration is employed. The last five stages of the LNA employ a 4th-order matching network, and lower-order matching networks are used for initial stages as a lower quality of passives severely degrade the noise

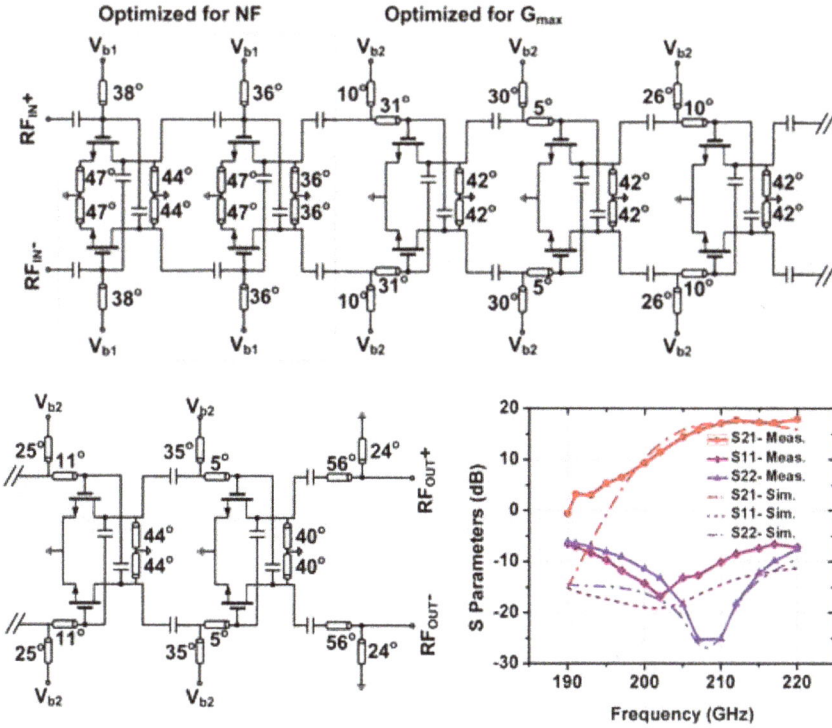

Fig. 1.41. A 210 LNA schematic and measurement result (© IEEE [26]).

figure [26]. The LNA achieves a peak gain of 18 dB with a 15 GHz bandwidth and the return loss is <-8 dB.

1.9. Modeling of RF Devices for 5G and Beyond

1.9.1. *Introduction to RF modeling*

With the recent technology demand and growth in the radio-frequency-based data wireless communication markets, demand for highly efficient, low-cost radio frequency solutions is on the rise. With the expeditious growth of CMOS and its shrinking technology nodes at lower cost, more complex and sophisticated RF solutions can be bound within a single system. This system-on-chip realization includes RF, analog, mixed signal and digital designs integrated on a single-chip system, which recently is more feasible and viable.

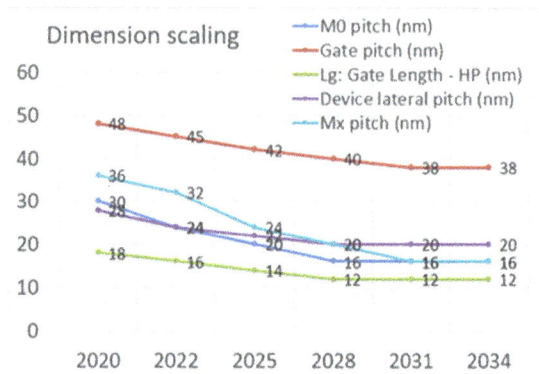

Fig. 1.42.   Technology node vs. year [35].

Figure 1.42 shows the evolution of technology nodes over the years [35].

All the technology scaling comes with the cause of increasing interconnect parasitic components. With the increasing speed of operation, low-cost manufacturing, and fast time to market, the need for proper modeling of the device is essential to obtain an optimal performance. As we advance to more and more radio-frequency-based data communication, the significance of high bandwidth and data rate becomes a complicating factor, and hence the modeling of device at higher frequencies like 5G and beyond gets much more complex [36, 37].

With the extreme reduction in technology nodes, CMOS parameters like short channel, gate leakage, mobility, etc., will start to have serious issues, hence, making it difficult to achieve actual specifications and requirements. Thus, a proper modeling of the RF components is required for optimal designs.

Some of the basic requirements of device modeling, equipment modeling, and characteristics are discussed in further topics.

1.9.2. *RF spectrum*

Figure 1.43 shows the wireless and radio frequency spectrum covering a wide range of applications like 2G, 3G, 4G, 5G, and beyond radios.

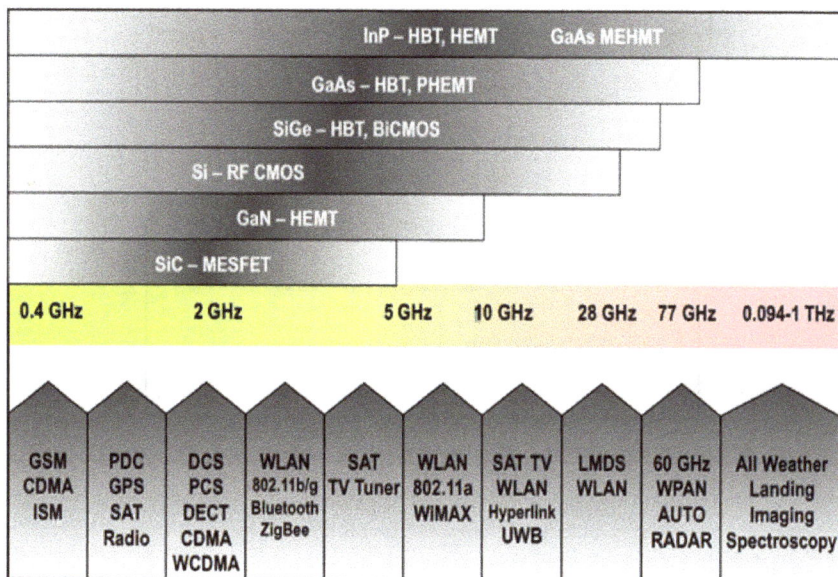

| 0.4 GHz | | 2 GHz | | 5 GHz | | 10 GHz | | 28 GHz | 77 GHz | 0.094-1 THz |

Fig. 1.43. Wireless application and spectrum [37].

The significance of wireless and mobile communication has a high impact in our daily lives as they empower us to communicate through voice, data images, and video. As seen in Fig. 1.43, many types of devices that can be operated at high frequencies were developed over the years [37].

1.9.3. *RF device parameters*

RF devices could be modeled by knowing some of its parameters like cut-off frequency f_T, the maximum oscillating frequency fmax, Noise Figure NFmin, etc.

1.9.3.1. *The cut-off frequency*

The cut-off frequency of f_T is defined as the transition frequency at which the small signal current gain starts dropping to unity and below. This parameter helps to choose the device based on the maximum frequency of a device when it is used as an amplifier. The f_T could be obtained by converting the S21 parameter to a H21

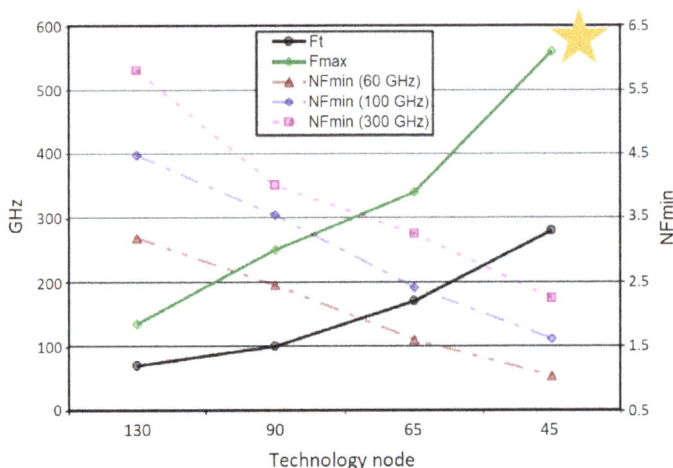

Fig. 1.44. Frequency vs. technology node [38].

parameter.

$$f_T = \frac{g_m}{2\pi C_{gs}}.$$

1.9.3.2. *Maximum oscillation frequency fmax*

The maximum oscillation frequency (or fmax) is the frequency at which unilateral power gain becomes unity. Figure 1.44 gives an indication of Technology nodes vs. f_T and Fmax.

1.9.3.3. *Noise figure NFmin*

In general, noise is defined as any unwanted signal. Noise figure is a measurement of the degradation of the signal-to-noise ratio as the signal passes through the system. Noise figure is an important factor for design; hence, improving the noise figure of a device is desirable. Figure 1.45 shows an example of an amplifier with signal, thermal noise, and additive noise [39].

1.9.4. *RF device modeling (5G and beyond)*

In general, modeling a device is a combination of physical and empirical methods to develop general equations and to describe its

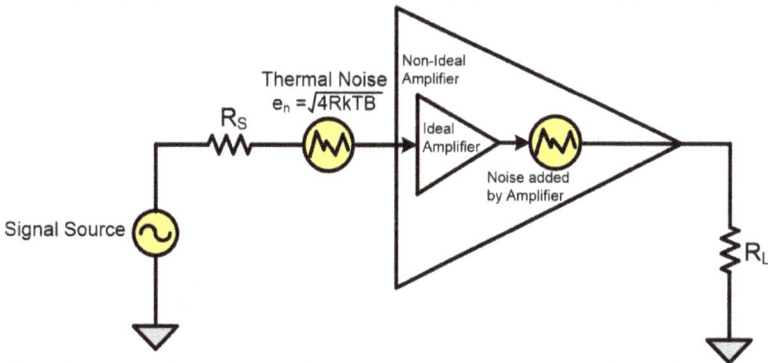

Fig. 1.45. Amplifier with noise [39].

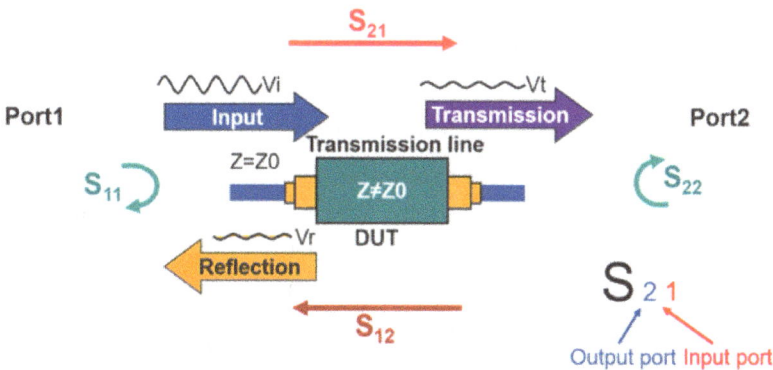

Fig. 1.46. S-parameter [40].

behavior. In any given technology, several parameters are embedded in it [40–44]. The models give a small- and large-scale signal analysis as well as noise analysis. By using the latest tools and technology, we can model the device accurately and efficiently over a wide range of frequencies. Figure 1.46 shows typical S-parameter characteristics [40].

The bandwidth of the device is limited by a transmission line. Hence, the S-parameter measures its accuracy and is very important for the high-speed design. Figure 1.47 gives a light wave analogy of the S-parameter.

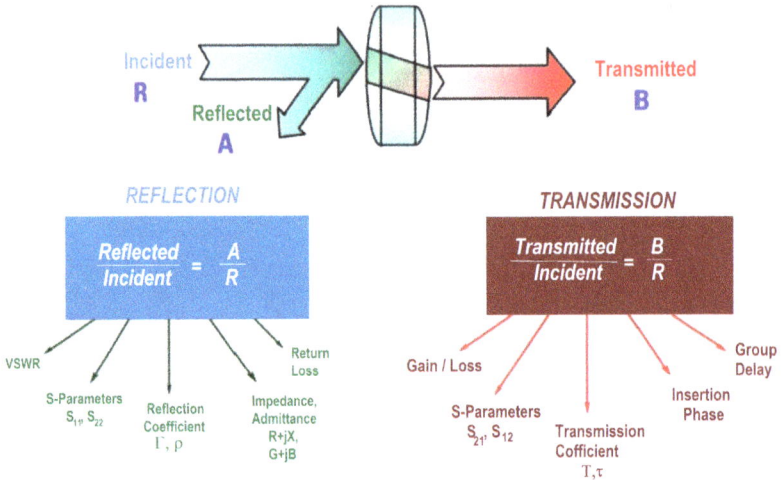

Fig. 1.47.   S-parameter — a lightwave analogy [40].

1.9.4.1. *High frequency modeling procedure*

An accurate RF measurement on wafer is difficult due to the parasitic components of the device and its external system. A typical MOS cross-section device parasitic is shown in Fig. 1.48.

The gate drain connection, usually in the form of a transmission line, connects the devices to the outside circuitry. At high frequency, the series resistance and inductance parasitic dominates. The parasitic changes with small and large signal frequency models. Figure 1.49 shows a small signal, high frequency equivalent circuit of a MOS transistor at high frequency.

1.9.5. *Calibration and de-embedding*

In order to obtain accurate RF modeling, a two-step correction procedure — calibration and de-embedding — is implanted [45, 46].

1.9.5.1. *Calibration*

The modeling measurement system must be calibrated by defining a reference plane for S-parameter measurements at the probe tip using some standard calibration procedure. Figure 1.50 shows a typical example of a 110-GHz (5G and beyond) wafer level calibration

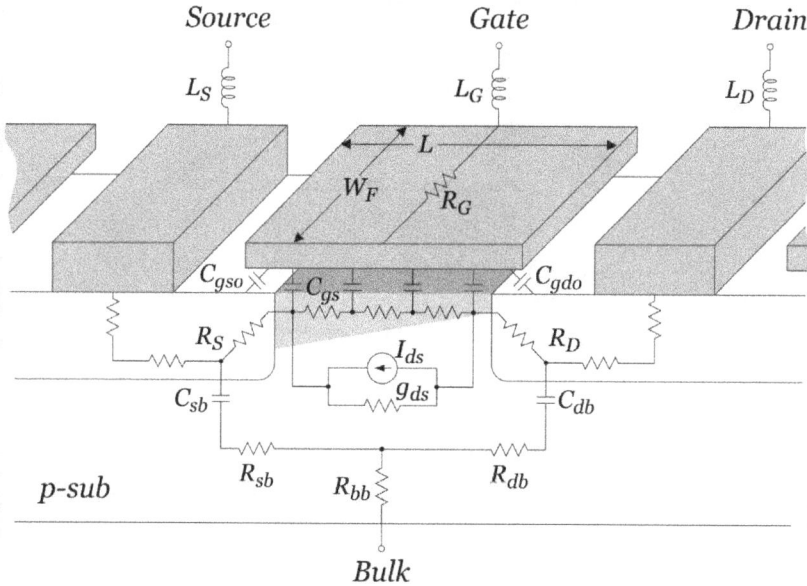

Fig. 1.48.   MOS cross-section with parasitic [37, 41].

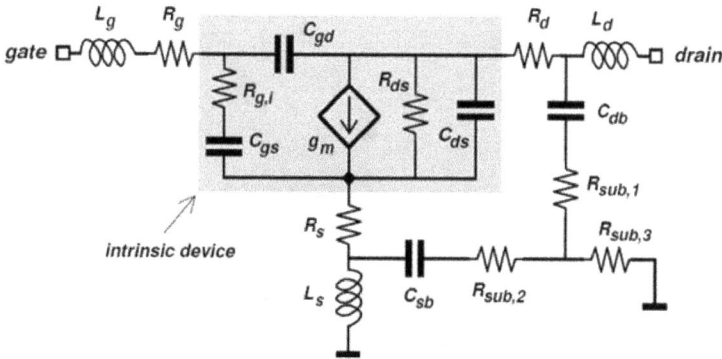

Fig. 1.49.   Small signal, high frequency model [38].

setup [42]. Frequency extenders consisting of up/down-convertors are used to up/down convert the frequency, due to the limitation of vector network analyzers of up to 67 GHz [47].

A proper calibration ensures good and accurate results. Usually, calibration substrates are used for the calibration procedure. Figure 1.51 shows a typical calibration substrate [48].

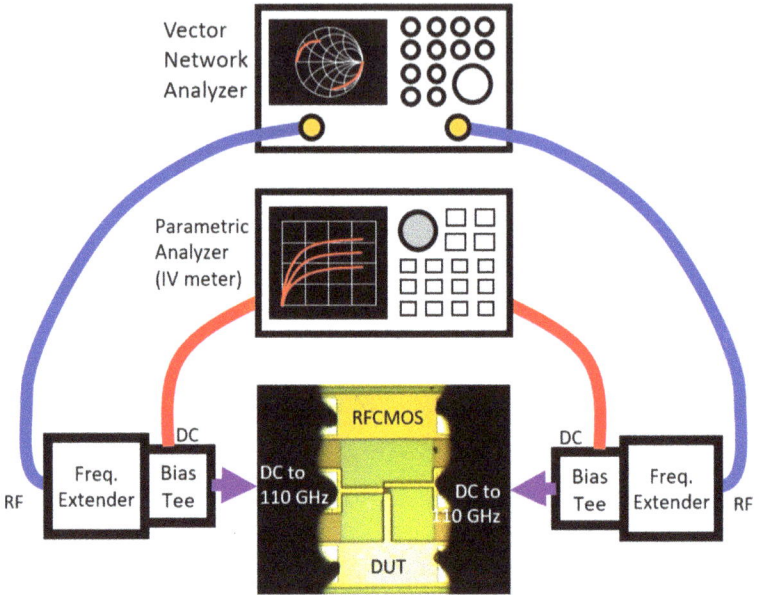

Fig. 1.50.   Wafer level calibration setup for up to 110 GHz [47].

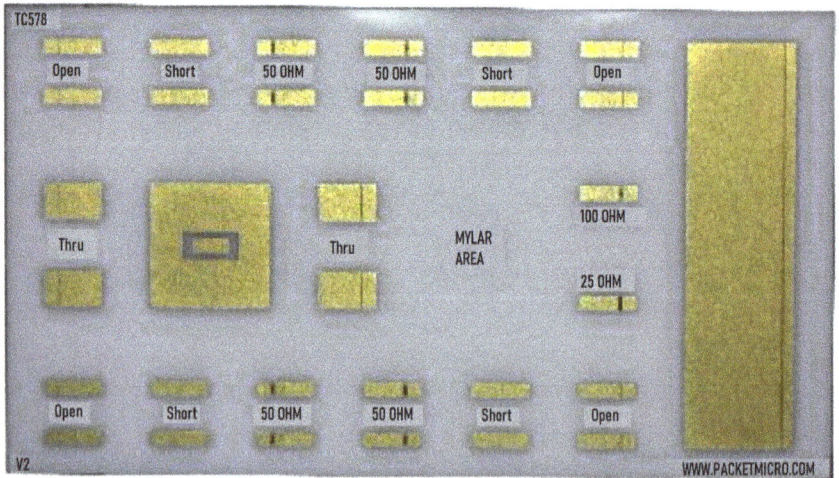

Fig. 1.51.   Typical calibration substrate [48].

1.9.5.2. *De-embedding*

In the measurement of high frequency systems, the effect of probing pads and extra leads are typically subtracted from the measurement through a de-embedding process. The parasitics include resistive, capacitance, and inductive. Hence, the losses due to these passive parasitics are significant. These parasitic components must be removed to evaluate intrinsic device characteristics at high frequency; hence, de-embedding, which is a method to remove these parasitic losses from the measurements. Various techniques are used for this process. The most common de-embedding is the open de-embedding and open-short de-embedding. Figure 1.52 shows the various test structures used for de-embedding.

As most interconnects to the external world go through the RF pads, a good model for the pad is critical; hence, de-embedding is implemented, including the pad models. Figure 1.53 shows a layout of a common RF GSG pad.

In the process of de-embedding, which is to remove the effects of the pads and the cables after the measurement structure, the measured Y parameter is subtracted from the measured device. Figure 1.54 shows an equivalent open de-embedding model.

As the application frequency becomes higher, open de-embedding may not be sufficient to de-embed all the parasitics; hence, open-short embedding is implemented. Figure 1.55 shows an equivalent model for open-short de-embedding.

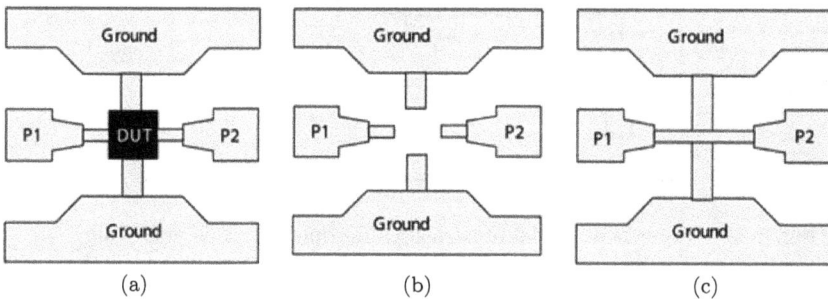

Fig. 1.52. (a) Design-under-test (DUT) on measurement, (b) the open test structure, and (c) the short test structure [38].

Fig. 1.53.   Layout of a common RF GSG pad [38].

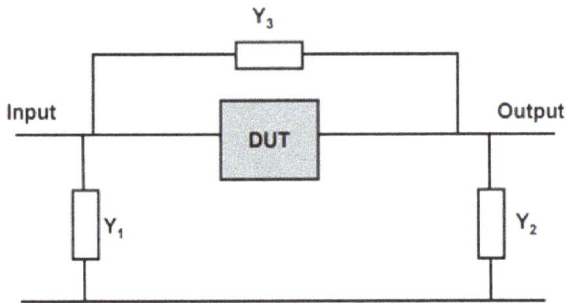

Fig. 1.54.   Equivalent models of parasitics for open de-embedding [38].

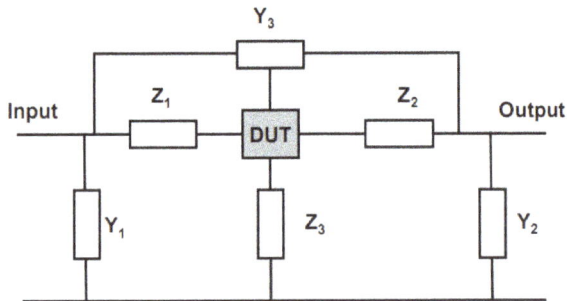

Fig. 1.55.   Equivalent model of parasitics for open-short de-embedding [38].

Once the de-embedding values are obtained, they are used to derive the S-parameter of the device only. As for systems design for higher frequencies like 5G and beyond, both the open and open-short methods start to have imperfect results, and, hence, these inaccuracies make the de-embedded results noisy and affect the accuracy of the models. Hence, several other high frequency de-embedding models are imposed. Some of the models are recursive, where most of the interconnects to the external world go through the RF pads and a good model for the pad becomes critical. Hence, de-embedding is implemented, including for the pad models.

1.9.6. *Noise figure measurement*

In an RF receiver system, the sensitivity of the receiver is very significant. This means the minimum amount of signal-to-noise ratio that an RF receiver can successfully detect. Hence, the sensitivity of the receiver depends on the noise figure of the entire receiver. Noise figure measurement modeling is implemented on the devices. There are several device parameters and parasitic, which contribute to the noise measurement. Figure 1.56 gives an idea of various noise contributions [38].

To attain the noise figure, a general two-port theory is implemented. This two-port theory provides a means to represent a noisy two-port in terms of a noiseless two-port and its corresponding two-noise sources. Modeling the noise of a two-port network is based on a generalized Thevenin's theorem. Figure 1.57 shows an equivalent two-port noise representation model.

Thermal noise is the main source of device noise in 5G and beyond frequencies. It is also the result of the kinetic energy in the particles. These thermally excited particles in a conductor undergo Brownian motion via collision with the lattice of the conductor. Among the various methods proposed for the MOS modeling, the Van Der Ziel model is the most widely accepted one [49]. Van Der Ziel modeled the FET noise as a voltage-modulated resistor, capacitively coupled to the gate, as depicted in Fig. 1.58.

The noise contribution determination from the various noise sources and the sensitivity of the noise performance is essential when

(a)

(b)

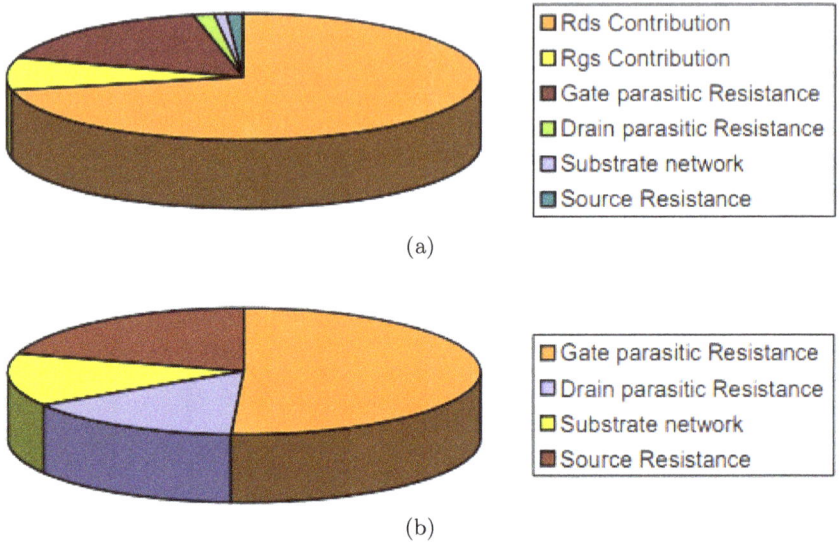

Fig. 1.56.   (a) Noise contribution of various noise sources for a round table device. (b) Noise sensitivity of the device to parasitic noise sources.

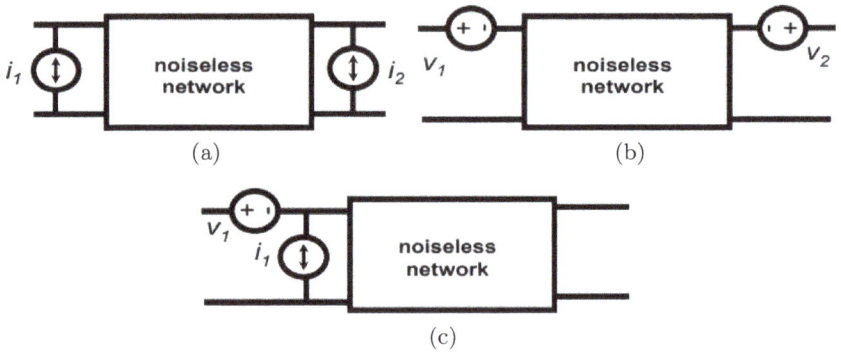

Fig. 1.57.   Equivalent two-port noise representation [38].

it comes to device optimization. The sensitivity of the noise figure increases with an increase in parasitic resistance. Hence, to get an optimum design, the gate resistance is kept very low, and most of the device noise would come from the drain side noise source. The parasitic gate resistance is the second-largest contribution to

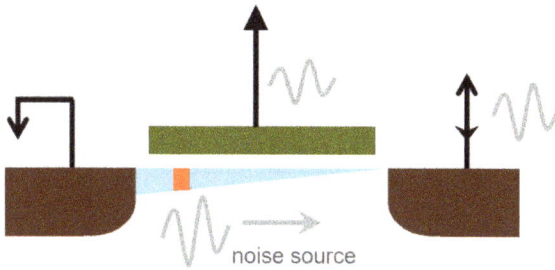

Fig. 1.58.   Generation of channel and induced gate noise in MOSFET [38].

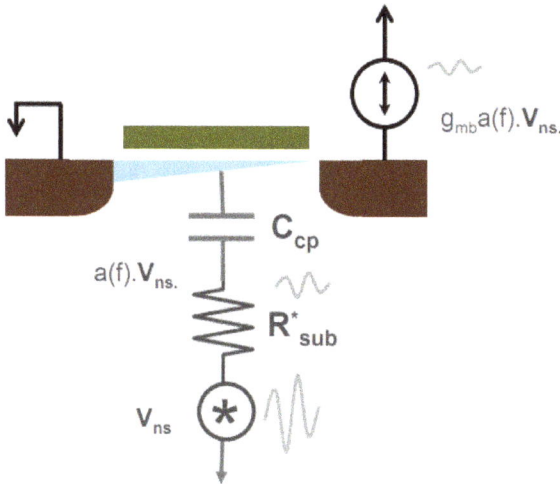

Fig. 1.59.   The substrate noise coupling to the channel [38].

the noise. The substrate resistance also contributes to the noise. Figure 1.59 shows the substrate noise coupling to the channel.

For the complete noise simulation, the small signal model, as shown in Fig. 1.60, is used. These noise measurements are performed for 5G and beyond frequencies. The noise data can be de-embedded using some of the techniques mentioned above.

A typical noise figure measurement setup is shown in Fig. 1.61.

Using the typical noise measurement setup, several noise parameters of the device are measured. These parameters are used to model the noise behavior of the transistor.

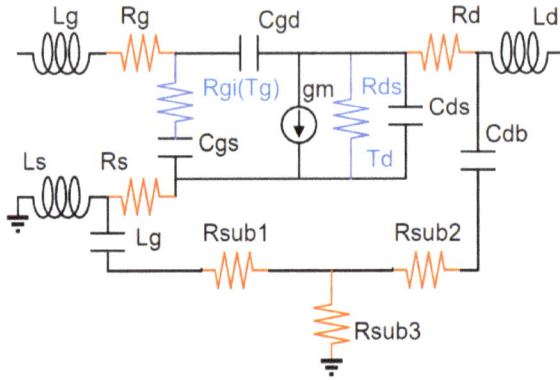

Fig. 1.60.   The employed model for noise analysis [38].

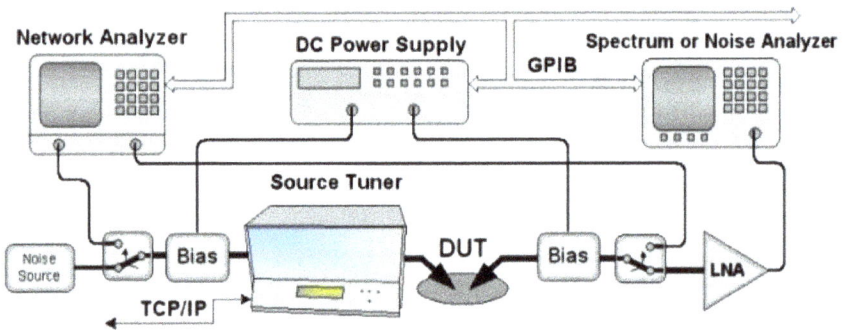

Fig. 1.61.   Noise figure measurement setup.

- Minimum noise figure (NFmin):

This is the given smallest noise figure that a device can reach at a given frequency and bias condition for an optimum design.

- Equivalent noise resistance (Rn):

This is a parameter that indicates how fast the noise figure increases with the inputs, which are mismatched.

- Optimum noise reflection factor (Γopt):

This parameter is used as optimum admittance Yopt.

The noise could be modeled using two independent noise sources at the source and drain side of the transistor with a proper choice of temperature. The optimal noise impedance at the source of the device approaches the optimal gain impedance as the frequency approaches f_T. On the other hand, the noise sensitivity of the device Rn reduces with frequency, making the overall noise figure less sensitive from the optimal noise impedance deviation. Hence, a proper noise modeling approach is implemented to achieve an optimal design.

1.9.7. *RF device modeling for advanced beyond 5G and terahertz designs*

Future advanced 5G and terahertz communication will utilize the frequency spectrum around 300 GHz [51]. At these frequencies, promising high data-rate communication can be achieved due to the wide bandwidth. Hence, to achieve this, a THz communication standard has been defined and the modeling of the device becomes very challenging. The advance device modeling of THz channels is characterized by numerous properties like lower transmission channel gain, specular power-angle spectrum, and signal attenuation due to molecular absorption [52].

Achieving the THz communication is not realizable as the current semiconductor integrated circuit technology is unable to be implemented on these high bandwidth designs with optimum transmission power. The high carrier frequency, which will result in a huge data path loss and short wavelength, causes a small communication distance, which implies that the channel is very sensitive to antenna displacement, making dynamic applications difficult to realize. The IEEE 802.15.3d Task Group (TG) has developed the IEEE standard 802.15.3d-20176 as the world's first THz communication standard [53].

1.9.7.1. *THz channel measurement and modeling*

An approach to THz modeling is shown in Fig. 1.62.

The modeling begins with a scenario. The various channel measurements are then carried out. This is followed by various channel

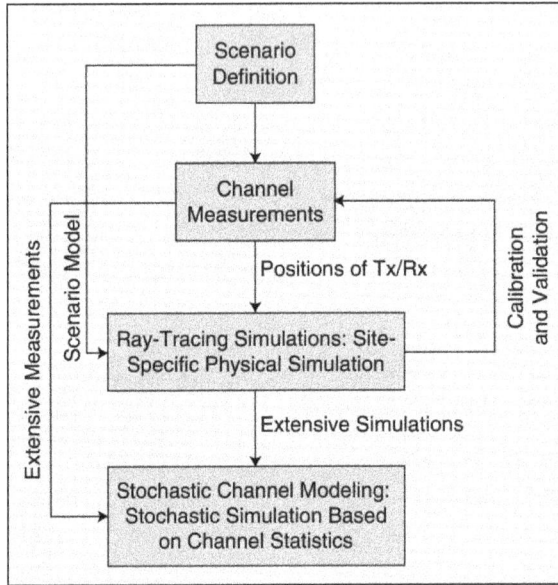

Fig. 1.62.   The canonical approach to T-channel modeling [51].

simulations (CS) such as ray or beam tracing. They are applied for predetermined channel modeling, which does not include a detailed scenario description and TX/RX position. There are two essential THz channel-measurement devices and time-domain CS. Both the devices will transmit a reference signal and measure the received signal at the Rx side. The proper channel information is obtained by processing the reference and the Rx signals. For each narrow band measurement, the channel transfer function (CTF) is described by a scalar quantity, and the broadband CTF is obtained by combining all narrow band CTFs. Various mathematical modeling is applied to the CTF. The RF signal is the convolution of the transmitted signal and the channel impulse response (CIR); hence, the correlation between the transmitted m-sequence and the Rx signals results in a CIR.

As the VNA measurements are composed of many narrowband tests, this mythology has an advantage based on narrow bandwidth on the channel measurement, such as low measuring noise and individual calibration for each frequency point, hence, compromising

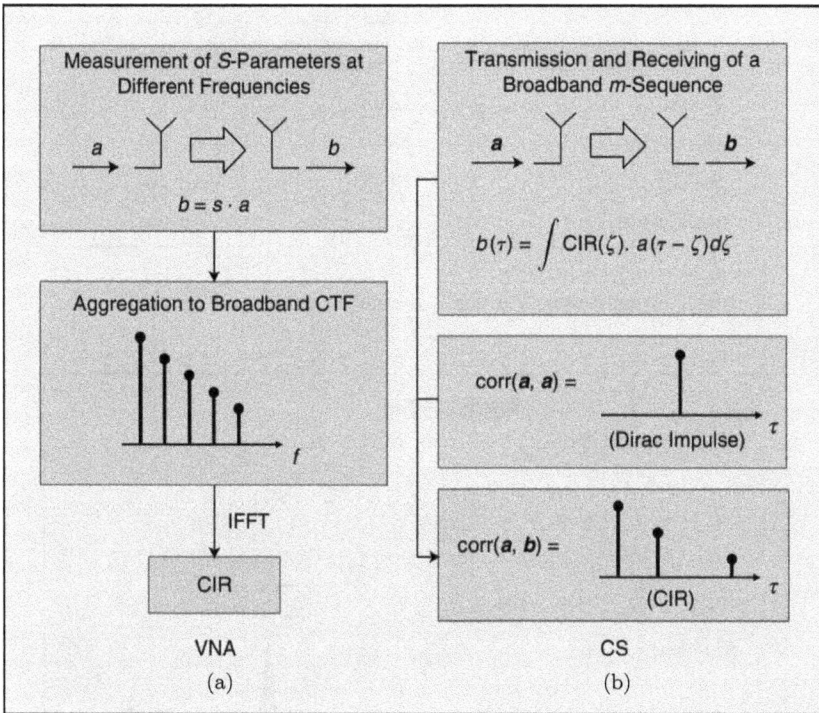

Fig. 1.63. The measurement methodologies of a (a) VNA and (b) CS [50].

on the measurement time. The CS operates in the time domain and takes real-time measurement, hence, allowing dynamic channel effects to be captured. But there is a strong thermal noise due to the large bandwidth limited by the device's technology limitation. Figure 1.63 details the VNA and CS measurements.

The comparison of the THz channel measurement techniques is mentioned in Fig. 1.64.

Once the channel measurements are done, a channel simulator is designed to reproduce the measurement results as a means to better understand the propagation mechanism. The ray-tracing simulation approach implements the path finding techniques according to the geometric optics and calculates the angle of departure and angle of arrival and, eventually, the propagation delay. Figure 1.65 shows the principle of ray-tracing channel simulation.

	VNA	**CS**
Measurement domain	Frequency	Time delay
Speed	Low	High
Dynamic	No	Yes
Precision	High	Medium
Bandwidth	High	Medium

Fig. 1.64.   A comparison of THz channel measurement techniques [51].

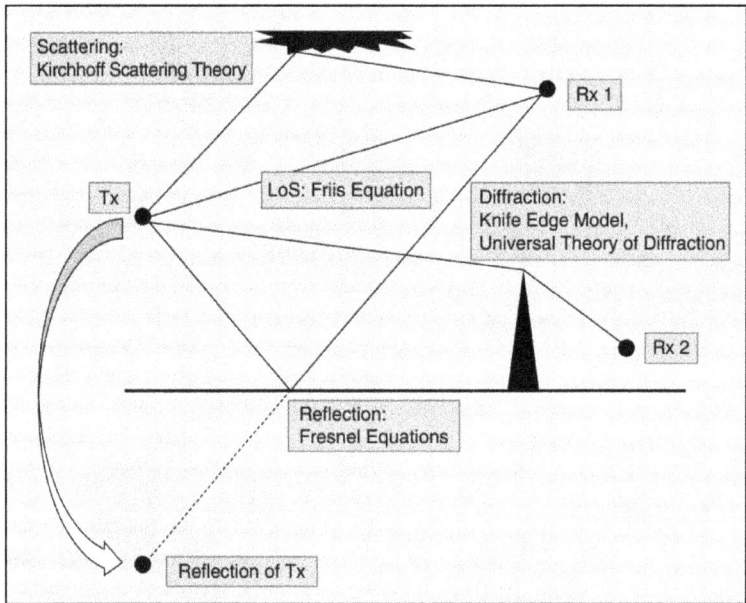

Fig. 1.65.   The principle of ray-tracing channel simulations [51].

1.9.7.2.   *THz channel characteristics*

Based on measurement and modeling, some of the THz channel characteristics need to be investigated.

- High path loss
  The THz channel suffers from an extremely high path loss. A 300-GHz free-space communication for 10 m can have up to 120 dB of

loss. They are mainly due to the strong thermal noise and limited Tx output power. A high gain antenna could be implemented as a counter measure for these losses.

- Power azimuth spectral (PAS)
  Due to the quasi-optical wave propagation, the most received signal power comes from certain discrete direction, which makes the THz channel very sensitive to the scenario geometry. Hence, a slight tilt of a reflector can make the multipath component (MPC) appear and disappear [54].

  Most of the MPCs cause inter-symbol interferences and will affect the performances. Hence, care should be taken to reduce MPCs, and a high-gain antenna is to be used. However, high-gain antennas are always directional, and antenna alignment is a must for THz communication [55].

- Frequency selectivity
  The THz channel has strong frequency selectivity due to its broadband nature and reasons like carrier frequency specular reflection and diffuse reflection [55, 56]. Figure 1.66 shows the frequency selectivity.

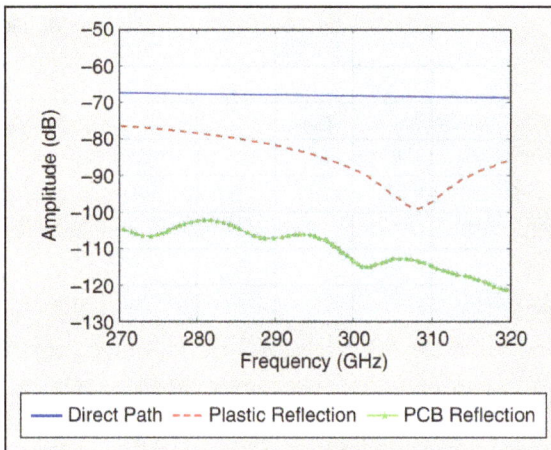

Fig. 1.66. The CTFs of a direct path, a path with one specular reflection on a plastic board, and a path with one diffuse reflection in a printed circuit board [50].

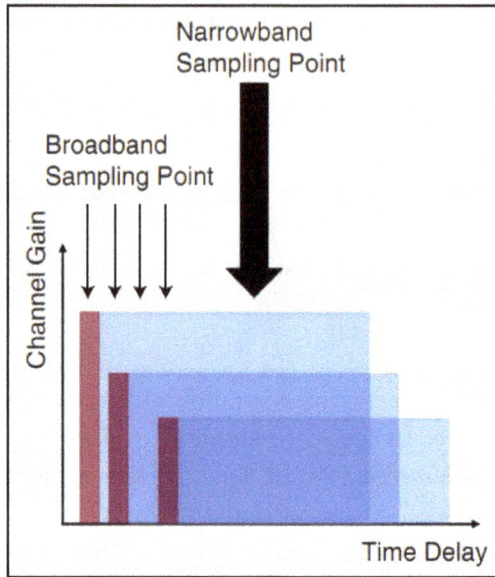

Fig. 1.67.   A comparison between narrowband (blue) and broadband (red) [50].

- Fast fading behavior

  This behavior is closely related to frequency selectivity. The multipath propagation often causes frequency selectivity instead of fast fading. Hence, time diversity cannot be used in THz communication. Figure 1.67 explains the comparison between narrowband and broadband.

- Atmospheric effects

  For outdoor applications, attenuations are caused by gaseous absorption, rain, cloud, and fog, which have an impact on the channel gain, hence, making the attention weather-dependent.

1.9.7.3.   *THz channel properties in various scenarios*

Figure 1.68 explains an overview of THz channel properties in four scenarios. We have hence investigated various RF modeling and technologies implemented up to the THz range. As more advanced technologies come in, the requirement for more bandwidth and, hence, faster devices would rise. This would make modeling much

Scenario	Kiosk Downloading	Intradevice Communication	Wireless Fronthaul/ Backhaul	Wireless Data Center
Description	Transmission of large multimedia files to user equipment or between electronic devices, such as a camera and laptop, within a very short time period	Communication link within a device, e.g., interchip and interboard communication, such that data cables are no longer necessary, and the device is more space efficient	Wireless transmissions between a base station and central network element (backhaul) or between a base station and remote radio head (fronthaul)	Wireless data exchange within or between racks in a data center to simplify data center deployment and maintenance and improve cooling
Data rate (Gb/s)	10–20	10–100	10–100	10–100
Dynamics	Limited	No	No	No
Path loss (dB)	50–60	50–60	100–160	55–100
Propagation	LoS	LoS, reflection, scattering	LoS	LoS, reflection
Peculiarity	User equipment position uncertainty	Near-field propagation	Rain and fog attenuation	Complex scenario

Fig. 1.68.   An overview of THz channel properties in various scenarios.

more complex [56, 57]. High-frequency research will continue to push to the ultra-THz range, which would lower the price of designs. A proper modeling and measurement technique would have to be implemented in order to have proper device optimization. On the other hand, when the technology nodes keep shrinking, the inter-parasitic becomes more complex and significant, which makes device modeling much more tedious.

References

[1] I. F. Akyildiz, A. Kak and S. Nie, "6G and beyond: The future of wireless communications systems," *IEEE Access*, vol. 8, pp. 133995–134030, Jul. 2020.

[2] S. Chen, Y.-C. Liang, S. Sun, S. Kang, W. Cheng and M. Peng, "Vision, requirements, and technology trend of 6G: How to tackle the challenges of system coverage, capacity, user data-rate and movement speed," *IEEE Wirel. Commun.*, vol. 27, no. 2, pp. 218–228, Apr. 2020.

[3] W. Jiang, B. Han, M. A. Habibi and H. D. Schotten, "The road towards 6G: A comprehensive survey," *IEEE Open J. Commun. Soc.*, vol. 2, pp. 334–366, Feb. 2021.

[4] H. Tataria, M. Shafi, A. F. Molisch, M. Dohler, H. Sjöland and F. Tufvesson, "6G wireless systems: Vision, requirements, challenges, insights, and opportunities," *Proc. IEEE*, vol. 109, no. 7, pp. 1166–1199, Jul. 2021.

[5] M. Alsabah, M. A. Naser, B. A. Mahmmod, *et al.*, "6G wireless communications networks: A comprehensive survey," *IEEE Access*, vol. 9, pp. 148191–148243, Nov. 2021.

[6] [Online]. Available: https://www.keysight.com/sg/en/assets/7121-1085/article-reprints/RF-Enabling-6G-Opportunities-and-Challenges-from-Technology-to-Spectrum.pdf

[7] H. Tataria, M. Shafi, M. Dohler and S. Sun, "Six critical challenges for 6G wireless systems: A summary and some solutions," *IEEE Veh. Technol. Mag.*, vol. 17, no. 1, pp. 16–26, Mar. 2022.

[8] P. de Maagt, P. H. Bolivar and C. Mann, "Terahertz science, engineering and systems — from space to earth applications," K. Chang (ed.), *Encyclopedia of RF and Microwave Engineering*, vol. 6, pp. 5175–5195. Wiley-Interscience, 2005.

[9] M. Fujishima, "Device characterization and modeling for terahertz CMOS design," *2015 IEEE MTT-S Int. Microw. RF Conf. (IMaRC)*, pp. 361–364, 2015.

[10] M. Fujishima, "Future of 300 GHz band wireless communications and their enabler, CMOS transceiver technologies," *Jpn. J. Appl. Phys.*, 60 SB0803, 2021.

[11] IEEE Standard for High Data Rate Wireless Multi-Media Networks, Amendment 2: 100 Gb/s Wireless Switched Point-to-Point Physical Layer IEEE Computer Society sponsored by the LAN/MAN Standards Committee, 2017. https://standards.ieee.org/standard/802_15_3d-2017.html

[12] "Sharing and compatibility studies between land-mobile, fixed and passive services in the frequency range 275–450 GHz," Report ITU-R SM.2450-0, Jun. 2019. www.itu.int/pub/R-REP-SM.2450-2019

[13] A. Pärssinen, M. Alouini, M. Berg, T. Kuerner, P. Kyösti, M. E. Leinonen, M. Matinmikko-Blue, E. McCune, U. Pfeiffer and P. Wambacq (eds.). White Paper on RF Enabling 6G — Opportunities and challenges from technology to spectrum. 6G Research Visions, No. 13, 2020.

[14] U. R. Pfeiffer, "Silicon CMOS/SiGe transceiver circuits for THz applications," *IEEE 12th Topical Meeting on Silicon Monolithic Integrated Circuits in RF Systems*, pp. 159–162, 2012.

[15] E. Ojefors, B. Heinemann and U. R. Pfeiffer, "Active 220- and 325-GHz frequency multiplier chains in an SiGe HBT technology," *IEEE Trans. Microw. Theor. Technol.*, vol. 59, no. 5, pp. 1311–1318, May 2011.

[16] O. Momeni and E. Afshari, "High power terahertz and millimeterwave oscillator design: A systematic approach," *IEEE J. Solid State Circuits*, vol. 46, no. 3, pp. 583–597, Mar. 2011.

[17] E. Ojefors, B. Heinemann and U. R. Pfeiffer, "A 220 GHz subharmonic receiver front end in a SiGe HBT technology," *IEEE Radio Freq. Integr. Circuits Symp. (RFIC)*, pp. 69–72, Jun. 2011.

[18] P. Rodríguez-Vázquez, J. Grzyb, B. Heinemann and U. R. Pfeiffer, "Performance evaluation of a 32-QAM 1-meter wireless link operating at 220–260 GHz with a data-rate of 90 Gbps," *IEEE Asia-Pacific Microw. Conf.*, pp. 723–725, Nov. 2018.

[19] N. Sarmah, *et al.*, "A fully integrated 240-GHz direct-conversion quadrature transmitter and receiver chipset in SiGe technology," *IEEE Trans. Microw. Technol.*, vol. 64, no. 2, pp. 562–574, Feb. 2016.

[20] N. Sarmah, P. R. Vazquez, J. Grzyb, W. Foerster, B. Heinemann and U. R. Pfeiffer, "A wideband fully integrated SiGe chipset for high data

rate communication at 240 GHz," *IEEE Europ. Microw. Int. Cir. Conf.*, pp. 181–184, Oct. 2016.

[21] M. Fujishima, "Device characterization and modeling for terahertz CMOS design," *IEEE MTT-S Int. Microw. RF Conf.*, pp. 361–364, Feb. 2016.

[22] K. Takano, *et al.*, "17.9 A 105 Gb/s 300 GHz CMOS transmitter," *IEEE Int. Solid-State Circuits Conf. (ISSCC) Dig. Tech. Papers*, pp. 308–309, Feb. 2017.

[23] J. Yun, D. Yoon, S. Jung, M. Kaynak, B. Tillack and J.-S. Rieh, "Two 320 GHz signal sources based on SiGe HBT technology," *IEEE Microw. Wireless Compon. Lett.*, vol. 25, no. 3, pp. 178–180, Mar. 2015.

[24] P.-Y. Chiang, Z. Wang, O. Momeni and P. Heydari, "A silicon-based 0.3 THz frequency synthesizer with wide locking range," *IEEE J. Solid State Circuits*, vol. 49, no. 12, pp. 2951–2963, Dec. 2014.

[25] C. Jiang, M. Aseeri, A. Cathelin and E. Afshari, "A fully on-chip frequency-stabilization mechanism for terahertz sources eliminating frequency reference and dividers," *IEEE Trans. Microw. Theory Techn.*, vol. 67, no. 7, pp. 2523–2536, Jul. 2019.

[26] Z. Wang, P.-Y. Chiang, P. Nazari, C.-C. Wang, Z. Chen and P. Heydari, "A 210 GHz fully integrated differential transceiver with fundamental-frequency VCO in 32 nm SOI CMOS," *IEEE Int. Solid-State Circuits Conf. (ISSCC) Dig. Tech. Papers*, pp. 136–137, Feb. 2013.

[27] X. Liu and H. C. Luong, "A fully integrated 0.27-THz injection-locked frequency synthesizer with frequency-tracking loop in 65-nm CMOS," *IEEE J. Solid State Circuits*, vol. 55, no. 4, pp. 1051–1063, Apr. 2020.

[28] Y. Zhao, *et al.*, "A 0.56 THz phase-locked frequency synthesizer in 65 nm CMOS technology," *IEEE J. Solid-State Circuits*, vol. 51, no. 12, pp. 3005–3019, Dec. 2016.

[29] N. Sharma, *et al.*, "200–280 GHz CMOS RF front-end of transmitter for rotational spectroscopy," *Proc. IEEE Symp. VLSI Technol.*, pp. 116–117, Jun. 2016.

[30] S. Mattisson, "An overview of 5G requirements and future wireless networks: Accommodating scaling technology," *IEEE Solid-State Circuits Mag.*, vol. 10, no. 3, pp. 54–60, Aug. 2018.

[31] S. Kim, J. Yun, D. Yoon, M. Kim, J.-S. Rieh, M. Urteaga, *et al.*, "300 GHz integrated heterodyne receiver and transmitter with on-chip fundamental local oscillator and mixers," *IEEE Trans. Terahertz Sci. Technol.*, vol. 5, no. 1, pp. 92–101, Jan. 2015.

[32] J. Al-Eryani, H. Knapp, J. Kammerer, K. Aufinger, H. Li and L. Maurer, "Fully integrated single-chip 305–375-GHz transceiver with on-chip antennas in SiGe BiCMOS," *IEEE Trans. Terahertz Sci. Technol.*, vol. 8, no. 3, pp. 329–339, May 2018.

[33] S. P. Singh, T. Rahkonen, M. E. Leinonen and A. Pärssinen, "A 290 GHz low noise amplifier operating above fmax/2 in 130 nm SiGe technology for sub-THz/THz receivers," *IEEE Radio Freq. Int. Cir. Symp. (RFIC)*, pp. 223–226, Jul. 2021.

[34] S. Hara, *et al.*, "A 76-Gbit/s 265-GHz CMOS receiver," *IEEE Asian Solid-State Cir. Conf.*, pp. 1–3, Dec. 2021.

[35] IEEE IRDS 2020 edition. Online: https://irds.ieee.org/editions/2020

[36] R. W. Dutton, *et al.*, "Device simulation for RF applications," *Int. Electron. Devices Meeting. IEDM Technical Digest*, pp. 301–304, 1997.

[37] H. S. Bennett, J. J. Pekarik and M. Huang, "ITRS CHAPTER: RF and A/MS technologies for wireless communication".

[38] B. Heydari, "CMOS circuits and devices beyond 100 GHz," Technical Report No. UCB/EECS-2008-121.

[39] Tektronix Noise Figure Application notes, White Paper 37W-30477-0.

[40] Keysight S-Parameter Measurements Application Notes 5991-3736EN.

[41] Y. Tsividis, *Operation and Modeling of the MOS Transistor*, 2nd edition. Oxford University Press, 2003.

[42] Semiconductor Industry Association, *et al.*, "International Technology Roadmap for Semiconductors (ITRS)," 2003 Edition.

[43] P. R. Gray and R. G. Meyer, "Future directions in silicon IC for RF personal communications," *Proc. IEEE Custom Integr. Circuits Conf.*, pp. 83–90, May 1995.

[44] A. Abidi, "Low-power radio-frequency IC's for portable communications," *Proc. IEEE*, vol. 83, no. 4, pp. 544–569, 1995.

[45] M. C. A. M. Koolen, J. A. M. Geelen and M. P. J. G. Versleijen, "An improvement deembedding technique for on-wafer high-frequency characterization," *Proc. IEEE Bipolar Circuit Technology Meeting*, pp. 191–194, 1991.

[46] C. H. Chen and M. J. Deen, "A general noise and S-parameter deembedding procedure for on-wafer high-frequency noise measurements of MOSFETs," *IEEE Trans. Microw. Theor. Technol.*, vol. 49, no. 5, pp. 1004–1005, May 2001.

[47] C. B. Sia, "Improving wafer-level S-parameters measurement accuracy and stability with probe-tip power calibration up to 110 GHz for 5G applications," *49th European Microwave Conference*.

[48] TCS70 Calibration Substrate, S-Parameter Calibration and TDR Impedance Validation PacketMicro.

[49] A. van der Ziel, *Noise in Solid State Devices and Circuits*. John Wiley Sons, New York, 1986, pp. 60, 62.

[50] B. Peng, K. Guan, A. Kuter, S. Rey, M. Patzold and T. Kuerner, "Channel modeling and system concepts for future terahertz communications: Getting ready for advances beyond 5G," *IEEE Veh. Technol. Mag.*, vol. 15, no. 2, pp. 136–143, Jun. 2020.

[51] P. Heymann, *et al.*, "Experimental evaluation of microwave field-effect transistor noise models," *IEEE Trans. Microw. Theory Technol.*, vol. 47, pp. 156–163, Feb. 1999.

[52] IEEE Standard for High Data Rate Wireless Multi-Media Networks — Amendment 2: 100 Gb/s Wireless Switched Point-to-Point Physical Layer, IEEE Standard 802.15.3d2017 (amendment to IEEE Standard 802.15.3).

[53] D. He, *et al.*, "Stochastic channel modeling for kiosk applications in the terahertz band," *IEEE Trans. Terahertz Sci. Technol.*, vol. 7, no. 5, pp. 502–513, Jul. 2017.

[54] S. Priebe, M. Jacob and T. Kürner, "Affection of THz indoor communication links by antenna misalignment," *Proc. 6th IEEE Euro. Conf. Antennas Propagat. (EUCAP)*, Mar. 2012, pp. 483–487.

[55] T. Schneider, A. Wiatrek, S. Preussler, M. Grigat and R.-P. Braun, "Link budget analysis for terahertz fixed wireless links," *IEEE Trans. Terahertz Sci. Technol.*, vol. 2, no. 2, Mar. 2012.

[56] D. P. Triantis, "Thermal noise modeling for short-channel MOSFETs," *IEEE Trans. Electron. Devices*, vol. 43, pp. 1950–1955. Nov. 1996.

[57] S. Koenig, D. Lopez-Diaz, J. Antes, *et al.*, "Wireless sub-THz communication system with high data rate," *Nat. Photon.*, pp. 977–981, 2013.

COMMUNICATION

We are stronger for our diversity and weaker for our exclusivity.

Kiat Seng Yeo

Chapter 2

Modified Shannon Capacity for Wireless Communications

Jianguo Ma

2.1. Introduction to Wireless Age and Shannon Theory

In 1895, Marconi did the first wireless communication experiment by communicating the letter "S" along a distance of 3 km in the form of a three-dot Morse code with the help of radio waves. On 2 June 1896, Marconi filed his first patent application for the Marconi's wireless transmitter and receiver with the patent number 12039, which, together with his famous "four 7's invention" granted on 26 April 1900 with the patent number 7777, opened the door to the wireless era. On 12 December 1901, Marconi did the first transatlantic wireless communication between England and Canada, followed by a public service opened in 1907 [1]. Since then, wireless communications have gradually become an important part of society with wireless technology as one of the key strategic technologies [2, 3]. Particularly, since the introduction of modern digital mobile communication (2G), wireless communications have become the must-have of present society, and Cisco predicted that: (i) total number of global mobile subscribers will grow from 5.1 billion (66% of the population) in 2018 to 5.7 billion (71% of the population) by 2023, (ii) global mobile devices will grow from 8.8 billion in 2018 to 13.1 billion by 2023 [4], and (iii) the number of mobile devices will almost double the global population by 2023. The fastest growing

Fig. 2.1. Evolution of wireless technologies in terms of data rates (cartoons are assembled from the internet).

mobile device category is M2M (machine-to-machine), followed by smartphones. The mobile M2M category is projected to grow at a 30% CAGR from 2018 to 2023, with smartphones growing at a 7% CAGR within the same period — 1.4 billion of those will be 5G capable [4]. In order to catch up with the fast-growing demands, future wireless communication networks must provide much higher communication data rates, resulting in much higher communication capacity for 5G and beyond [5]. Figure 2.1 shows the evolving generations of wireless technologies in terms of data rates (the cartoons are assembled from the internet).

It can be imagined wireless technologies have drawn much attention and are still being paid much attention both in industries and academia. Wireless communications were and are one of the most popular research fields, resulting in numerous publications in journals and conferences. There are 667,014 publications in total by using the keywords of "wireless" or "mobile" in IEEE Xplore from 1897 to the end of 2021, as illustrated in Fig. 2.2, where conference papers make up 80% of total publications — latest achievements and results can be published quickly in conferences, which are most welcome by R&D engineers in the industries, who prefer to report their latest results to reflect the fast-growing technologies.

The first conference paper appeared in 1979, and due to the mobile/wireless boom, the number of conference and journal papers on mobile/wireless technologies and applications is increasing dramatically with 2G/3G/4G/5G implementations. The number of publications would increase further due to 6G, which will be under discussion thereafter.

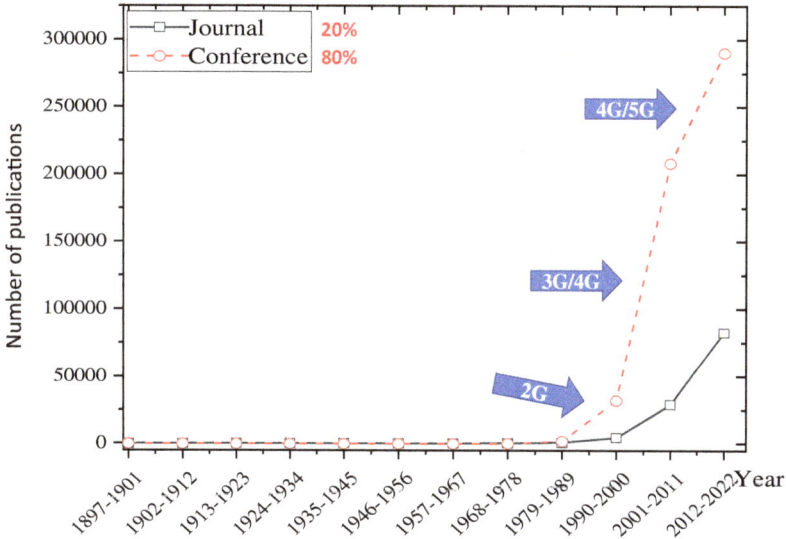

Fig. 2.2. Number of publications with the keywords "wireless" or "mobile" in IEEE Xplore from 1897 to 2021.

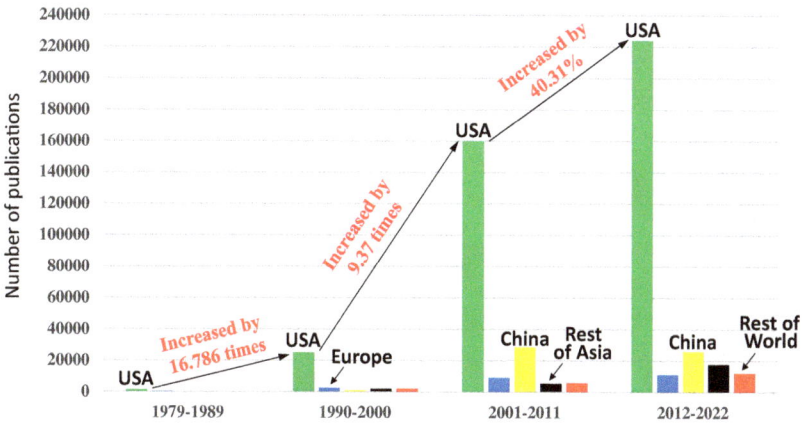

Fig. 2.3. Numbers of publications in the US, Europe, and China.

Figure 2.3 shows the distribution of conference papers with respect to key regions of the world. It is straightforward that the United States (US) has played a leading role in the past and will continue to lead technology development in the next 10 years, even though China has grown tremendously in the last two decades. Before 1989, China had no conference publications on mobile/wireless, but it made a big step in the 1990s, resulting in fantastic progress in wireless communications over the last two decades. This is reflected in China publishing many papers and overtaking Europe to become the second major player. However, it is interesting to note that China's number of conference papers has decreased in the last decade compared to the first decade of the century. On the other hand, the US's number is still growing, possibly reflecting more core technologies and new ideas generated by US organizations, although China has become one of the major players in wireless communications. It might mainly focus on the applications due to its huge market. Before China's publications reach similar numbers of publications in wireless technologies, it is hard to say whether China will be a dominant technology player in wireless communications worldwide.

The distributions of Top 10 organizations with more papers in wireless technologies are shown in Fig. 2.4. It coincides with Fig. 2.3, which shows that the US played a dominant role before 2000. Table 2.1 illustrates the Top 15 most cited papers from IEEE

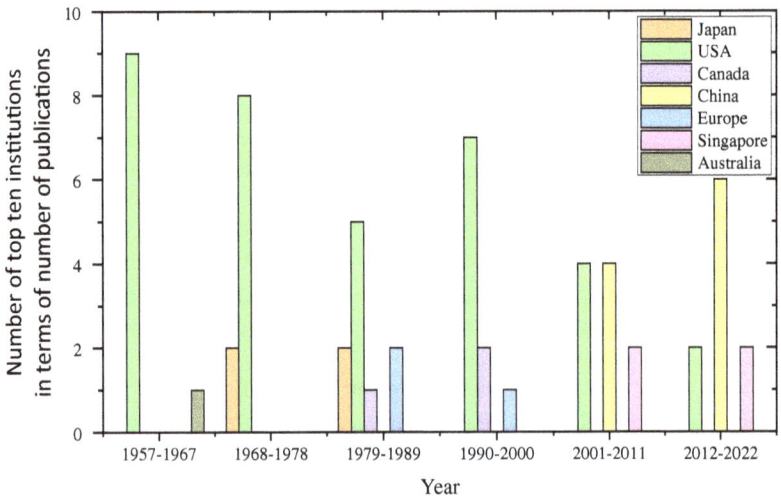

Fig. 2.4.   Numbers of Top 10 organizations with the most publications on mobile/wireless technologies.

Table 2.1. The Top 15 most cited papers from IEEE Xplore.

No.	Paper Information	Citations
1	J. N. Laneman, D. N. C. Tse and G. W. Wornell, "Cooperative diversity in wireless networks: Efficient protocols and outage behavior," in IEEE Transactions on Information Theory, vol. 50, no. 12, pp. 3062–3080, Dec. 2004, US.	9,672
2	S. M. Alamouti, "A simple transmit diversity technique for wireless communications," in IEEE Journal on Selected Areas in Communications, vol. 16, no. 8, pp. 1451–1458, Oct. 1998, US.	9,386
3	I. F. Akyildiz, W. Su, Y. Sankarasubramaniam and E. Cayirci, "A survey on sensor networks," in IEEE Communications Magazine, vol. 40, no. 8, pp. 102–114, Aug. 2002, US.	8,680
4	S. Haykin, "Cognitive radio: Brain-empowered wireless communications," in IEEE Journal on Selected Areas in Communications, vol. 23, no. 2, pp. 201–220, Feb. 2005, Canada.	8,626
5	C. Szegedy, V. Vanhoucke, S. Ioffe, J. Shlens and Z. Wojna, "Rethinking the inception architecture for computer vision," 2016 IEEE Conference on Computer Vision and Pattern Recognition (CVPR), 2016, pp. 2818–2826, US.	6,501
6	W. B. Heinzelman, A. P. Chandrakasan and H. Balakrishnan, "An application-specific protocol architecture for wireless microsensor networks," in IEEE Transactions on Wireless Communications, vol. 1, no. 4, pp. 660–670, Oct. 2002, US.	6,443
7	G. Bianchi, "Performance analysis of the IEEE 802.11 distributed coordination function," in IEEE Journal on Selected Areas in Communications, vol. 18, no. 3, pp. 535–547, Mar. 2000, Italy.	5,964
8	A. Jadbabaie, Jie Lin and A. S. Morse, "Coordination of groups of mobile autonomous agents using nearest neighbor rules," in IEEE Transactions on Automatic Control, vol. 48, no. 6, pp. 988–1001, Jun. 2003, US.	5,768
9	P. Gupta and P. R. Kumar, "The capacity of wireless networks," in IEEE Transactions on Information Theory, vol. 46, no. 2, pp. 388–404, Mar. 2000, US.	5,656
10	V. Tarokh, N. Seshadri and A. R. Calderbank, "Space-time codes for high data rate wireless communication: Performance criterion and code construction," in IEEE Transactions on Information Theory, vol. 44, no. 2, pp. 744–765, Mar. 1998, US.	5274

(Continued)

Table 2.1.   (*Continued*)

No.	Paper Information	Citations
11	J. G. Andrews *et al.*, "What will 5G be?" in IEEE Journal on Selected Areas in Communications, vol. 32, no. 6, pp. 1065–1082, Jun. 2014, US.	5,226
12	W. R. Heinzelman, A. Chandrakasan and H. Balakrishnan, "Energy-efficient communication protocol for wireless microsensor networks," Proceedings of the 33rd Annual Hawaii International Conference on System Sciences, 2000, vol. 2, p. 10, US.	5,193
13	A. Sendonaris, E. Erkip and B. Aazhang, "User cooperation diversity. Part I. System description," in IEEE Transactions on Communications, vol. 51, no. 11, pp. 1927–1938, Nov. 2003, US.	5,134
14	V. Tarokh, H. Jafarkhani and A. R. Calderbank, "Space-time block codes from orthogonal designs," in IEEE Transactions on Information Theory, vol. 45, no. 5, pp. 1456–1467, Jul. 1999, US.	5,055
15	R. R. Yager, "On ordered weighted averaging aggregation operators in multicriteria decisionmaking," in IEEE Transactions on Systems, Man, and Cybernetics, vol. 18, no. 1, pp. 183–190, Jan.–Feb. 1988, US.	4,686

Xplore, which reflects the interests of the wireless community. In 13 papers, the first authors were from US organizations, which demonstrates that the US produces the most useful work in wireless technologies than any other countries, coincident with Fig. 2.3. Table 2.2 presents the Top 15 most popular papers (i.e., most numbers of views/downloads), which reflects the hotter areas in wireless communications.

It is well-known that communication is called an "information highway", and "capacity" is the most critical parameter of planning/ designing highways. Therefore, communication capacity is also a very critical parameter for wireless communications. It is desirable to predict the maximum communication capacity theoretically. Shannon published his famous Shannon formula to predict the maximum communication capacity (or Shannon limit) in 1949 [5]. Since then, wireless communications have or are still trying to reach the Shannon

Table 2.2. The Top 15 papers with the most number of views/downloads from IEEE Xplore.

No.	Paper Information	Citations
1	A. Gupta and R. K. Jha, "A Survey of 5G network: Architecture and emerging technologies," in IEEE Access, vol. 3, pp. 1206–1232, 2015, India	306,336
2	T. S. Rappaport *et al.*, "Millimeter wave mobile communications for 5G cellular: It will work!" in IEEE Access, vol. 1, pp. 335–349, 2013, US.	301,734
3	S. M. R. Islam, D. Kwak, M. H. Kabir, M. Hossain and K. -S. Kwak, "The Internet of Things for health care: A comprehensive survey," in IEEE Access, vol. 3, pp. 678–708, 2015, Korea.	291,777
4	A. Al-Fuqaha, M. Guizani, M. Mohammadi, M. Aledhari and M. Ayyash, "Internet of Things: A survey on enabling technologies, protocols, and applications," in IEEE Communications Surveys & Tutorials, vol. 17, no. 4, pp. 2347–2376, 2015, US.	142,782
5	B. Chen, C. Zhu, W. Li, J. Wei, V. C. M. Leung and L. T. Yang, "Original symbol phase rotated secure transmission against powerful massive MIMO eavesdropper," in IEEE Access, vol. 4, pp. 3016–3025, 2016, China.	131,640
6	T. N. Pham, M. Tsai, D. B. Nguyen, C. Dow and D. Deng, "A cloud-based smart-parking system based on Internet-of-Things technologies," in IEEE Access, vol. 3, pp. 1581–1591, 2015, Taibei, China.	109,642
7	Y. Gao, R. Ma, Y. Wang, Q. Zhang and C. Parini, "Stacked patch antenna with dual-polarization and low mutual coupling for massive MIMO," in IEEE Transactions on Antennas and Propagation, vol. 64, no. 10, pp. 4544–4549, Oct. 2016, UK.	102,191
8	E. G. Larsson, O. Edfors, F. Tufvesson and T. L. Marzetta, "Massive MIMO for next generation wireless systems," in IEEE Communications Magazine, vol. 52, no. 2, pp. 186–195, Feb. 2014, Sweden.	102,049
9	J. G. Andrews *et al.*, "What will 5G be?" in IEEE Journal on Selected Areas in Communications, vol. 32, no. 6, pp. 1065–1082, Jun. 2014, US.	90,001
10	D. Fang, Y. Qian and R. Q. Hu, "Security for 5G mobile wireless networks," in IEEE Access, vol. 6, pp. 4850–4874, 2018, US.	70,460

(*Continued*)

Table 2.2.   (*Continued*)

No.	Paper Information	Citations
11	M. Agiwal, A. Roy and N. Saxena, "Next generation 5G wireless networks: A comprehensive survey," in IEEE Communications Surveys & Tutorials, vol. 18, no. 3, pp. 1617–1655, third quarter 2016, Korea.	67,792
12	S. Chen, H. Xu, D. Liu, B. Hu and H. Wang, "A vision of IoT: Applications, challenges, and opportunities with China perspective," in IEEE Internet of Things Journal, vol. 1, no. 4, pp. 349–359, Aug. 2014, China.	66,266
13	L. D. Xu, W. He and S. Li, "Internet of Things in industries: A survey," in IEEE Transactions on Industrial Informatics, vol. 10, no. 4, pp. 2233–2243, Nov. 2014, China.	64,825
14	H. Kaushal and G. Kaddoum, "Underwater optical wireless communication," in IEEE Access, vol. 4, pp. 1518–1547, 2016, India.	54,705
15	V. Chamola, V. Hassija, V. Gupta and M. Guizani, "A comprehensive review of the COVID-19 pandemic and the role of IoT, drones, AI, blockchain, and 5G in managing its impact," in IEEE Access, vol. 8, pp. 90225–90265, 2020, India.	55,198

limits with various efforts (see, for example, [4] as well as papers listed in Tables 2.1 and 2.2 and references therein). As a result, many publications shown in Fig. 2.2 are on the topic of communication capacity.

Figure 2.5 presents the number of publications on "channel capacity" among the total publications given in Fig. 2.2. It is obvious that publications in both journals and conferences were very limited before 1980. Due to the second generation of mobile (2G) and satellite communications, more papers focusing on channel capacity had been published in the late 1980s. Thanks to the third generation of mobile communications and Bluetooth, research publications have increased tremendously, and about 4,000 conference papers, as well as around 1,000 journal papers on channel capacity, have been published in the first decade of the new century. However, due to the capacity of the practical wireless systems being very close to the Shannon limit,

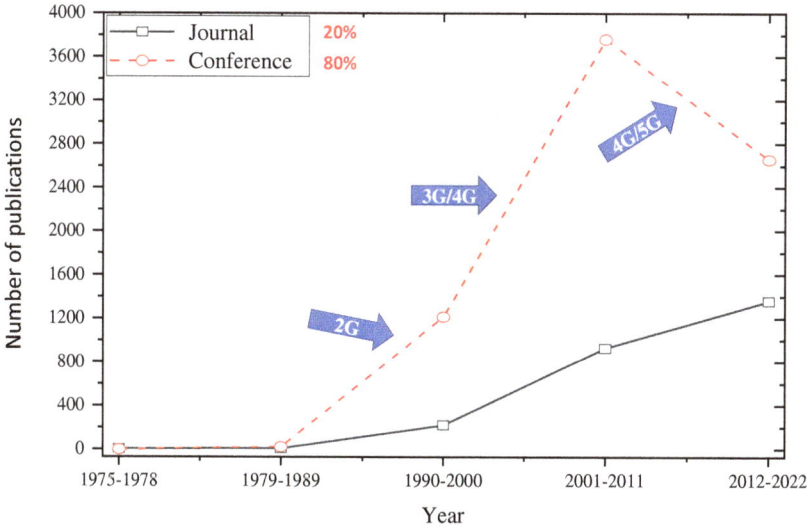

Fig. 2.5. Number of publications with the key word "channel capacity" among publications in Fig. 2.2.

fewer papers have been published in the last decade compared to 2001–2011.

Figure 2.6 shows the paper distributions on channel capacity for different regions, and the US is always the dominant leader in publications on channel capacity. Compared to the first decade of the century, papers on channel capacity from the US, European countries, and China decreased 37%, 33%, and 21%, respectively, but increased in other regions. That might reflect the technology interest delays in developing countries/regions.

Figure 2.7 presents the distributions of the Top 10 organizations with the most number of publications on channel capacity worldwide. Before 2000, North America was dominant. China stepped up in the last decade of the last century and progressed aggressively in the first decade of the new century while the US was declining. That reflects the technology migration from the US to China on wireless, but in terms of the number of the publications, the US is still absolutely dominant, meaning that the US is still the absolute dominant player worldwide in wireless communications.

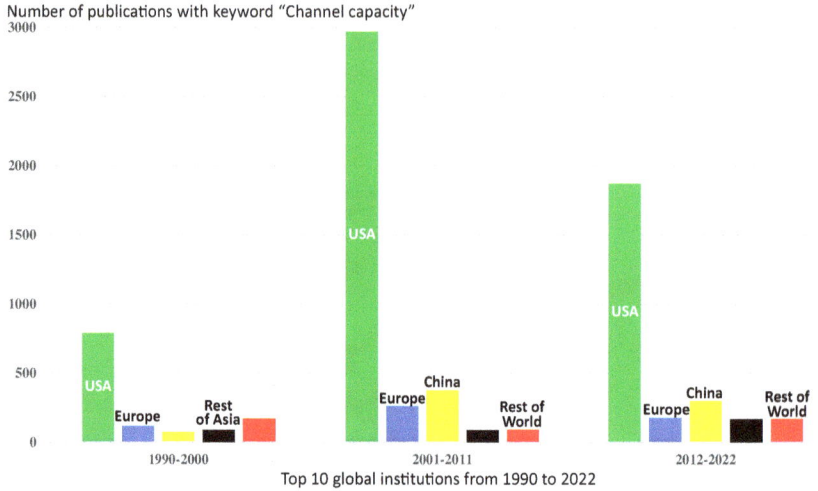

Fig. 2.6.   Publication distributions in regions with the keyword "channel capacity".

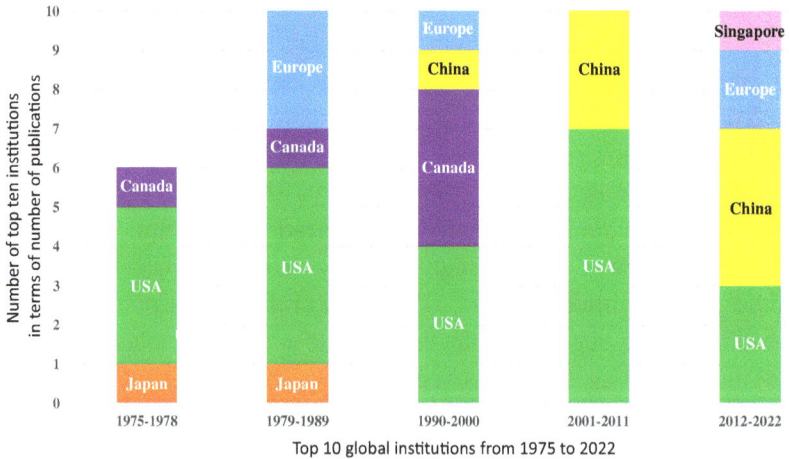

Fig. 2.7.   Distributions by regions of the Top 10 organizations' most publications on "channel capacity".

Figure 2.8 shows that there has been a rapid increase in both conference and journal papers on the Shannon theory due to 3G and 4G deployment in 2000. Although the publication of conference papers has stabilized, the publication of journal papers continues

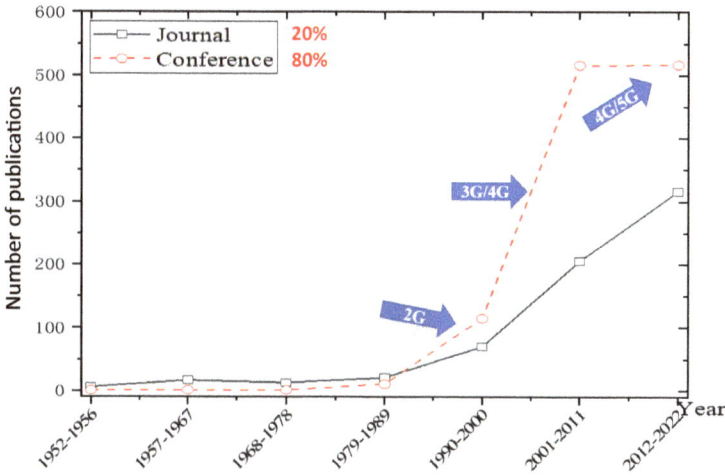

Fig. 2.8.   Publications on "Shannon capacity".

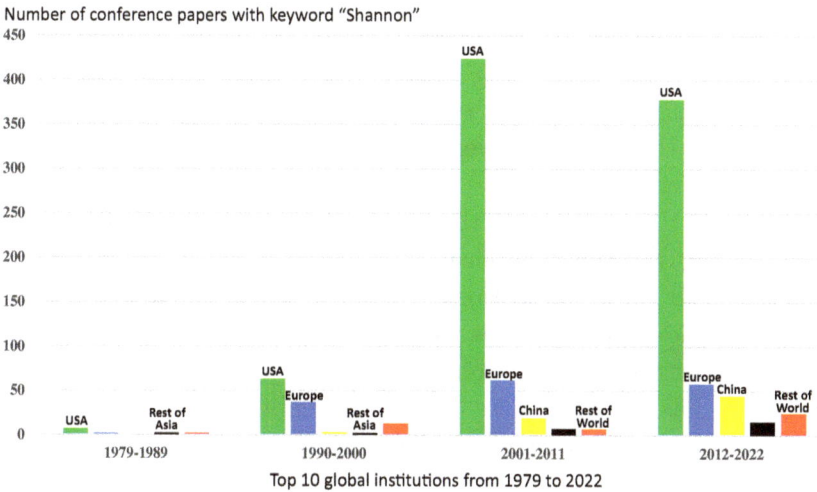

Fig. 2.9.   Publication distributions in regions with the keyword "Shannon".

to rise as the carriers began trailing fixed wireless 5G services as a replacement for wired broadband home internet in 2017 and beyond.

Figure 2.9 discusses the publication distributions in regions on the Shannon theory. Clearly, the US has published the most number of

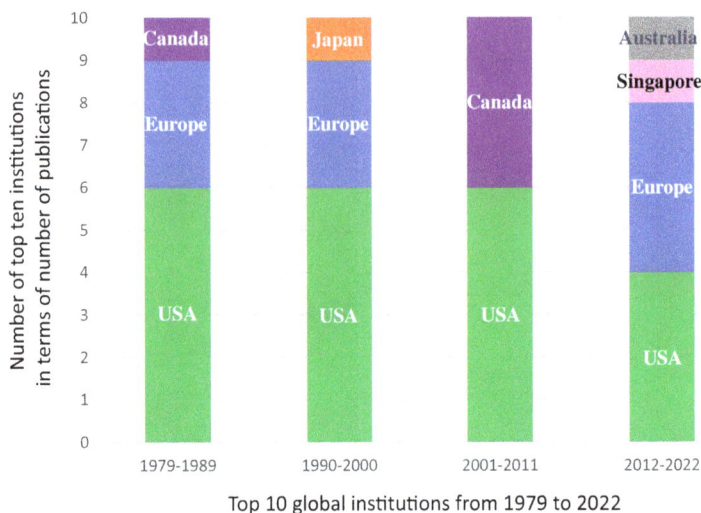

Fig. 2.10. Distribution by regions of Top 10 organizations with the most publications on "Shannon".

papers, much more than the sum of all the other regions/countries. That means the US is also very strong in its fundamental research on wireless communications.

Figure 2.10 summarizes the distributions of Top 10 organizations having the most publications with the keyword "Shannon" amongst their total publications in wireless/mobile technologies and communications. That means the US and European countries have been spending more effort on fundamental research. In addition, Singapore has progressed well in fundamental research in the last decade; however, China had no place on the list.

2.2. Modified Shannon's Capacity for Wireless Communication

2.2.1. *The classic Shannon formula*

As said, communication capacity is the most important parameter of a communication system — it is desirable to understand what the upper limit theoretically is for the communication system.

C.E. Shannon published his classic work in 1948 entitled "A mathematical theory of communication" in two parts in the July and October Issues of the *Bell System Technical Journal* [6]. This paper founded the discipline of information theory [8] and laid down the fundamental cornerstone of modern digital communications. In [6], a channel and the channel capacity C have been defined:

"The channel is merely the medium used to transmit the signal from transmitter to receiver. It may be a pair of wires, a coaxial cable, a band of radio frequencies, a beam of light, etc."

"In the more general case with different lengths of symbols and constraints on the allowed sequences, we make the following definition:

The capacity C of a discrete channel is given by

$$C = \lim_{T \to \infty} \left(\frac{\log N(T)}{T} \right), \tag{2.1}$$

where N(T) is the number of allowed signals of duration T."

In [7], Shannon further elaborated on his ideas and concepts presented in [6] and proposed the classic Shannon theorem:

"Theorem 2: Let P be the average transmitter power, and suppose the noise is white thermal noise of power N in the band W. By sufficiently complicated encoding systems, it is possible to transmit binary digits at a rate

$$C = W \log_2 \left(\frac{P + N}{N} \right) \tag{2.2}$$

with as small a frequency of errors as desired. It is not possible by any encoding method to send at a higher rate and have an arbitrarily low frequency of errors."

Shannon further explained Theorem 2 clearly as:

"This shows that the rate $W \log_2 (P+N)/N$ measures in a sharply defined way the capacity of the channel for transmitting information. It is a rather surprising result, since

one would expect that reducing the frequency of errors would require reducing the rate of transmission, and that the rate must approach zero as the error frequency does. Actually, we can send at the rate C but reduce errors by using more involved encoding and longer delays at the transmitter and receiver. The transmitter will take long sequences of binary digits and represent this entire sequence by a particular signal function of long duration. The delay is required because the transmitter must wait for the full sequence before the signal is determined. Similarly, the receiver must wait for the full signal function before decoding into binary digits."

2.2.2. *Modified Shannon formula*

Shannon defined P in Eq. (2.2) (originally Eq. (2.19) in [8]) as the *"average transmitter power"*; the original Shannon formula can be interpreted as the maximum channel capacity of a transmitter. Practically, what we care about is the total number of messages received rather than the messages sent for any communication system. Of course, if the communication system is a wired system, like telephony, the channel capacity will be identical for both the transmitter and receiver if the transmitting losses of the wires can be ignored.

For any wireless communication systems, a message is carried by radio frequency (RF) waves, radiated from the transmitting antenna of the transmitter and received by the receiving antenna of the receiver. The RF waves propagate in the space, and this forms the wireless channel as depicted in Fig. 2.11.

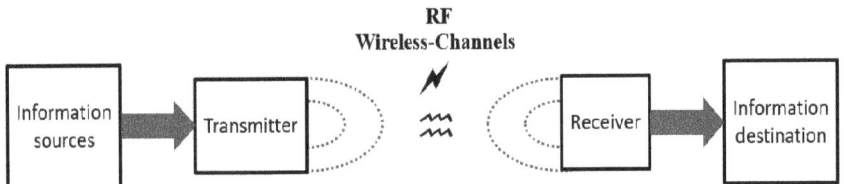

Fig. 2.11.   A general point-to-point wireless communication system.

In analyzing, designing, and implementing wireless communication systems, it is normally assumed:

- The transmitting and antennas are "point sources"
- The receiving antenna is located in the far-field region of the transmitting antenna

The far-field assumption indicates that the radio waves transmitted by the transmitting antenna is a scale wave and is no longer a vector wave governed by the Poynting vector.

A point source implies the radiation is omnidirectional, and the radiation wave is in the form of a spheric wave, as shown in Fig. 2.12, and the energy is radiated uniformly in all directions. If the total power radiated from the antenna is fixed as P, the energy density is proportional to $1/r^2$, where r is the distance to the point source. That means if the distance to the point source is doubled, the power density will reduce to a quarter (shown in Fig. 2.12). The received power of a receiver is proportional to the power density. That gives the so-called propagation-loss L in the free space with the operating

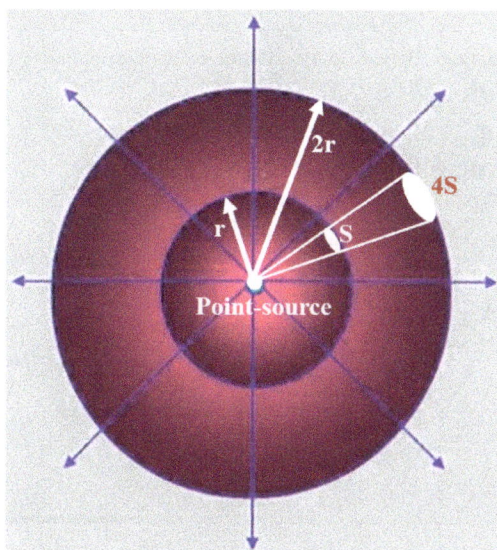

Fig. 2.12. Radiation pattern of a point source in the free space.

wavelength λ [9]:

$$L = \left(\frac{\lambda}{4\pi r}\right)^2 . \tag{2.3}$$

Equation (2.3) tells us that only **a very small portion of the total power P radiated by the transmitter will be received by the receiving antenna** in free space. In any practical environment, the actual path losses of the power will be much higher than that given by Eq. (2.3). The further away the receiver is from the transmitter, the less power the receiver receives. The effectiveness of a wireless communication system is how much of the messages is received by the receiver rather than the messages sent by the transmitter; otherwise, it will be meaningless. The most critical parameter of any wireless receiver is its sensitivity, which *"is defined as the minimum signal level that a receiver can detect with 'acceptable quality'. In the presence of excessive noise, the detected signal becomes unintelligible and carries little information"* [10]. The data rate strongly depends on how high the received power level for the receiver is with respect to the minimum detectable signal power level.

An antenna can be equivalent as a voltage source series with a resistor as shown in Fig. 2.13 [10, 11]:

This resistor Rant will generate thermal noise in terms of the power spectral density (PSD) [10]:

$$N_{ant} = 4kTWR_{ant} \tag{2.4}$$

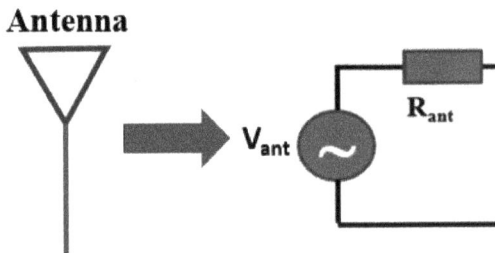

Fig. 2.13.   Equivalent circuit model of an antenna.

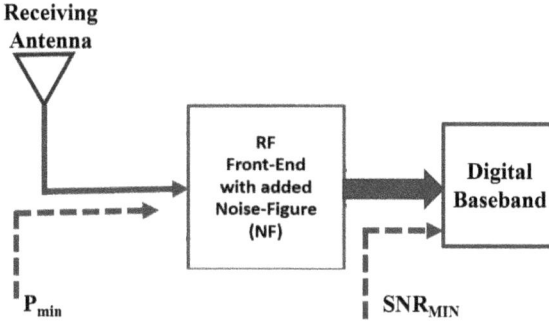

Fig. 2.14. Block diagram of a general wireless receiver.

Where T is the temperature in Kelvin, W is the bandwidth of the RF (here we use the same symbol to represent the frequency bandwidth as Shannon's original Eq. (2.19) in [8]), and k is the Boltzmann constant $k = 1.380649 \times 10^{-23}$ J/K.

Figure 2.14 illustrates a general receiver block diagram where the receiving antenna receives the RF signal in the air and transmits the received RF signal to the RF front-end; the RF front-end down-converts the RF signal with amplifications to the analog base band followed by an analog digital converter (ADC) with a certain signal-to-noise ratio (SNR), which must be higher than the minimum SNR (SNR_{MIN}) required by the digital baseband. Those devices and circuitries of the RF front-end contribute noise, which is described by the added noise-figure (NF) [10, 11]; thus, the minimum detectable power level P_{min} can be expressed as [10]:

$$P_{min}(mW) = kTW \times NF \times SNR_{MIN} \qquad (2.5)$$

or

$$P_{min}(dBm) = \log(kTW) + NF(dB) + SNR_{MIN}(dB)$$
$$\approx -174(dBm) + 20\log(W) + NF(dB) + SNR_{MIN}(dB). \qquad (2.6)$$

Equations (2.5) and (2.6) are the most important formula to design any RF front-end systems in order to plan and design the entire RF front-end. As long as the received power level P_R is higher than the

P_{min} defined in Eqs. (2.5) and (2.6), the communication between the transmitter and receiver can be established. The higher the P_R, the more messages sent from the transmitter will be received with fewer errors. Equations (2.5) and (2.6) provide the rule for the trade-off between the expected bit error rate (BER) and the difficulties of implementing the RF front-end hardware. For instance, the minimum detectable power level (sensitivity) of -107 dBm is set for GSM and IMT2000 to guarantee the minimum required SNR if the acceptable BER is equal and less than 2×10^{-2}. As mentioned above, if the received signal power level is below P_{min}, the wireless receiving system cannot detect any meaningful information because the BER will not be acceptable.

By replacing P and N with P_R and P_{min}, the original Shannon formula (2.2) can be re-written as:

$$C_R = W \log_2 \left(1 + \frac{P_R}{P_{min}} \right). \tag{2.7}$$

Here C_R is dedicated to the maximum channel capacity theoretically for a wireless receiver, while the original Shannon formula (2.2) is the maximum channel capacity for a transmitter.

As illustrated in Fig. 2.2, the received power P_R is only a portion of the transmitted power P. Let G_T be the gain of the transmitting antenna and G_R be the gain of the receiving antenna (for omnidirectional antenna $G = G_T = G_R = 1$). Then [12]:

$$P_R = PG_T G_R L, \tag{2.8}$$

where L is the pass-loss, for free space L is defined in Eq. (2.3). Re-writing (2.7) as

$$
\begin{aligned}
C_R &= W \times \log_2 \left(1 + \frac{PG_T G_R L}{kTW \times NF \times SNR_{MIN}} \right) \\
&= W \times \log_2 \left(1 + \frac{PG_T G_R}{kTW \times NF \times SNR_{MIN}} \left(\frac{\lambda}{4\pi r} \right)^2 \right),
\end{aligned}
\tag{2.9}
$$

where r is the distance from the transmitter to the receiver shown in Fig. 2.12. Equation (2.9) indicates that if **all parameters are fixed** except for the operating wavelength λ, the higher the radio frequency (shorter wavelength) is, the lower the receiver channel capacity is.

$$\frac{C_R(f)}{C_R(f_0)} = \frac{W(f)}{W(f_0)} \frac{\log_2\left[1 + \frac{PG_TG_R}{kTW(f)\times NF\times SNR_{MIN}}\left(\frac{\lambda}{4\pi r}\right)^2\right]}{\log_2\left[1 + \frac{PG_TG_R}{kTW(f_0)\times NF\times SNR_{MIN}}\left(\frac{\lambda_0}{4\pi r}\right)^2\right]}$$

$$= \frac{W(f)}{W(f_0)} \frac{\log_2\left[1 + \frac{PG_TG_R}{kTW(f_0)\times NF\times SNR_{MIN}}\left(\frac{\lambda_0}{4\pi r}\right)^2 \frac{W(f_0)}{W(f)}\left(\frac{\lambda}{\lambda_0}\right)^2\right]}{\log_2\left[1 + \frac{PG_TG_R}{kTW(f_0)\times NF\times SNR_{MIN}}\left(\frac{\lambda_0}{4\pi r}\right)^2\right]}$$

$$\frac{C_R(f)}{C_R(f_0)} = \frac{W(f)}{W(f_0)} \frac{\log_2\left[1 + A\frac{W(f_0)}{W(f)}\left(\frac{f_0}{f}\right)^2\right]}{\log_2[1 + A]}. \tag{2.10}$$

Equation (2.10) illustrates the channel capacity ratio between a lower radio frequency (f_0) and higher radio frequency (f) if all other parameters are the same (here denoted as constant A) for both frequencies.

For instance, the first-millimeter wave band assigned to wireless by ITU is a frequency range of 24.25–27.5 GHz with a center frequency (f_0) of 25.875 GHz and a total bandwidth of 3.25 GHz, while the first terahertz band for wireless is 275–296 GHz with the center frequency of 285.5 GHz and the total bandwidth of 21 GHz. Now

$$\frac{C_R(285.5\text{GHz})}{C_R(25.875\text{GHz})} = \frac{21}{3.25} \frac{\log_2\left[1 + \frac{A}{6.4615}\left(\frac{1}{11}\right)^2\right]}{\log_2[1 + A]}$$

$$= 6.4615\frac{\log_2\left[1 + \frac{A}{781.842}\right]}{\log_2[1 + A]}. \tag{2.11}$$

It is straightforward that if A is less than 1,000 (which is already not practical), the channel capacity in the terahertz band (275–296 GHz) is **much** less than the channel capacity in the millimeter-wave band (24.25–27.5 GHz). **The result shows that the channel capacity is controversial to the conventional belief where**

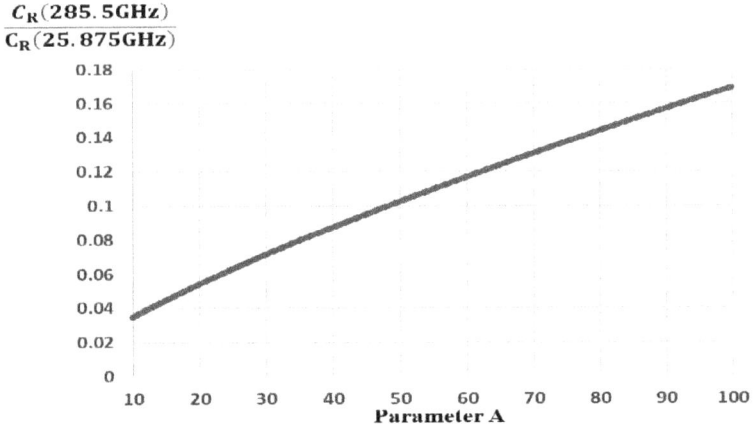

Fig. 2.15.　Comparisons of the channel capacity for millimeter-wave and terahertz bands.

higher RF with large bandwidth would have large channel capacity. Figure 2.15 presents the ratio of channel capacities in 25.875 and 285.5 GHz — it highlights that the channel capacity of using a terahertz band is less than that of using a millimeter-wave. While it is not practical for A = 100, A is less than 100 in most of the practical cases. For the same distance of communications, **higher RF will result in less channel capacity.**

Re-writing Eq. (2.9) as:

$$C_R = W \times \log_2 \left(1 + \frac{PG_TG_RL}{kTW \times NF \times SNR_{MIN}} \right)$$

$$= W \times \log_2 \left(1 + \frac{B}{SNR_{MIN}} \right). \qquad (2.12)$$

Equation (2.12) shows that for all given parameters except the minimum SNR required by the digital baseband (referring to Fig. 2.14), the receiving channel capacity for a wireless network is reversely proportional to SNR_{MIN}.

The bit error rate (BER) is required for every wireless system, and different modulation schemes provide different BERs, as illustrated in Fig. 2.16. Clearly, for the same BER, a higher order modulation scheme is required to achieve a better SNR. Referring to Fig. 2.16 again, Eq. (2.12) concludes that **the higher the SNR_{MIN} is, the**

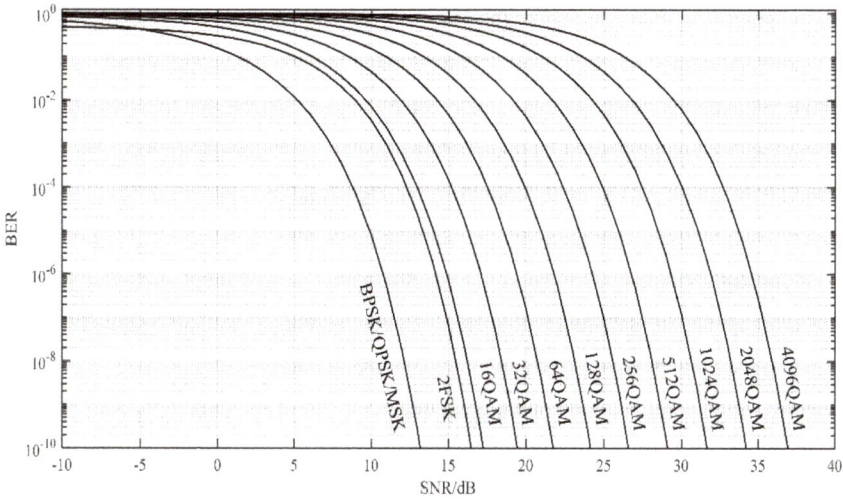

Fig. 2.16. The bit error rate (BER) vs. different SNRs.

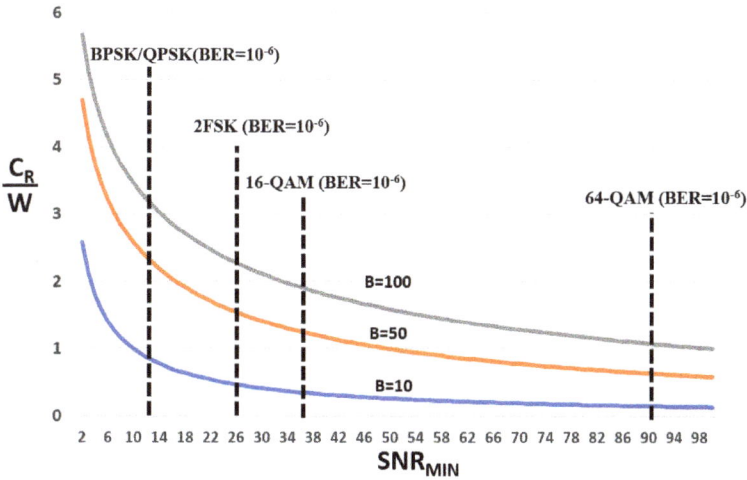

Fig. 2.17. Channel capacity vs. SNR$_{\mathrm{MIN}}$.

less the channel capacity, in contrast to the conventional statements in the literature.

The channel capacity versus the minimum required SNR is shown in Fig. 2.17 with all parameters fixed. For comparisons, different modulations with the same BER $= 10^{-6}$ have been marked in the graph. It indicates, for example, that the channel capacity for BPSK/QPSK is more than three times higher than the channel capacity for 64-QAM if all parameters are fixed (B $= 100$).

From Eq. (2.9), we can understand that there is no way to reduce the pass-losses because L is determined by physics. For a given hardware implementation, P_{min} is fixed too, and the only way to increase the channel capacity is to increase the gains for both transmitting and receiving antenna. Therefore, using directional antennas to replace omnidirectional ones can increase the channel capacity. It is known that antenna arrays can have higher gain, with antenna-array-based MIMO (multiple-input-multiple-output) becoming one of the key technologies for higher data-rate wireless communications.

Traditionally, it is assumed that a receiving antenna is located in the far-field region of the transmitting antenna as shown in Fig. 2.18 [12, 13]:

$$r > 2\frac{D^2}{\lambda}, \qquad (2.13)$$

where D is the antenna size and λ is the wavelength. Suppose we are using Wi-Fi in the office to connect the PC to the internet, and the PC is receiving the signal by its receiving antenna, which is assumed

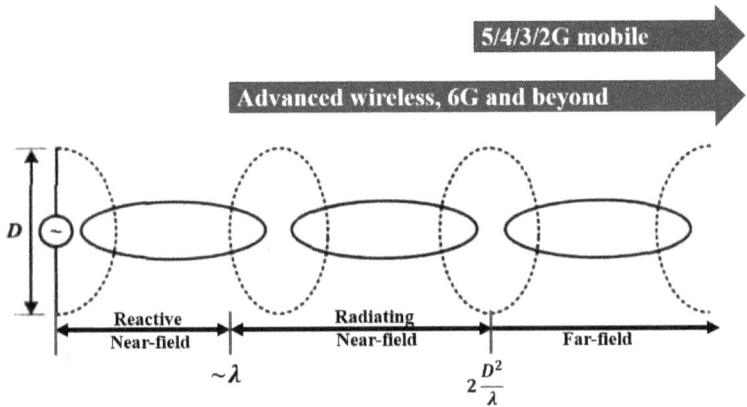

Fig. 2.18.   Fields for an antenna.

to be a point source. The point source is around 1.5 m above the floor, while the floor can be assumed to be a perfect ground. So, the radiation field above the up-half space can be calculated by a dipole, which is formed by the point source and its imaging point source. The distance between the point source and the imaging point source is 3 m now, which is antenna size D. The wavelength for Wi-Fi is about 0.12 m; thus, the far-field condition is greater than 150 m away from the transmitting antenna. Normally, a Wi-Fi hot-point is located roughly a few meters to a couple of ten meters away from the PC. It is definitely not 150 m away. Now, the new challenge for 6G wireless is to define the channel capacity for covering the region of radiative near-field, together with the traditional far-field.

In the far-field region, electromagnetic waves are scale waves with real wave impedance and quantity of energy. However, in the radiative near field, electromagnetic waves are still vector waves, and the energy carried by both electric field $\overrightarrow{\mathbf{E}}$ and magnetic field $\overrightarrow{\mathbf{H}}$ is a complex quantity because the electric/magnetic fields have a 90° shift, both in the time-domain and space-domain. For any antenna, it is a metallic component — the received electric field will induce the voltage, and the received magnetic field will induce the current too. In any case, the induced voltage and current will convert to the energy received in a complex form. We can also define the minimum detectable energy level for the receiver, so another kind of modified Shannon formula can be proposed for covering the region of the radiative near field [2]:

$$C_{\mathrm{R}} = W \log_2 \left(1 + \frac{\|\mathrm{P_R}\|}{\|\mathrm{P_{min}}\|} \right), \qquad (2.14)$$

where $\| \bullet \|$ represents the modulus of the complex quantity.

2.3. Conclusion

From a publication point of view, the US is still playing the dominant role in modern wireless communication research, and it holds the most critical technologies of modern wireless communications. European countries and Japan are declining in publications, which reflects that these regions are not paying much attention to wireless technologies as they did a few decades ago. China is a rising star in modern wireless communication research, which might benefit from its huge domestic market in wireless communications.

The classic Shannon formula gives the maximum available channel capacity without differentiating transmitting and receiving. This work introduces pass losses and antenna gains as well as the hardware parameters of the RF front-end into the channel capacity formula so that the channel capacity can be optimized according to the hardware performance, while the classic Shannon formula has no link with the hardware parameters, which makes the channel capacity analysis and hardware design separately.

The modified Shannon formula proves that higher modulation schemes will lead to lower channel capacity, which is controversial to conventional thinking. The modified Shannon formula shows it will not give more channel capacity by using higher RF compared to lower frequencies. An example has been given to comparing the channel capacity of a band of 25.875 GHz with a band of 285.5 GHz.

Finally, the radiative near-field case has been briefly discussed with a new modified Shannon formula, which might be critical for future advanced wireless technologies such as 6G and beyond.

References

[1] "Anniversary of Marconi's first patent," *Nature*, vol. 137, p. 940, Jun. 1936.
[2] J. Ma, "The challenging of radio access technology for 5G," *2019 IEEE MTT-S International Wireless Symposium (IWS)*, 2019, pp. 1–4. doi:10.1109/IEEE-IWS.2019.8803884
[3] J. Ma, "From MHz to THz: systems and applications," *IEEE Trans. Microwave Theor. Tech.*, vol. 70, no. 3, pp. 1459—460, Mar. 2022. doi: 10.1109/TMTT.2022.3151504
[4] "Cisco annual internet report (2018–2023)," White Paper. https://www.cisco.com/c/en/us/solutions/collateral/service-provider/visual-networking-index-vni/white-paper-c11-738429.html
[5] M. Agiwal, A. Roy and N. Saxena, "Next generation 5G wireless networks: A comprehensive survey," *IEEE Commun. Surv. Tutor.*, vol. 18, no. 3, pp. 1617–1655, 2016. doi: 10.1109/COMST.2016.2532458
[6] C. E. Shannon, "A mathematical theory of communication," *Bell Syst. Tech. J.*, vol. 27, no. 3, pp. 623–656, Oct. 1948.

[7] A. D. Wyner and S. Shamai, "Introduction to 'communication in the presence of noise'," *Proc. IEEE*, vol. 86, no. 2, pp. 442–446, Feb. 1998. doi: 10.1109/JPROC.1998.659496.

[8] C. E. Shannon, "Communication in the presence of noise," *Proc. Inst. Radio Eng.*, vol. 86, no. 1, pp. 10–21, Jan. 1949.

[9] "Recommendation ITU-R P.525-4: Calculation of free-space attenuation," International Telecommunication Union (ITU), Aug. 2019.

[10] B. Razavi, *RF Microelectronics*, 2nd edition. Prentice Hall, 2012.

[11] J. Ma, "Modified Shannon's capacity for wireless communication [Speaker's corner]," *IEEE Microw. Mag.*, vol. 22, no. 9, pp. 97–100, Sept. 2021. doi: 10.1109/MMM.2021.3086386.

[12] R. Johnson, *Antenna Engineering Handbook*, 2nd edition. McGraw-Hill, Inc., New York, 1984, pp. 1–12.

[13] J. Ma, *Third Generation Communication Systems: Future Developments and Advanced Topics*. Springer-Verlag, Heidelberg, Germany, 2003.

CONNECTIVITY

Any organizational transformation should adopt a balance approach. Specifically, we should find new equilibrium between Technology and Talent (T), between Idea and Innovation (I), between Competition and Collaboration (C), between Knowledge and Know-how (K), and between Science and Society (S) — TICKS. They are checkboxes to ensure successful transformation.

If TICKS does not work well, we must use STICK — between Speed and Scale (S), between Today and Tomorrow (T), between Individual and Institution (I), between Change and Cost (C), and between Knowledge creation and Knowledge transfer (K).

Kiat Seng Yeo

Chapter 3

Random Access: Connection-Free or Connection-Based?

Yayu Gao, Wen Zhan and Lin Dai

3.1. Introduction

As a fundamental type of multiple access, random access has been widely adopted in various wireless communication networks, including Wi-Fi and cellular networks [1]. With random access, each node determines when to access in a distributed manner. Due to the lack of coordination among nodes, transmission failures may occur when there are multiple concurrent transmissions.

According to whether a connection is established before data transmissions, random access schemes can be roughly categorized into two groups: connection-free or connection-based. As Fig. 3.1(a) illustrates, in conventional connection-free random access schemes, every data packet in nodes' queues needs to contend for channel access. The time wasted in each transmission failure is then determined by the length of a data packet, which could be quite large. In contrast, for connection-based random access, before each data packet transmission, each node first sends a short request to the receiver. The data packet is transmitted only after the request is successfully received, i.e., a connection is established, as Fig. 3.1(b) illustrates. As the length of a request is typically smaller than that of a data packet, the time wasted in transmission failures can

(a)

(b)

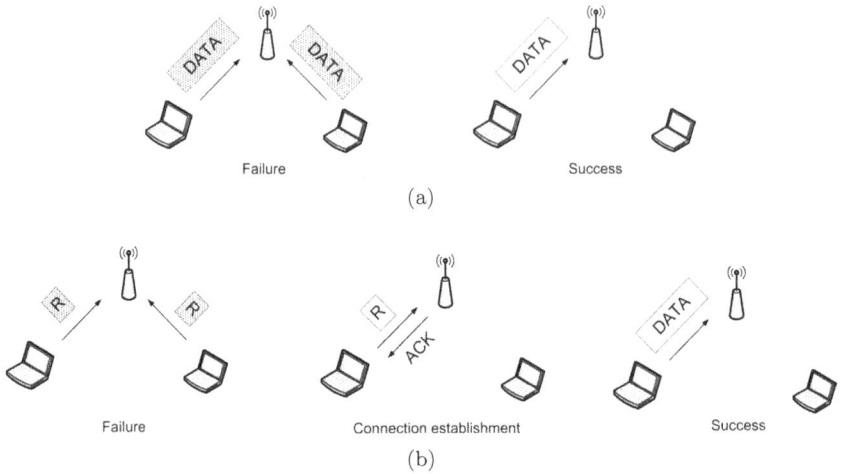

Fig. 3.1. (a) Connection-free random access. (b) Connection-based random access.

be reduced, though at the cost of extra overhead caused by the connection establishment process.

Both connection-free random access and connection-based random access have found wide applications in practical wireless communication networks. For instance, in Wi-Fi networks, the two access mechanisms of the distributed coordination function (DCF) [2], basic access and request-to-send/clear-to-send (RTS/CTS) access, are examples of connection-free random access and connection-based random access, respectively. With RTS/CTS access, before transmitting each data packet, each node would first send an RTS frame to establish a connection with the access point. By doing so, transmission failures only involve RTS frames that are much shorter than data packets. Nevertheless, the benefits brought by the reduction of transmission failure time may not always overweigh the overhead of connection establishment, especially when the transmission time of each data packet is small [3].

With the increasing popularity of Internet-of-Things, Machine-to-Machine (M2M) communications are expected to play a dominant role in next-generation wireless communication networks, where a typical scenario is that a massive number of machine-type devices

(MTDs) send short packets [4]. For M2M communications, there have been growing concerns over the necessity of connection establishment, with many advocating that connection-free (i.e., grant-free) random access protocols should be adopted [5, 6]. In cellular networks, for instance, a connection-based random access scheme has long been adopted [7]. That is, each device with packets to transmit first sends an access request to the Base Station (BS) via a four-way-handshake random access procedure to establish a connection, and then the BS assigns resource blocks for the device to clear its data queue. When supporting the M2M communications, the four-way-handshake random access procedure has been widely criticized for its low efficiency. Therefore, in the 5G New Radio (NR) standard [8], a connection-free random access scheme was further introduced, where each device sends one small packet within the random access procedure.

Intuitively, there exists a critical threshold of the transmission time of each data packet, only above which establishing a connection is beneficial. Characterization of such a threshold is of great practical importance, which, nevertheless, has received little attention. Except for a few early studies on the RTS threshold for Wi-Fi networks [3, 9–12], in general, characterizing the critical conditions for beneficial connection establishment in random-access networks remains largely unexplored.

For performance comparison of connection-free random access and connection-based random access, the key challenge lies in the proper modeling of random-access networks, with the overhead taken into consideration. Most existing theoretical models focus on connection-free random access, where the overhead is usually ignored by assuming instant acknowledgment [13–25]. For practical networks where the overhead is taken into account when evaluating the network performance, analytical models are usually customized for specific system configurations, and the differences in protocol details render it difficult to generalize the results from one to another [3, 9–12, 26–31].

In this chapter, we will provide a thorough study on modeling and throughput optimization of both connection-free random access and connection-based random access. We will start from two modeling methodologies, channel-centric modeling and node-centric modeling, of random-access networks in Section 3.2 and derive the maximum network throughput in Section 3.3. The optimal throughput performance of connection-based and connection-free random access is compared in Section 3.4. The analysis is further applied to Wi-Fi and 5G networks in Section 3.5. Finally, conclusions are summarized in Section 3.6.

3.2. Modeling: Channel-Centric vs. Node-Centric

3.2.1. *Network assumptions*

Consider a homogeneous slotted random-access network where nodes transmit packets to a common receiver over lossless channels, where each node is equipped with a buffer of infinite size and the maximum number of retransmission attempts for each packet is infinite. Suppose that the time axis is divided into multiple slots. All the nodes are synchronized and can start a transmission only at the beginning of a time slot. The classical collision model is assumed. That is, when multiple nodes transmit their packets simultaneously, a collision occurs, and none of them can be successfully decoded. A packet transmission is successful only if there are no concurrent transmissions.

3.2.2. *Modeling methodologies*

As Fig. 3.2 illustrates, a random-access network can be regarded as a multi-queue-single-server system. Based on whether the modeling focus is on the aggregate channel or each node's queue, the analytical

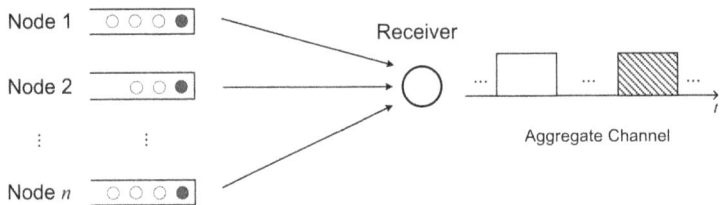

Fig. 3.2. A random-access network can be regarded as a multi-queue-single-server system.

models can be roughly divided into two categories: channel-centric and node-centric.

1) *Channel-centric models*: Channel-centric modeling can be traced back to Abramson's landmark paper [13] for Aloha networks and has been extended to more sophisticated CSMA networks [14]. Such approach captures the essence of contention among nodes by characterizing the aggregate traffic of nodes, and greatly simplifies the throughput analysis. It, nevertheless, ignores the behavior of each node's queue, which makes it hard to analyze the effect of backoff parameters on network performance, and therefore sheds little light on the optimal setting of node-level backoff parameters.

2) *Node-centric models*: To further characterize the queueing performance of random-access networks, a more elaborate model should be established for each node's queue in the network. With interactions among nodes' queues taken into consideration, nevertheless, the node-centric approach may lead to high modeling complexity. To establish a scalable node-centric model, a great deal of effort has been made to tackle the coupled queues using approximate techniques, which gradually come to a consensus: each node's queue can be approximately considered as an independent queueing system with identically distributed service time if the number of nodes is large. For random-access networks, the key to obtaining the service time distribution then lies in proper characterization of the state transition of the head-of-line (HOL) packet of each node's queue [25].

In the following sections, we will introduce a channel-centric model and a node-centric model for random-access networks, respectively.

3.2.3. *Channel-centric model*

In this section, we focus on channel-centric modeling. Specifically, with the collision model at the receiver side, the aggregate channel has three states: successful transmission (State S), collision (State C) and idle (State I). The state transition of the aggregate channel can be modeled as a discrete-time Markov renewal process $(\mathbf{X}, \mathbf{V}) = \{(X_j, V_j), j = 0, 1, \ldots\}$, where X_j denotes the channel state at the

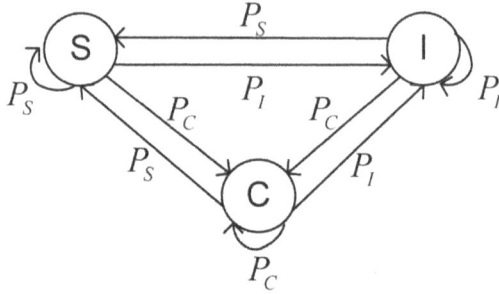

Fig. 3.3.   Embedded Markov chain $\{X_j\}$ of the state transition process of the aggregate channel.

j-th transition and V_j denotes the epoch at which the j-th transition occurs.

The embedded Markov chain $\mathbf{X} = \{X_j\}$ is shown in Fig. 3.3, where P_S, P_C and P_I denote the transition probabilities that the aggregate channel enters States S, C, and I, respectively. Assume that at each time slot the aggregate traffic, i.e., the total number of transmitted packets, is a Poisson random variable with rate G. The probabilities that the aggregate channel enters State I, State S, and State C can then be obtained as

$$P_I = e^{-G}, P_S = Ge^{-G}, P_C = 1 - P_I - P_S = 1 - e^{-G} - Ge^{-G}, \quad (3.1)$$

respectively.

Let τ_I, τ_S, and τ_C denote the holding time at State I, State S, and State C, respectively, in the unit of time slots. The limiting state probabilities of the Markov renewal process $(\mathbf{X}, \mathbf{V})$ can then be written as

$$\tilde{\pi}_i = \frac{\pi_i \cdot \tau_i}{\sum\limits_{j \in \{I,S,C\}} \pi_j \cdot \tau_j}, i \in \{I, S, C\}, \quad (3.2)$$

where $\{\pi_i\}_{i \in \{I,S,C\}}$ denotes the limiting state probability distribution of the embedded Markov chain shown in Fig. 3.3, which can be easily obtained as $\pi_S = P_S$, $\pi_I = P_I$, and $\pi_C = P_C$. Without loss of generality, we normalize the holding time at the idle state, i.e., State I, as $\tau_I = 1$ time slot. The probability of the aggregate channel

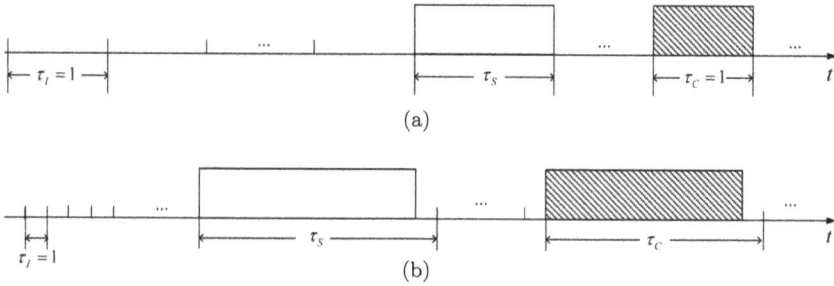

Fig. 3.4.   States of the aggregate channel. (a) Without sensing (Aloha). (b) With sensing (CSMA).

being at the successful transmission state, i.e., State S, can then be obtained by combining (3.1) and (3.2) as

$$\tilde{\pi}_S = \frac{\tau_S \cdot Ge^{-G}}{(\tau_S - \tau_C) \cdot Ge^{-G} - (\tau_C - 1) \cdot e^{-G} + \tau_C}. \tag{3.3}$$

We can see from (3.3) that $\tilde{\pi}_S$ is determined by the holding time of the aggregate channel at the collision state τ_C and success state τ_S, which further depends on whether nodes sense the channel or not, i.e., the random access scheme is Aloha or CSMA:

- For Aloha, nodes do not sense the channel before their transmissions, as Fig. 3.4(a) illustrates. As a result, they cannot distinguish between collision and idle states of the channel. The holding time of the aggregate channel at the collision state τ_C has to be equal to the holding time at the idle state $\tau_I = 1$ time slot.
- For CSMA, as sensing is available, the length of an idle time slot is determined by the sensing time, which is typically much smaller than the collision time. In that case, the holding time at the collision state τ_C can be much larger than the holding time at the idle state $\tau_I = 1$ time slot, as Fig. 3.4(b) illustrates.

3.2.4. *Node-centric model*

Due to ignoring the queueing behavior of each node, the channel-centric models, in general, shed little light on the optimal setting of node-level backoff parameters. In general, for a multi-queue

(a)

(b)

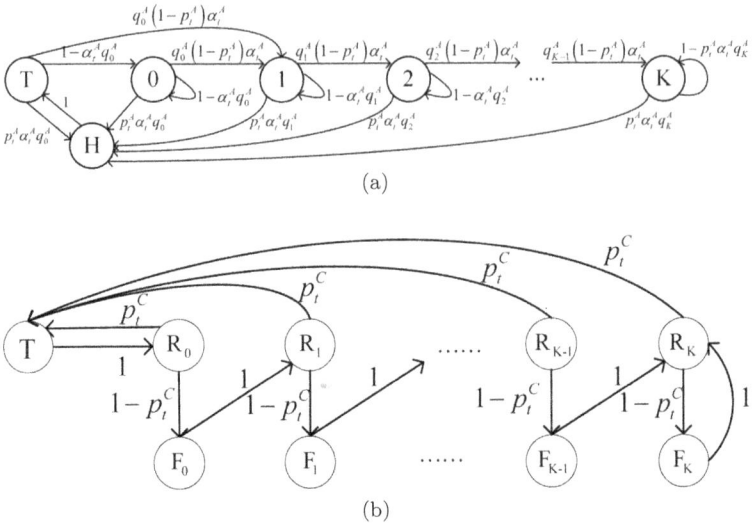

Fig. 3.5.   Embedded Markov chain of the state transition process of HOL packets with (a) Aloha and (b) CSMA [25].

single-server system, the service rate is determined by the aggregate activities of all the HOL packets.

Specifically, for both Aloha and CSMA networks, a discrete-time Markov renewal process $(\boldsymbol{X}, \boldsymbol{V}) = \{(X_j, V_j), j = 0, 1, \ldots\}$ can be established to model the behavior of each HOL packet, where X_j denotes the state of a tagged HOL packet at the j-th transition and V_j denotes the epoch at which the j-th transition occurs [25]. The difference, however, lies in the state characterization. Figure 3.5(a) and 3.5(b) show the embedded Markov chain $\boldsymbol{X}^A = \{X_j^A\}$ and $\boldsymbol{X}^C = \{X_j^C\}$ for Aloha and CSMA, respectively.

In both cases, State T denotes the successful transmission state. For Aloha, a fresh HOL packet is initially in State T, and moves to State H if it transmits successfully, where State H denotes the data transmission state and α_t^A denotes the probability that no node is in State H at slot t. Otherwise, it moves to State 0 if it does not transmit, or State 1 if it transmits but unsuccessfully. The HOL packet moves to State $i+1$ every time when it experiences a transmission failure until i reaches the cutoff phase K, as Fig. 3.5(a) shows. For CSMA, the sensing states need to further be considered and distinguished from transmission states. Specifically, as Fig. 3.5(b) illustrates, a fresh HOL packet is initially in State R_0, and moves to State T if it transmits successfully, or to State F_0 if its transmission fails. By comparing Fig. 3.5(a) with Fig. 3.5(b), we can clearly see that the retransmission state $i \in \{0, \ldots, K\}$ in the Aloha case is split into two states in the CSMA case, with R_i for sensing and waiting to request, and F_i for unsuccessful transmission.

In Fig. 3.5, p_t^A denotes the probability of successful transmission of HOL packets with Aloha at time slot t, and p_t^C denotes the probability of successful transmission of HOL packets with CSMA at mini slot t, given that the channel is idle at mini slot $t - 1$, respectively. Let $\lim_{t\to\infty} p_t^A = p^A$ and $\lim_{t\to\infty} p_t^C = p^C$ denote the steady-state probability of successful transmission of HOL packets with Aloha and CSMA, respectively. Moreover, let q_i^A and q_i^C denote the transmission probability of a State-i HOL packet with Aloha and a State-R_i HOL packet with CSMA, respectively, $i = 0, \ldots, K$. The steady-state probability distribution $\{\pi_i^A\}$ and $\{\pi_i^C\}$ of the embedded Markov chains $\boldsymbol{X}^A$ and $\boldsymbol{X}^C$ can be obtained as

$$
\begin{cases}
\pi_0^A = \dfrac{(1 - q_0^A \alpha_A)}{q_0^A \alpha_A} \pi_T^A, \\[2mm]
\pi_H^A = \pi_T^A, \\[2mm]
\pi_i^A = \dfrac{(1 - p^A)^i}{q_i^A \alpha_A} \pi_T^A, i = 1, \ldots, K - 1, \\[2mm]
\pi_K^A = \dfrac{(1 - p^A)^K}{p^A q_K^A \alpha_A} \pi_T^A,
\end{cases}
\tag{3.4}
$$

where

$$
\alpha_A = \frac{1}{1 - (\tau_S - 1) p^A \ln p^A},
\tag{3.5}
$$

and

$$
\begin{cases}
\pi_{R_i}^C = (1 - p^C)^i \pi_T^C, i = 0, \ldots, K - 1, \\[2mm]
\pi_{R_K}^C = \dfrac{(1 - p^C)^K}{p^C} \pi_T^C, \\[2mm]
\pi_{F_i}^C = (1 - p^C) \pi_{R_K}^C, i = 0, \ldots, K.
\end{cases}
\tag{3.6}
$$

Let τ_i denote the mean holding time in State i, respectively, $i \in S$, in the unit of time slots. Note that for a Markov renewal process $(\boldsymbol{X},\boldsymbol{V})$, the limiting state probabilities can be obtained as

$$\tilde{\pi}_i = \frac{\pi_j \cdot \tau_j}{\sum_{j \in S} \pi_j \cdot \tau_j}, \quad i \in S, \tag{3.7}$$

where S is the state space of $\boldsymbol{X}$. For Aloha, the holding time of State i, $i \in \{T, 0, \ldots, K\}$,

$$\tau_i^A = 1, \ i = T, 0, \ldots, K, \tag{3.8}$$

while the holding time of State H

$$\tau_H^A = \tau_S - 1. \tag{3.9}$$

Note that τ_S is the holding time of each successful transmission. When $\tau_S = 1$, States T and H in Fig. 3.5(a) merge into one state, i.e., State T.

For CSMA, the mean holding time of State R_i, $i \in \{0, \ldots, K\}$, is further determined by the steady-state probability of channel being idle, α_C, which can be written as [25]

$$\tau_{R_i}^C = \frac{a}{\alpha_C q_i^C}, i = 0, \ldots, K, \tag{3.10}$$

where a is the length of a mini-slot and

$$\alpha_C = \frac{1}{1 + \tau_F^C - \tau_F^C p^C - (\tau_T^C - \tau_F^C) p^C \ln p^C}. \tag{3.11}$$

Note that for Aloha, each node's queue has a successful output if and only if the HOL packet stays at State T or State H. Therefore, the service rate of each node's queue is $\tilde{\pi}_T^A + \tilde{\pi}_H^A$, where $\tilde{\pi}_T^A$ and $\tilde{\pi}_H^A$ are the probabilities of the HOL packet being at State T and State H, respectively. By combining (3.4)–(3.5) and (3.7)–(3.9), we have the service rate

$$\tilde{\pi}_T^A + \tilde{\pi}_H^A = \frac{\tau_S^A}{\tau_S^A - 1 + \sum_{i=0}^{K-1} \frac{(1-p^A)^i}{q_i^A \alpha_A} + \frac{(1-p^A)^K}{p^A q_K^A \alpha_A}}, \tag{3.12}$$

where p^A denotes the limiting probability of successful transmission of HOL packets with Aloha.

On the other hand, for CSMA, each node's queue has a successful output if and only if the HOL packet stays at State T. Therefore, the probability $\tilde{\pi}_T^C$ of the HOL packet being at State T is the service rate of each node's queue, which can be written as

$$\tilde{\pi}_T^C = \frac{\tau_T^C}{\tau_T^C + \tau_F^C \frac{1-p^C}{p^C} + \frac{a}{\alpha_C} \cdot \left(\sum_{i=0}^{K-1} \frac{(1-p^C)^i}{q_i^C} + \frac{(1-p^C)^K}{p^C q_K^C} \right)}, \tag{3.13}$$

by combining (3.6)–(3.7) and (3.10)–(3.11), where p^C denotes the limiting probability of successful transmission of HOL packets with CSMA, given that the channel is idle.

3.3. Throughput Optimization

Due to uncoordinated transmissions of nodes, the number of successfully decoded packets varies from time to time. In random-access networks, the average number of successfully decoded packets per time slot is an important performance metric, which is referred to as the network throughput, denoted as $\hat{\lambda}_{out}$. With the collision model, one packet at most can be successfully decoded in each time slot. Therefore, the network throughput is also the percentage of the time that a random-access network has a successful packet transmission, which reflects the access efficiency.

It has been shown in our previous works [25] that the throughput performance and the network steady-state operating point of random access networks is determined by whether the network is in unsaturated or saturated conditions. Specifically, if the network is not saturated, the network throughput $\hat{\lambda}_{out}$ is always equal to the aggregate input rate and independent of the backoff parameters. On the other hand, if the aggregate input rate increases and all the nodes are busy with non-empty queues, the network becomes saturated. In this case, the network throughput $\hat{\lambda}_{out}$ is determined by the aggregate service rate, which is closely dependent on the backoff parameters of nodes.

The focus of this chapter is on the maximum throughput of random-access networks. Therefore, the saturated condition is of more interest, with which the network throughput is pushed to the limit. In the following, the maximum throughput will be derived via the two modeling approaches introduced in Section 3.2.2, channel-centric and node-centric, respectively. It will be shown that with the channel-centric model, the throughput analysis is greatly simplified, yet by excluding each node's queue, it is difficult to obtain the optimal setting of backoff parameters for achieving the maximum network throughput. In contrast, with the node-centric model, the maximum network throughput and the corresponding optimal backoff parameters can be obtained.

3.3.1. *Throughput optimization via channel-centric model*

Denote the throughput of Aloha network and CSMA network as $\hat{\lambda}_{out}^A$ and $\hat{\lambda}_{out}^C$, respectively. With the channel-centric model shown in Section 3.2.3, the aggregate channel is in State S when there is a successful transmission. Therefore, the network throughput $\hat{\lambda}_{out}^A$ with Aloha is equal to the steady-state probability $\tilde{\pi}_S$ that the aggregate channel is at State S, which is given by (3.3) as

$$\hat{\lambda}_{out}^A = \frac{\tau_S \cdot Ge^{-G}}{(\tau_S - 1) \cdot Ge^{-G} + 1}. \tag{3.14}$$

With CSMA, as nodes can sense the channel and withhold their transmissions when the channel is sensed busy, there is no state transition for sure at the time slot following a successful transmission or a collision. Therefore, as Fig. 3.4(b) shows, when the aggregate channel is at State S, only $\tau_S - 1$ out of τ_S time slots are with successful transmission. The network throughput $\hat{\lambda}_{out}^C$ is then given by $\frac{\tau_S-1}{\tau_S} \cdot \tilde{\pi}_S$, which can be written as

$$\hat{\lambda}_{out}^C = \frac{(\tau_S - 1) \cdot Ge^{-G}}{(\tau_S - \tau_C) \cdot Ge^{-G} - (\tau_C - 1) \cdot e^{-G} + \tau_C}, \tag{3.15}$$

according to (3.3). In both cases, the network throughput is a function of the aggregate traffic rate G. Define the maximum network throughput as $\hat{\lambda}_{max} = \max_G \hat{\lambda}_{out}$. It can be obtained from (3.14) and (3.15) that

$$\hat{\lambda}_{max}^A = \frac{\tau_S}{\tau_S - 1 + e}, \tag{3.16}$$

which is achieved when $G^{*,A} = 1$, and

$$\hat{\lambda}_{max}^C = \frac{\tau_S - 1}{\tau_S - \tau_C - (\tau_C - 1)\mathbb{W}_0^{-1}\left(-\frac{\tau_C-1}{\tau_C} \cdot e^{-1}\right)}, \tag{3.17}$$

which is achieved when $G^{*,C} = 1 + \mathbb{W}_0\left(-\frac{\tau_C-1}{\tau_C} \cdot e^{-1}\right)$.

A closer look at (3.16) and (3.17) shows that the maximum network throughput $\hat{\lambda}_{max}$ is only determined by the holding time of each successful transmission τ_S for Aloha and the holding times of each successful transmission and collision τ_S and τ_C for CSMA, respectively. To achieve $\hat{\lambda}_{max}$, the aggregate traffic G should be optimally set as 1 for Aloha and $1 + \mathbb{W}_0\left(-\frac{\tau_C-1}{\tau_C} \cdot e^{-1}\right)$ for CSMA, respectively. Due to the missing connection between G and the node-level backoff parameters, nevertheless, tuning the system parameters to achieve it remains unknown. In the following subsection, we will show how to further derive the optimal backoff parameters for achieving the maximum network throughput based on the node-centric model.

3.3.2. *Throughput optimization via node-centric model*

For each node, its throughput is equal to the service rate in saturated conditions. Based on the node-centric model in Section 3.2.4, for

Aloha, we have $\hat{\lambda}_{out}^A = n_A \left(\tilde{\pi}_T^A + \tilde{\pi}_H^A \right)$, which can be obtained as

$$\hat{\lambda}_{out}^A = -\frac{\tau_T^A p^A \ln p^A}{1 - (\tau_S - 1)p^A \ln p^A}, \tag{3.18}$$

according to (3.12), where the steady-state probability of successful transmission of HOL packets in saturated conditions p^A is the non-zero root of the following fixed-point equation [32]:

$$p^A = \exp \left\{ -\frac{n_A}{\sum_{i=0}^{K-1} \frac{p^A (1-p^A)^i}{q_i^A} + \frac{(1-p^A)^K}{q_K^A}} \right\}. \tag{3.19}$$

For CSMA, we have $\hat{\lambda}_{out}^C = n_C \tilde{\pi}_T^C$, which can be obtained as

$$\hat{\lambda}_{out}^C = \frac{-\tau_T^C p^C \ln p^C}{1 + \tau_F^C - \tau_F^C p^C - (\tau_T^C - \tau_F^C)p^C \ln p^C}, \tag{3.20}$$

according to (3.13), where the steady-state probability of successful transmissions of HOL packets in saturated conditions p^C is the non-zero root of the following fixed-point equation [25]:

$$p^C = \exp \left\{ -\frac{n_C}{\sum_{i=0}^{K-1} \frac{p^C (1-p^C)^i}{q_i^C} + \frac{(1-p^C)^K}{q_K^C}} \right\}. \tag{3.21}$$

In both cases, the network throughput is determined by the network steady-state point p^A or p^C. Define the maximum network throughput as $\hat{\lambda}_{\max} = \max_p \hat{\lambda}_{out}$. It can be obtained from (3.18) and (3.20) as

$$\hat{\lambda}_{\max}^A = \frac{\tau_S}{\tau_S - 1 + e}, \tag{3.22}$$

which is achieved when

$$p^A = p^{*,A} = e^{-1}, \tag{3.23}$$

and

$$\hat{\lambda}_{\max}^C = \frac{-\tau_T^C \mathbb{W}_0\left(-\frac{1}{e(1+1/\tau_F^C)}\right)}{\tau_F^C - (\tau_T^C - \tau_F^C)\mathbb{W}_0\left(-\frac{1}{e(1+1/\tau_F^C)}\right)}, \tag{3.24}$$

which is achieved when

$$p^C = p^{*,C} = -(1 + 1/\tau_F^C)\mathbb{W}_0\left(-\frac{1}{e(1+1/\tau_F^C)}\right), \tag{3.25}$$

respectively.

By comparing (3.16) and (3.17), (3.22) and (3.24), we can see that the maximum throughputs derived via the channel-centric model and the node-centric model are the same. To achieve the maximum network throughput, (3.23) and (3.25) show the corresponding optimal network steady-state point p^* for Aloha and CSMA, respectively. Note that the network steady-state point p has been derived as a function of the transmission probabilities $\{q_i\}$ in (3.19) and (3.21) for Aloha and CSMA, respectively. By combining (3.19) with (3.23) and (3.21) with (3.25), the optimal transmission probability $\{q_i^{*,A}\}$ and $\{q_i^{*,C}\}$ for achieving the maximum network throughput of Aloha and CSMA can be obtained.

Specifically, by combining (3.19) and (3.23), it can be obtained that to achieve $\hat{\lambda}_{\max}^A$, the optimal transmission probabilities $\{q_i^{*,A}\}$ for each node in Aloha should satisfy

$$\frac{n_A}{\sum_{i=0}^{K-1}\frac{(1-e^{-1})^i e^{-1}}{q_i^{*,A}} + \frac{(1-e^{-1})^K}{q_K^{*,A}}} = 1. \tag{3.26}$$

When $K = 0$, the optimal transmission probability $q_0^{*,A} = \frac{1}{n_A}$.

On the other hand, it can be obtained that to achieve $\hat{\lambda}_{\max}^C$, the optimal transmission probabilities $\{q_i^{*,C}\}$ for each node in CSMA should satisfy

$$\sum_{i=0}^{K-1}\frac{p^{*,C}(1-p^{*,C})^i}{q_i^{*,C}} + \frac{(1-p^{*,C})^K}{q_K^{*,C}} = \frac{n_C}{-\ln p^{*,C}}, \tag{3.27}$$

according to (3.21), where $p^{*,C}$ is given in (3.25). When $K = 0$, the optimal transmission probability $q_0^{*,C} = \frac{-\ln p^{*,C}}{n_C}$.

3.4. Maximum Effective Throughput

Note that the network throughput evaluates how efficient the time is used for successful transmissions. However, in each successful transmission, there exists certain overhead that do not carry any information. With connection-free random access, for instance, as Fig. 3.6(a) illustrates, the receiver broadcasts an acknowledgment (ACK) to indicate that the transmitted packet is correctly decoded. If nodes wait for a certain amount of time without receiving an ACK, they infer that the preceding transmission has failed; with connection-based random access, on the other hand, as Fig. 3.6(b) illustrates, each node sends a request to the receiver first and transmits its data packet only after the request is successfully received, i.e., a connection is established. Compared to connection-free random access, the collision time becomes shorter as it is determined by the length of the request, which is usually much smaller than that of a data packet.

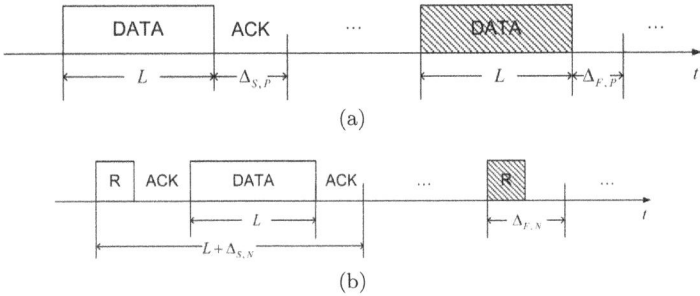

Fig. 3.6.   Graphic illustration of (a) connection-free and (b) connection-based random-access networks.

To compare the throughput performance, we need to further take the overhead into account. Let L, Δ_S and Δ_F denote the transmission time of a data packet, the overhead time for each successful or failed transmission, respectively, all in the unit of seconds. Δ_S is usually larger than Δ_F, both of which are much smaller than L. Define the effective throughput η_{out} as the fraction of time that is spent on the data payload transmission, which can be written as $\eta_{out} = \frac{L}{L+\Delta_S} \cdot \hat{\lambda}_{out}$. It is clear that the maximum effective throughput is given by

$$\eta_{\max} = \frac{L}{L + \Delta_S} \cdot \hat{\lambda}_{\max}. \tag{3.28}$$

3.4.1. *Connection-free random access*

For connection-free random access, every data packet needs to contend. If sensing is not available, then the holding time of the aggregate channel in the successful transmission state τ_S is equal to that in the collision state τ_C, with $\tau_S = \tau_C = \tau_I = 1$ time slot. The maximum effective throughput of connection-free Aloha can easily be obtained from (3.16) and (3.28) as

$$\eta_{\max}^{A,P} = \frac{\gamma_P}{1 + \gamma_P} \cdot e^{-1}, \tag{3.29}$$

where $\gamma_P = \frac{L}{\Delta_{S,P}}$.

On the other hand, if nodes can sense the channel before transmission, then the length of each time slot is determined by the sensing time, which is denoted as σ_C (in the unit of seconds). In this case, we can obtain from Fig. 3.4(b) and Fig. 3.6(a) that the holding time (in the unit of time slots) of the aggregate channel in the successful transmission state is $\tau_S = \frac{L + \Delta_{S,P}}{\sigma_C} + 1$, and the holding time (in the unit of time slots) of the aggregate channel in the collision state is $\tau_C = \frac{L + \Delta_{F,P}}{\sigma_C} + 1$. The maximum effective throughput of connection-free CSMA can then be obtained from (3.17) and (3.28) as

$$\eta_{\max}^{C,P} = \frac{-\gamma_P \cdot \mathbb{W}_0 \left(-\frac{\gamma_P + \delta_P}{e \cdot \left(\gamma_P + \delta_P + \frac{\sigma_C}{\Delta_{S,P}} \right)} \right)}{\gamma_P + \delta_P - (1 - \delta_P) \mathbb{W}_0 \left(\frac{-\gamma_P - \delta_P}{e \cdot \left(\gamma_P + \delta_P + \frac{\sigma_C}{\Delta_{S,P}} \right)} \right)}, \tag{3.30}$$

where $\delta_P = \frac{\Delta_{F,P}}{\Delta_{S,P}}$.

3.4.2. *Connection-based random access*

With connection-based random access, each node would send a short request to establish a connection with the receiver before transmitting the data packet. By doing so, the holding time of the aggregate channel in the successful transmission state τ_S can be much larger than that in the collision state τ_C. Without sensing, we have $\tau_C = \tau_I = 1$ time slot. We can further obtain from Fig. 3.4(a) and Fig. 3.6(b) that $\tau_S = \frac{L+\Delta_{S,N}}{\Delta_{F,N}}$. The maximum effective throughput of connection-based Aloha can then be obtained from (3.16) and (3.28) as

$$\eta_{\max}^{A,N} = \frac{\gamma_N}{\gamma_N + 1 + (e-1)\delta_N}, \tag{3.31}$$

where $\gamma_N = \frac{L}{\Delta_{S,N}}$ and $\delta_N = \frac{\Delta_{F,N}}{\Delta_{S,N}}$.

With sensing, we can obtain from Fig. 3.4(b) and Fig. 3.6(b) that the holding time of the aggregate channel in the successful transmission state is $\tau_S = \frac{L+\Delta_{S,N}}{\sigma_C} + 1$, and the holding time of the aggregate channel in the collision state is $\tau_C = \frac{\Delta_{F,N}}{\sigma_C} + 1$. The maximum effective throughput of connection-based CSMA can then be obtained from (3.17) and (3.28) as

$$\eta_{\max}^{C,N} = \frac{-\gamma_N \cdot \mathbb{W}_0\left(-\dfrac{\delta_N}{e\left(\delta_N + \frac{\sigma_C}{\Delta_{S,N}}\right)}\right)}{\delta_N - (\gamma_N + 1 - \delta_N)\,\mathbb{W}_0\left(-\dfrac{\delta_N}{e\left(\delta_N + \frac{\sigma_C}{\Delta_{S,N}}\right)}\right)}. \tag{3.32}$$

3.4.3. *Connection-establishment threshold*

In Sections 3.4.1 and 3.4.2, the maximum effective throughput in both connection-free random access and connection-based random access cases has been derived as a function of the data packet transmission time L and the overhead time for each successful transmission Δ_S and failed transmission Δ_F. Intuitively, the overhead time for each successful transmission of connection-based random

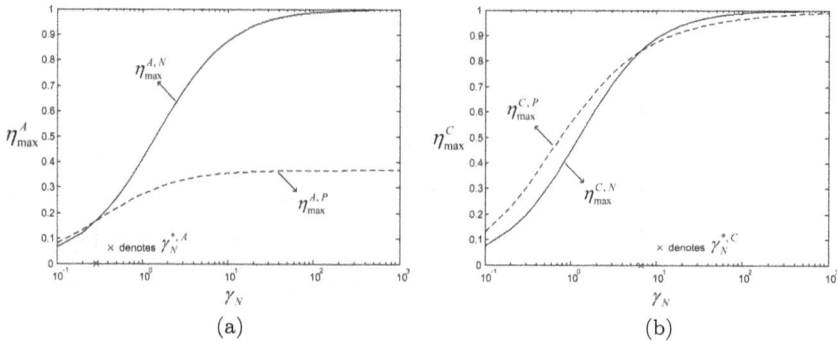

Fig. 3.7. (a) Maximum effective throughput $\eta_{\max}^A$ of Aloha networks and (b) maximum effective throughput $\eta_{\max}^C$ of CSMA networks versus $\gamma_N = \frac{L}{\Delta_{S,N}} \cdot \delta_P = 1/2$. $\delta_N = 1/4$. $\Delta_{S,N}/\sigma_C = 20$. $\Delta_{S,N}/\Delta_{S,P} = 2$.

access is larger than that of connection-free random access due to the connection establishment process, i.e., $\Delta_{S,N} > \Delta_{S,P}$. The collision time of connection-based random access, on the other hand, is shorter than that of connection-free random access, i.e., $\Delta_{F,N} < L + \Delta_{F,P}$. Whether the overhead of establishing a connection overweighs its benefits clearly depends on the data packet transmission time L.

As Fig. 3.7 shows, for both Aloha and CSMA networks, the connection-based maximum effective throughput exceeds the connection-free maximum effective throughput only when $\gamma_N = \frac{L}{\Delta_{S,N}}$ is sufficiently large. For large γ_N, in contrast to Aloha networks, where the maximum effective throughput can be significantly increased by connection establishment, only a slight improvement is observed in CSMA networks. Intuitively, if each node can perfectly sense the channel, the collision probability can be effectively diminished by reducing the slot length σ_C, i.e., the sensing time. The throughput gain of connection-based CSMA over connection-free CSMA, which comes from the reduction of collision time, thus becomes marginal.

To find out when connection-based random access outperforms connection-free random access, let us define γ_N^* as the threshold for beneficial connection establishment, i.e., $\eta_{\max}^N \geq \eta_{\max}^P$ if $\gamma_N \geq \gamma_N^*$. By combining (3.29) and (3.31), (3.30) and (3.32), the connection-establishment threshold $\gamma_N^{*,A}$ and $\gamma_N^{*,C}$ for Aloha and CSMA networks can be obtained as

$$\gamma_N^{*,A} = \frac{1 - e \cdot \frac{\Delta_{S,P}}{\Delta_{S,N}}}{e - 1} + \delta_N, \tag{3.33}$$

and

$$\gamma_N^{*,C} = \frac{\left(\frac{-\sigma_C}{\Delta_{S,N}} - (b - \frac{\sigma_C}{\Delta_{S,N}})\mathbb{W}_0\left(\frac{-b}{b - \frac{\sigma_C}{\Delta_{S,N}}}\exp\left\{\frac{-b}{b - \frac{\sigma_C}{\Delta_{S,N}}}\right\}\right)\right)}{\left(1 + (1 - \frac{\sigma_C}{\Delta_{S,N} \cdot b})\mathbb{W}_0\left(\frac{-b}{b - \frac{\sigma_C}{\Delta_{S,N}}}\exp\left\{\frac{-b}{b - \frac{\sigma_C}{\Delta_{S,N}}}\right\}\right)\right)}$$

$$-\delta_P \cdot \frac{\Delta_{S,P}}{\Delta_{S,N}}, \tag{3.34}$$

where $b = -\dfrac{\delta_N}{\mathbb{W}_0\left(-\dfrac{\delta_N}{e(\delta_N + \frac{\sigma_C}{\Delta_{S,N}})}\right)} + 1 - \dfrac{\Delta_{S,P}}{\Delta_{S,N}} - \delta_N$, respectively.

Figure 3.8 illustrates how the thresholds $\gamma_N^{*,A}$ and $\gamma_N^{*,C}$ vary with $\frac{\Delta_{S,N}}{\Delta_{S,P}}$, the ratio of the overhead time for each successful transmission in the connection-based case to that in the connection-free case. With an increasing $\frac{\Delta_{S,N}}{\Delta_{S,P}}$, that is, a higher cost of establishing a connection, a larger data packet transmission time is required

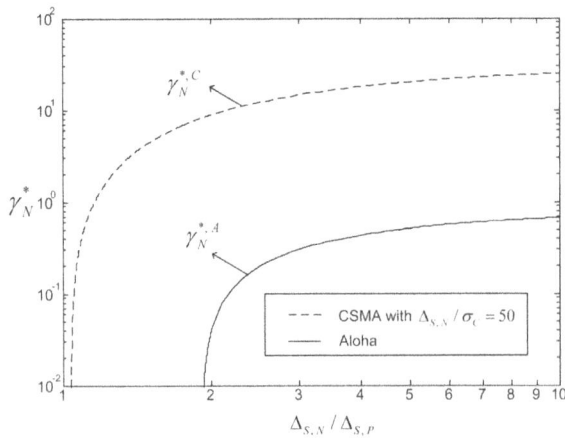

Fig. 3.8.   Connection-establishment threshold γ_N^* versus $\Delta_{S,N}/\Delta_{S,P} \cdot \delta_P = 1/2$. $\delta_N = 1/4$.

for connection-based random access to outperform connection-free random access. As a result, both thresholds $\gamma_N^{*,A}$ and $\gamma_N^{*,C}$ increase with $\frac{\Delta_{S,N}}{\Delta_{S,P}}$. A closer look at Fig. 3.8 shows that $\gamma_N^{*,A}$ for Aloha networks is much smaller than $\gamma_N^{*,C}$ for CSMA networks, corroborating with the finding that the benefit from connection establishment is more significant when sensing is absent.

So far, we have characterized the maximum effective throughput performance with Aloha and CSMA in both connection-free and connection-based cases, and demonstrated that sensing can substantially improve the throughput performance when each packet needs to contend for the channel. The improvement, nevertheless, becomes marginal if a connection is established before each data packet transmission. The analysis also shows that with sensing, the throughput gain from connection establishment could be quite insignificant.

3.5. Case Studies

In this section, we will demonstrate how to apply the preceding analysis to provide practical insights for access design in Wi-Fi and 5G networks, respectively.

3.5.1. *Wi-Fi*

The MAC protocol of IEEE 802.11 networks, distributed coordination function (DCF) is based on CSMA with two access mechanisms, including basic access and request-to-send/clear-to-send (RTS/CTS) access [2]. The basic access mechanism is an example of connection-free CSMA. As Fig. 3.9(a) illustrates, after a node transmits its data packet, it waits for an ACK from the receiver, i.e., the access point (AP). If the ACK is received within a time interval of SIFS, then the node knows that its data packet transmission has been successful. If no ACK is received within a time interval of DIFS, then the node enters the backoff process.

The RTS/CTS access mechanism, on the other hand, is an example of connection-based CSMA. As Fig. 3.9(b) illustrates, before a node transmits a data packet, an RTS frame is first sent out to

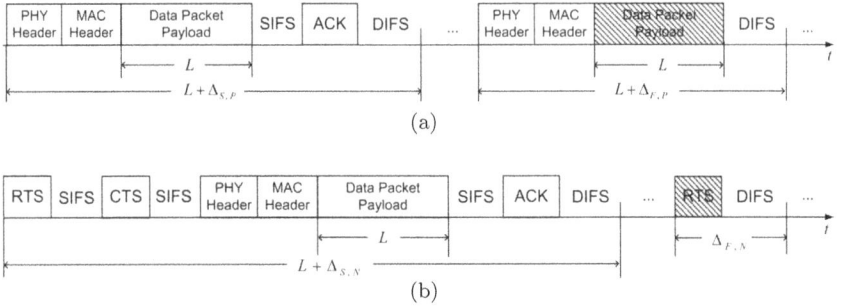

| PHY Header | MAC Header | Data Packet Payload | SIFS | ACK | DIFS | ... | PHY Header | MAC Header | Data Packet Payload | DIFS | ... |

(a)

| RTS | SIFS | CTS | SIFS | PHY Header | MAC Header | Data Packet Payload | SIFS | ACK | DIFS | ... | RTS | DIFS | ... |

(b)

Fig. 3.9.   Graphic illustration of IEEE 802.11 DCF networks with (a) basic access mechanism and (b) RTS/CTS access mechanism.

request and establish a connection with the AP. If a CTS frame is received within a time interval of Short Inter-frame Spacing (SIFS), then the node starts its data packet transmission. If no CTS is received within a time interval of Distributed Inter-frame Spacing (DIFS), then the node enters the backoff process.

We can see from Fig. 3.9 that the overhead time for each successful transmission $\Delta_{S,P}$ with the basic access mechanism includes the time for PHY header, MAC header, SIFS, ACK, and DIFS, and $\Delta_{S,N}$ with the RTS/CTS mechanism includes the time for PHY header, MAC header, RTS, CTS, ACK, DIFS, and three SIFSs. The overhead time for each failed transmission $\Delta_{F,P}$ with the basic access mechanism includes the time for PHY header, MAC header, and DIFS, and $\Delta_{F,N}$ with the RTS/CTS mechanism includes the time for RTS and DIFS.

3.5.1.1. *An illustrative example*

Let us take an example to demonstrate the maximum effective throughput performance of the basic access and RTS/CTS access mechanisms. Table 3.1 lists the typical values of key system parameters of the IEEE 802.11ac standard [33]. It should be noted that ACK, RTS, and CTS are transmitted using the basic rate R_B, while the MAC header and the data packet payload are transmitted using the data rate R_D. Given the parameter setting in Table 3.1, the overhead time for each successful and failed transmission in the basic access

Table 3.1. System parameter setting in the 802.11ac standard [33].

PHY Header	20 μs
MAC Header	288 bits
ACK	112 bits+PHY header
RTS	160 bits+PHY header
CTS	112 bits+PHY header
DIFS	34 μs
SIFS	16 μs
Slot Time σ_C	9 μs
Data Packet Payload Length PL	Up to 2^{23} bits
Data Rate R_D	7.2–96.3 Mbps
Basic Rate R_B	6–54 Mbps

and RTS/CTS access mechanisms can be obtained as:

$$\Delta_{S,P} = \frac{288}{R_D} + \frac{112}{R_B} + 90, \tag{3.35}$$

$$\Delta_{F,P} = \frac{288}{R_D} + 54, \tag{3.36}$$

$$\Delta_{S,N} = \frac{288}{R_D} + \frac{384}{R_B} + 162, \tag{3.37}$$

$$\Delta_{F,N} = \frac{160}{R_B} + 54, \tag{3.38}$$

and the data packet transmission time can be written as

$$L = \frac{PL}{R_D}, \tag{3.39}$$

all in the unit of μs. The data rate R_D and basic rate R_B in (3.35)–(3.39) are in the unit of Mbps, and the data packet payload length PL is in the unit of bits. The maximum effective throughput with the basic access and RTS/CTS access mechanisms can be calculated by combining (3.35)–(3.39) with (3.30) and (3.32), respectively.

Figure 3.10 illustrates how the maximum effective throughput varies with the packet payload length PL under various values of data rate R_D and basic rate R_B. It can be clearly seen from Fig. 3.10 that the RTS/CTS access mechanism outperforms the basic access

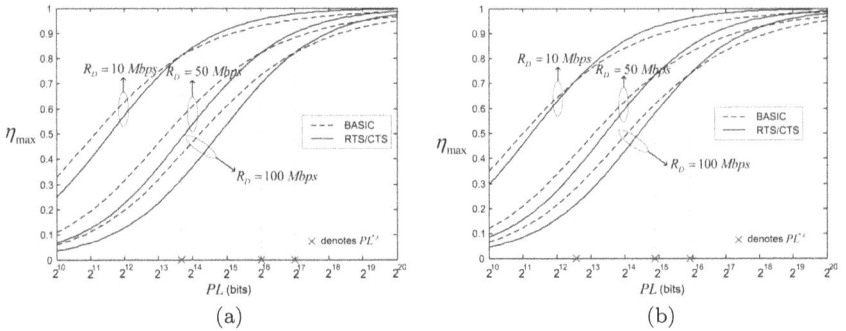

Fig. 3.10.   Maximum effective throughput $\eta_{\max}$ versus packet payload length PL of IEEE 802.11 DCF networks with the basic access and RTS/CTS access mechanisms. (a) $R_B = 6$ Mbps. (b) $R_B = 54$ Mbps.

mechanism in terms of the maximum effective throughput only when the packet payload length PL is sufficiently large. In Section 3.4.3, the connection-establishment threshold $\gamma_N^{*,C}$ for CSMA networks has been derived. Define $PL^{*,C}$ as the threshold of packet payload length for the RTS/CTS access mechanism to outperform the basic access mechanism, i.e., $\eta_{\max}^{RTS} \geq \eta_{\max}^{BASIC}$ if $PL \geq PL^{*,C}$. By combining (3.35)–(3.39) with (3.34), we have

$$PL^{*,C} = g(R_B) \cdot R_D - 288, \qquad (3.40)$$

where $g(R_B) = \dfrac{m(9-m)\mathbb{W}_0\left(\frac{m}{9-m}\exp\left\{\frac{m}{9-m}\right\}\right) - 9m}{m - (9-m)\mathbb{W}_0\left(\frac{m}{9-m}\exp\left\{\frac{m}{9-m}\right\}\right)} - 54$ with $m = \dfrac{160+54R_B}{-R_B \cdot \mathbb{W}_0\left(\frac{-160-54R_B}{e(160+63R_B)}\right)} + \dfrac{112}{R_B} + 18$. R_D and R_B are in the unit of Mbps, and $PL^{*,C}$ is in the unit of bits.

Note that in [3, 9–12], the RTS threshold was also derived and found to be decreasing as the network size increases or the initial backoff window size decreases. By contrast, the threshold $PL^{*,C}$ is independent of the network size and the initial backoff window size because the maximum effective throughput, achieved by optimally tuning the initial backoff window size, is independent of the network size. Table 3.2 lists the threshold[1] for typical values of data rate R_D

---

[1]Note that the threshold values shown in Table 3.2 have been rounded up to a power of 2.

Table 3.2. Threshold of packet payload length $PL^{*,C}$ for typical values of data rate R_D and basic rate R_B.

R_B (Mbps) \ R_B (Mbps)	7.2	14.4	21.7	28.9	43.3	57.8	65	72.2	96.3
6	2^{14}	2^{15}	2^{15}	2^{16}	2^{16}	2^{17}	2^{17}	2^{17}	2^{17}
9	2^{13}	2^{14}	2^{15}	2^{15}	2^{16}	2^{16}	2^{17}	2^{17}	2^{17}
12	2^{13}	2^{14}	2^{15}	2^{15}	2^{16}	2^{16}	2^{16}	2^{16}	2^{17}
18	2^{13}	2^{14}	2^{15}	2^{15}	2^{16}	2^{16}	2^{16}	2^{16}	2^{17}
24	2^{13}	2^{14}	2^{14}	2^{15}	2^{15}	2^{16}	2^{16}	2^{16}	2^{17}
36	2^{13}	2^{14}	2^{14}	2^{15}	2^{15}	2^{16}	2^{16}	2^{16}	2^{16}
48	2^{13}	2^{14}	2^{14}	2^{15}	2^{15}	2^{16}	2^{16}	2^{16}	2^{16}
54	2^{13}	2^{14}	2^{14}	2^{15}	2^{15}	2^{16}	2^{16}	2^{16}	2^{16}

and basic rate R_B. It can be seen from (3.40) and Table 3.2 that the threshold $PL^{*,C}$ increases as the data rate R_D increases or the basic rate R_B decreases. As Fig. 3.10 shows, more significant throughput gains can be achieved by the RTS/CTS access mechanism over the basic access mechanism when the data rate R_D is smaller, or, the basic rate R_B is larger, with which the overhead time for establishing a connection is relatively small compared to the packet transmission time.

Note that the threshold of packet payload length $PL^{*,C}$ is corresponding to the attribute *RTSThreshold* defined in the IEEE 802.11 standards. The default value of *RTSThreshold* is usually set to 2,346 or 2,347 bytes (approximately 2^{14} bits) [2]. As we can see from Table 3.2, with a large data rate R_D, the threshold should be much higher than the default value. In practice, the attribute *RTSThreshold* should be adaptively adjusted according to the basic rate R_B and the data rate R_B based on Eq. (3.40).

3.5.1.2. *Simulation results*

In this section, simulation results are presented to verify the preceding analysis on the maximum effective throughput for IEEE 802.11 DCF networks with the basic access and RTS/CTS mechanisms. The simulations are conducted based on the *ns*-2 simulator, and the values of system parameters are in accordance with Table 3.1.

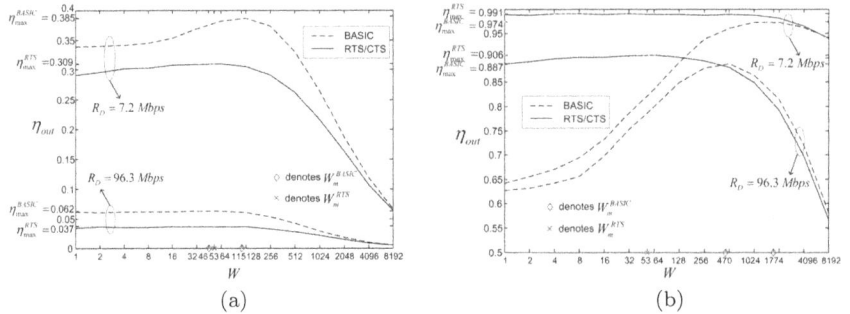

Fig. 3.11.   Simulated effective throughput η_{out} versus initial backoff window size W of IEEE 802.11 DCF networks with the basic access and RTS/CTS access mechanisms. $n = 20$. $R_B = 6$ Mbps. (a) $PL = 2^{10}$ bits. (b) $PL = 2^{18}$ bits.

The running time of each simulation is 100 seconds (1.11×10^7 time slots). We consider the saturated scenario where each node always has a packet to transmit and assume no packet dropping, i.e., each HOL packet stays in the queue until it is successfully transmitted.

Figure 3.11 presents the simulation results of the effective throughput under various values of initial backoff window size, i.e., $W = 2^i$, $i = 0, 1, \ldots, 13$. The maximum effective throughput with the basic access and RTS/CTS access mechanisms, $\eta_{\max}^{BASIC}$ and $\eta_{\max}^{RTS}$, which can be calculated by combining (3.35)–(3.39) with (3.30) and (3.32), respectively, are labeled in Fig. 3.11 and verified by simulation results. Note that to achieve the maximum effective throughput, the initial backoff window size W of each node should be carefully selected, as Fig. 3.11 illustrates. For IEEE 802.11 DCF networks, Binary Exponential Backoff (BEB) is adopted. That is, for a HOL packet with the i-th collision, $i = 0, 1, \ldots$, it chooses a random value from $\{0, 1, \ldots, W \cdot 2^{\min\{i,K\}} - 1\}$ to count down, where W and K denote the initial backoff window size and the cutoff phase, respectively. In [34], the optimal initial backoff window size W_m for achieving the maximum network throughput when the cutoff phase $K = \infty$ is obtained as

$$W_m = \frac{-2n\left(2\frac{1+\tau_F}{\tau_f}\mathbb{W}_0\left(-\frac{\tau_F}{e(1+\tau_F)}\right) + 1\right)}{\frac{1+\tau_F}{\tau_F}\mathbb{W}_0\left(-\frac{\tau_F}{e(1+\tau_F)}\right)\ln\left(-\frac{1+\tau_F}{\tau_F}\mathbb{W}_0\left(-\frac{\tau_F}{e(1+\tau_F)}\right)\right)}, \quad (3.41)$$

where $\tau_F = \tau_C - 1$, which is given by $(L + \Delta_{F,P})/\sigma_C$ for the basic access mechanism and $\Delta_{F,N}/\sigma_C$ for the RTS/CTS mechanism. The values of W_m are also labeled in Fig. 3.11 and verified by simulation results.

Figure 3.11 again corroborates that the RTS/CTS access mechanism achieves higher throughput than the basic access mechanism only when the packet payload length PL is sufficiently large. In that case, its throughput is also less sensitive to the change of the initial backoff window size W, indicating that the RTS/CTS access is more robust as well.

3.5.2. *5G*

In cellular systems, including 4G/5G, the Conventional Random Access (CRA) scheme has been adopted for a long time, in which each device has to perform the four-way-handshake random access procedure for establishing a connection with the BS prior to its data transmission, as shown in Fig. 3.12(a). As each device does not perform sensing, CRA is an example of connection-based Aloha.

The CRA scheme was originally designed to support traditional Human-Type Communications (HTC), where the number of devices is small, but each device may transmit a significant amount of data. The signaling overhead for connection establishment is usually negligible as the data transmission time is long. However, for the emerging massive Machine Type Communications (mMTC) traffic, where a large number of MTDs transmit short packets sporadically, establishing a connection prior to data transmission could be inefficient because of the comparatively heavy signaling overhead along with additional latency and energy consumption. Therefore, the Two-Step Random Access (TSRA) scheme was introduced in 5G systems, where MTDs can transmit one small data packet along with preamble transmission without connection establishment [8], as shown in Fig. 3.12(b). The TSRA scheme is an example of the connection-free Aloha.

We can see from Fig. 3.12(a) that with the CRA scheme, the overhead time for each successful transmission, $\Delta_{S,N}$, includes the time for preamble, *ra-ResponseWindow*, connection request,

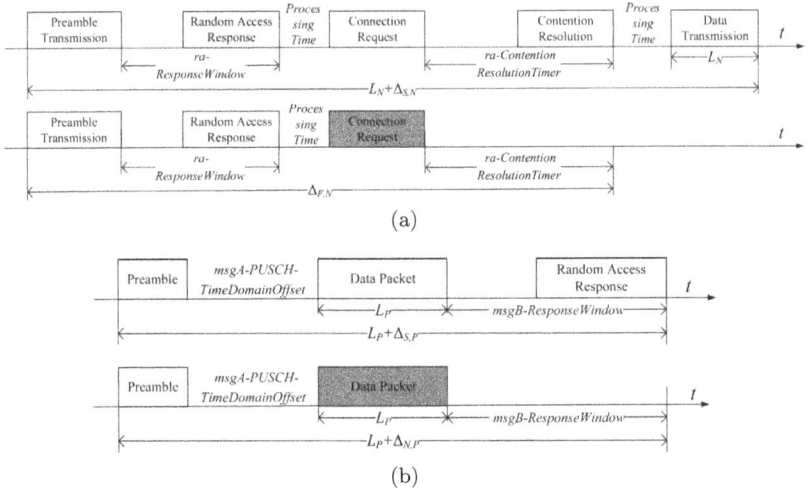

Fig. 3.12.   (a) Conventional Random Access (CRA) scheme in 5G. (b) Two-Step Random Access (TSRA) scheme in 5G.

and *ra-ContentionResolutionTimer*[2] and two processing time intervals. The overhead time for each failed transmission $\Delta_{F,N}$ includes preamble, *ra-ResponseWindow*, connection request, and *ra-ContentionResolutionTimer* and one processing time interval. On the other hand, as shown in Fig. 3.12(b), with the TSRA scheme, the overhead time for each successful transmission and that for each failed transmission, i.e., $\Delta_{S,P}$ and $\Delta_{F,P}$, include the time for preamble, *msgA-PUSCHTimeDomainOffset* and *msgB-ResponseWindow*.

3.5.2.1. *An illustrative example*

Let us take an example to demonstrate the maximum effective throughput performance of the CRA and TSRA mechanisms. Table 3.3 lists the typical values of key system parameters in 5G

---

[2]The *ra-ResponseWindow* and the *ra-ContentionResolutionTimer* are long enough for receiving the random access response and contention resolution message, respectively. If the device does not receive the contention resolution message when the *ra-ContentionResolutionTimer* runs out, then the random access procedure is considered to have failed.

Table 3.3. Typical system parameter setting in 5G [8, 35].

Preamble Transmission	1 slot
ra-ResponseWindow	10 slots
Connection request Transmission	1 slot
ra-ContentionResolutionTimer	16 slots
Processing time interval	1 slot
msgA-PUSCHTimeDomainOffset	1 slot
L_P	1 slot
msgB-ResponseWindow	4 slots
Slot length	0.25 milliseconds

[8, 35], based on which the overhead time for each successful and failed transmission in the CRA and TSRA mechanisms can be obtained as

$$\Delta_{S,N} = 0.0075, \tag{3.42}$$

$$\Delta_{F,N} = 0.00725, \tag{3.43}$$

$$\Delta_{S,P} = 0.0015, \tag{3.44}$$

$$\Delta_{F,P} = 0.0015, \tag{3.45}$$

all in the unit of seconds.

In the CRA scheme, the device can request uplink time-frequency resource from the BS as long as it has packets in the buffer. Let PL denote the number of information bits that each device transmits. The data transmission time can then be written as

$$L_N = \frac{PL}{R_{5G}}, \tag{3.46}$$

where R_{5G} is the data transmission rate in unit of bps. The maximum effective throughput with the CRA scheme in 5G can be calculated by combining (3.42)–(3.43) and (3.31) as

$$\eta_{\max}^{A,N} = \frac{PL}{PL + (0.00725e + 0.00025)R_{5G}}. \tag{3.47}$$

In the TSRA scheme, due to limited time-frequency resources, the device can transmit only one small packet of size 100 bits in one slot, i.e., $L_P = 1$ slot. The maximum effective throughput in 5G can be calculated by further combining (3.29) and (3.44) as

$$\eta_{\max}^{A,P} = 0.0526. \tag{3.48}$$

Define $PL^{*,A}$ as the threshold of packet payload length for the CRA to outperform TSRA, i.e., $\eta_{\max}^{A,N} \geq \eta_{\max}^{A,P}$. By combining (3.47)–(3.48), we have

$$PL^{*,A} = 0.0011 R_{5G}, \tag{3.49}$$

in the unit of bits, which reveals that the threshold $PL^{*,A}$ linearly increases with the data transmission rate R_{5G}. For instance, with $R_{5G} = 10$ Mbps, we have $PL^{*,A} = 11{,}000$ bits, which is much larger than the typical packet payload length in the mMTC case, i.e., 100 bits. It indicates that the newly introduced TSRA scheme is preferred in the context of sporadic small packet transmissions.

3.5.2.2. *Simulation results*

Let us further present simulation results to verify the preceding analysis on the maximum effective throughput for 5G with CRA and TSRA mechanisms. Figure 3.13 presents the simulated effective throughput under various values of the Access Class Barring (ACB) factor[3] q_0 with $PL \in \{2^{13}, 2^{14}, 2^{15}\}$ bits. The analysis in Section 3.3.1 has revealed that to achieve the maximum effective throughput in Aloha, the transmission probabilities should be carefully selected such that the network steady-state point $p = p^{*,A} = e^{-1}$, with which we can have the optimal transmission probability q_0^* in (3.26). The values of q_0^* are labeled in Fig. 3.13, and verified by simulation results.

Moreover, as shown in Fig. 3.13, the maximum effective throughput performance of the CRA scheme steadily increases as the packet payload length PL grows, while that of the TSRA scheme is insensitive to the variation of PL. The throughput performance of the CRA scheme can outperform that of the TSRA scheme only when the packet payload length PL is larger than the threshold $PL^{*,A} = 11{,}000$ bits.

---

[3]In the random access procedure of 5G, the transmission probability of the access request is referred to as the ACB factor in standards [36].

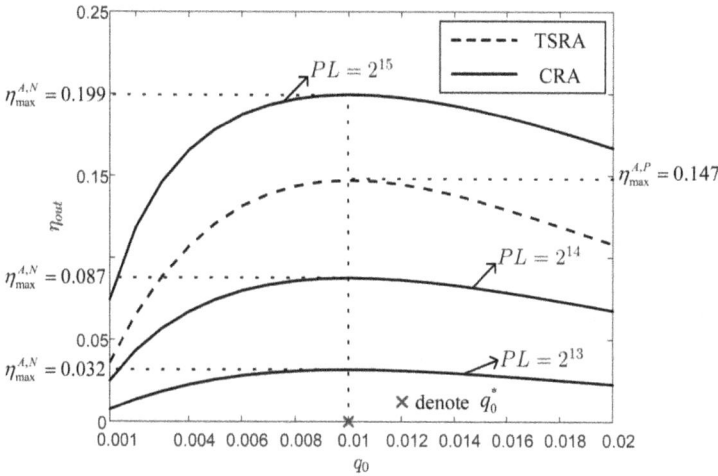

Fig. 3.13. Simulated effective throughput η_{out} versus ACB factor q_0 with the CRA and TSRA mechanisms. $n = 100$. $R_{5G} = 10$ Mbps. $PL \in \{2^{13}, 2^{14}, 2^{15}\}$ bits.

3.6. Summary

This chapter presents a comparative study of the optimal throughput performance of connection-free random access and connection-based random access. A channel-centric model and a node-centric model are established for random access networks, respectively. Based on that, explicit expressions of the maximum effective throughput are obtained for both connection-free and connection-based Aloha and CSMA. The analysis shows that both modeling methodologies lead to the same maximum network throughput, yet the corresponding optimal backoff parameters can only be obtained via the node-centric model.

By comparing the maximum effective throughput of connection-free and connection-based random access, thresholds of data packet transmission time for beneficial connection establishment in both Aloha and CSMA cases are obtained as functions of overhead parameters. The former is found to be much lower, which indicates that the throughput gain brought by connection establishment is more significant when sensing is absent.

The analysis sheds plenty of insights for practical network design. For Wi-Fi networks where CSMA-based DCF is adopted, the threshold of packet payload length for the connection-based RTS/CTS access mechanism to outperform the connection-free basic access mechanism in terms of the maximum effective throughput is shown to be much higher than the default value of RTS threshold given in the IEEE 802.11 standards. For 5G networks where Aloha-based CRA and TSRA are adopted, the threshold of packet payload length for the connection-based CRA scheme is also found to be much larger than the typical packet size in M2M communications, implying that the connection-free TSRA mechanism might be a preferable option for short packet transmissions.

References

[1] J. F. Kurose and K. W. Ross, *Computer Networking: A Top-Down Approach*, 5th edition. Addison-Wesley, 2009.

[2] IEEE Standard 802.11-2007, *IEEE Standard for information technology — Telecommunications and information exchange between systems — Locl and metropolitan area networks — Specific requirements — Part 11: Wireless LAN Medium Access Control (MAC) and Physical Layer (PHY) specifications,* 2007.

[3] G. Bianchi, "Performance analysis of the IEEE 802.11 distributed coordination function," *IEEE J. Sel. Areas Commun.*, vol. 18, no. 3, pp. 535–547, 2009.

[4] S. Chen, *et al.*, "Machineto-machine communications in ultra-dense networks — A survey," *IEEE Commun. Surveys Tuts.*, vol. 19, no. 3, pp. 1478–1503, 2017.

[5] C. Bockelmann, *et al.*, "Massive machine-type communications in 5G: Physical and MAC-layer solutions," *IEEE Commun. Mag.*, vol. 54, no. 9, pp. 59–65, 2016.

[6] L. Liu, *et al.*, "Sparse signal processing for grant-free massive connectivity: A future paradigm for random access protocols in the Internet of Things," *IEEE Signal Process. Mag.*, vol. 35, no. 5, pp. 88–99, 2018.

[7] C. Cox, *An Introduction to LTE: LTE, LTE-Advanced, SAE and 4G Mobile Communications.* Wiley, Hoboken, NJ, USA, 2012.

[8] 3GPP TS 38.321 V16.6.0, *5G;NR; Medium Access Control (MAC) protocol specification,* 2021.

[9] P. Chatzimisios, *et al.*, "Optimisation of RTS/CTS handshake in IEEE 802.11 wireless LANs for maximum performance," *IEEE GLOBECOM Workshops*, pp. 270–275, 2004.

[10] Z. Kong, *et al.*, "Adaptive RTS/CTS mechanism for IEEE 802.11 WLANs to achieve optimal performance," *IEEE ICC*, vol. 1, pp. 185–190, 2004.

[11] I. Tinnirello, *et al.*, "Revisit of RTS/CTS exchange in high-speed IEEE 802.11 networks," *IEEE WoWMoM*, pp. 240–248, 2005.

[12] J. Wang, *et al.*, "The impact of RTS threshold on the performance of multi-rate network," *IEEE CCWMC*, pp. 651–656, 2009.

[13] N. Abramson, "The Aloha system: Another alternative for computer communications," *Proc. Fall Joint Comput. Conf.*, pp. 281–285, 1970.

[14] L. Kleinrock, *et al.*, "Packet switching in radio channels: part I — carrier sense multiple-access modes and their throughput-delay characteristics," *IEEE Trans. Commun.*, vol. 23, no. 12, pp. 1400–1416, 1975.

[15] F. Tobagi, *et al.*, "Performance analysis of carrier sense multiple access with collision detection," *Computer Netw.*, vol. 4, no. 5, pp. 245–259, 1980.

[16] H. Takagi, *et al.*, "Throughput analysis for persistent CSMA systems," *IEEE Trans. Commun.*, vol. 33, no. 7, pp. 627–638, 1985.

[17] R. MacKenzie, *et al.*, "Throughput and delay analysis for p-persistent CSMA with heterogeneous traffic," *IEEE Trans. Commun.*, vol. 58, no. 10, pp. 2881–2891, 2010.

[18] P. Wong, *et al.*, "Analysis of non-persistent CSMA protocols with exponential backoff scheduling," *IEEE Trans. Commun.*, vol. 59, no. 8, pp. 2206–2214, 2011.

[19] B. Tsybakov, *et al.*, "Ergodicity of slotted Aloha system," *Probl. Inf. Transmission*, vol. 15, pp. 73–87, 1979.

[20] R. Rao, *et al.*, "On the stability of interacting queues in a multiple access system," *IEEE Trans. Inf. Theory*, vol. 34, no. 5, pp. 918–930, 1988.

[21] V. Anantharam, "The stability region of the finite-user slotted Aloha protocol," *IEEE Trans. Inf. Theory*, vol. 37, no. 3, pp. 535–540, 1991.

[22] W. Szpankowski, "Stability conditions for some multiqueue distributed systems: Buffered random access systems," *Adv. Appl. Prob.*, vol. 26, no. 2, pp. 498–515, 1994.

[23] W. Luo, *et al.*, "Stability of n interacting queues in random access systems," *IEEE Trans. Inf. Theory*, vol. 45, no. 5, pp. 1579–1587, 1999.

[24] J. Luo, *et al.*, "On the throughput, capacity, and stability regions of random multiple access," *IEEE Trans. Inf. Theory*, vol. 52, no. 6, pp. 2593–2607, 2006.

[25] L. Dai, "Toward a coherent theory of CSMA and Aloha," *IEEE Trans. Wireless Commun.*, vol. 12, no. 7, pp. 3428–3444, 2013.

[26] S. Andreev, *et al.*, "Efficient small data access for machine-type communications in LTE," *Proc. IEEE ICC*, pp. 3569–3574, 2013.

[27] O. Arouk, *et al.*, "General model for RACH procedure performance analysis," *IEEE Commun. Lett.*, vol. 20, no. 2, pp. 372–375, 2015.

[28] Z. Wang, *et al.*, "Optimal access class barring for stationary machine type communication devices with timing advance information," *IEEE Trans. Wireless Commun.*, vol. 14, no. 10, pp. 5374–5387, 2015.

[29] M. Koseoglu, "Lower bounds on the LTE-A average random access delay under massive M2M arrivals," *IEEE Trans. Wireless Commun.*, vol. 64, no. 5, pp. 2104–2115, 2016.

[30] Y. Beyene, *et al.*, "Random access scheme for sporadic users in 5G," *IEEE Trans. Wireless Commun.*, vol. 16, no. 3, pp. 1823–1833, 2017.

[31] M. Centenaro, *et al.*, "Comparison of collision-free and contention-based radio access protocols for the Internet of Things," *IEEE Trans. Commun.*, vol. 65, no. 9, pp. 3832–3846, 2017.

[32] W. Zhan and L. Dai, "Massive random access of machine-to-machine communications in LTE networks: Throughput optimization with a finite data transmission rate," *IEEE Trans. Wireless Commun.*, vol. 18, no. 12, pp. 5749–5763, 2019.

[33] IEEE Standard 802.11ac, *IEEE Standard for information technology– Telecommunications and information exchange between systems — Local and metropolitan area networks — Specific requirements — Part 11: Wireless LAN Medium Access Control (MAC) and Physical Layer (PHY) specifications — Amendment 4: Enhancements for very high throughput for operation in bands below 6 GHz*, 2013.

[34] L. Dai and X. Sun, "A unified analysis of IEEE 802.11 DCF networks: Stability, throughput and delay," *IEEE Trans. Mobile Comput.*, vol. 12, no. 8, pp. 1558–1572, 2013.

[35] 3GPP TS 38.533 V16.6.0, *5G;NR; Radio User Equipment (UE) conformance specification; Part 1: Common test environment*, 2021.

[36] 3GPP TS 38.331 V16.4.1, *5G;NR; Radio Resource Control (RRC); Protocol specification*, 2021.

COMPONENT

When we compete, we see things or people as relative. But when we collaborate, we see things or people as related. Although competition makes things better, collaboration makes a better person. Additionally, when we compete, we may create enemies. But when we collaborate, we become friends.

Kiat Seng Yeo

Chapter 4

Microwave and Millimeter-Wave Phase Change Material Devices for Future Communications

Tejinder Singh and Raafat R. Mansour

4.1. History of Phase Change Materials

While Ovshinsky [1] is generally credited as the inventor of phase change materials (PCM) for information storage, the discovery of phase changing electrical characteristics dates back to the early 1900s in the little-known and seldom-cited pioneering work of Alan Tower Waterman. While studying the thermionic emission of certain hot salts [2], Waterman observed a large negative coefficient of resistance of the molybdenum disulfide (MoS_2) with respect to temperature. More significantly, he observed a breakdown in resistivity characteristics when the device under test was heated by means of an electric current. He pointed out that MoS_2 may exist in two forms: α of high resistance and β of low resistance. The breakdown phenomenon and progressive conductivity changes are prominent of phase-change behavior in chalcogenide materials. Waterman also pointed out that the transition from α form to β form can be initiated by heat, electric field, or light. However, without the use of modern physical analysis tools such as transmission electron microscopy and x-ray crystallography, he was not able to observe any physical changes of MoS_2 along with the conductivity changes. He did, however, notice an increase in hardness with the increase in conductivity [3].

Since 1960, Ovshinsky has been working with amorphous chalcogenides. He developed both electrically controlled threshold and memory-switching devices and first reported his findings in a paper published in *Physical Review Letters* in 1968 [1], which detailed the operations of reversible switching in memory devices composed of 48% tellurium, 30% arsenic, 12% silicon, and 10% germanium. The most significant contribution of Ovshinsky's work is that he demonstrated practicality of the switching phenomenon in continuous successful switching operations of multiple devices over periods of many months. Interest in phase change memory was effectively initiated by Ovshinsky's groundbreaking paper, which remains the most cited literature in this field [3].

In the early 1970s, phase change memory drew a lot of interest in the industry and academia with the rapid expansion of applications of computers. The most notable work was the development of a 256-bit array, comprising a 16×16 matrix of phase change memory cells by R. Neale and D. Nelson of Energy Conversion Devices, along with Gordon E. Moore of Intel [4]. Their memory cell consisted of a storage element, coined as the Ovonic amorphous semiconductor switching device, in series with a silicon p-n junction diode. The Ovonic memory element comprised a thin film of PCM sandwiched between two molybdenum electrodes as a non-volatile bi-stable resistor. The high-to-low resistance ratio was about 10^3. Another attempt to build a phase change memory device was reported by R. Shanks and C. Davis of the Burroughs Corp.; they published their data of a 1,024-bit phase change memory in 1978 [5].

In early 2000s, the ability of having media, that is, being rewritable, was also an important requirement [6]. In the late 1980s, a landmark was achieved by Matsushita/Panasonic, which developed phase-change optical disc technology that remained stable over a million use cycles [7]. This technology became the mainstream in optical disc production, leading to the commercialization of 4.7-GB digital versatile disc random access memory (DVD-RAM) in the late 1990s. During the development process, various materials were examined, and the best performing PCM in terms of speed and stability were found to be $Ge_2Sb_2Te_5$ (GST) for DVD-RAM and an

AgInSbTe alloy for DVD-RW applications [8]. The phenomenology of phase-change optical recording is simple. An initially amorphous as-deposited GST layer is crystallized by exposure to a laser beam of intensity sufficient to heat the material to a temperature slightly above the glass-transition temperature. A subsequent exposure to an intense and short laser pulse melts the GST, which is then converted into the amorphous state on quenching. A recorded bit is an amorphized mark against the crystalline background. The reversibility of the crystallization–amorphization process allows the fabrication of rewritable memory [6].

A chalcogenide PCM GeTe-based reconfigurable switch was first presented by E. Chua in 2010 [9]. The switch exhibited low ON-state resistance of 180 Ω and a large dynamic range of (7e3). This paper also reported a partial crystallization and partial re-amorphization model to explain the differences between the measured and calculated device in OFF- and ON-state resistances, respectively. Although this paper reports the switching properties of PCM by using a phase change, no RF parameters are reported.

A thermally driven RF switch using the properties of metal insulator transition in VO_2 was reported by XLIM in 2010 [10] and Harvard University in 2013 [11]. Teledyne reported a high-performance RF switch based on MIT materials in 2015 [12]. The switch reportedly shows promising results for millimeter wave frequencies till 50 GHz. The heating mechanism is implemented using integrated chip heaters to provide local thermal control. The switch reported 0.13 dB of insertion loss and 20 dB isolation till 50 GHz. The dynamic range for this switch is (4.4e4), with a low resistance of 3.3 Ω at 68°C.

Early research on using directly and indirectly heated GeTe-based RF switches was presented by the University of Michigan [13], Northrop Grumman Corp. [14], HRL Laboratories [15], University of Limoges [16], and the University of Waterloo [17–19]. Most of these switches have demonstrated excellent RF performance, encouraging the development of a wide range of reconfigurable PCM-based RF components for both microwave and millimeter-meter wave applications.

4.2. Basic Principle of Phase Change Materials

Phase Change Material (PCM) has the unique property of reversible switching between amorphous and crystalline states upon specific heat treatment by means of electrical pulses. The state where atoms are arranged in a disorderly manner (short-range order) is called the amorphous state, whereas the state where atoms are organized in an orderly manner (long-range order) is called the crystalline state. The disordered amorphous state has a lower mean free path of conduction for electrons that impedes current flow due to electron scattering, thus resulting in a higher resistance when compared to the crystalline state. For non-volatile memory applications, the conventional principle of PCM is illustrated in Fig. 4.1. The large contrast in resistance between the states is used as a form of non-volatile memory to represent two states [20].

A relatively low amplitude and long duration (typically $> 1\mu s$) SET electrical pulse is used for crystallization during a transition to the ON state. Energy from the SET pulse heats the material for sufficient time to crystallize the material and provides adequate time for the atoms to reorganize to an orderly arrangement, thus transforming from an amorphous state to a crystalline state. The SET pulse is illustrated by the blue arrow in Fig. 4.1.

A short duration (typically < 100 ns) and high amplitude RESET electrical pulse is used for re-amorphization. The RESET pulse

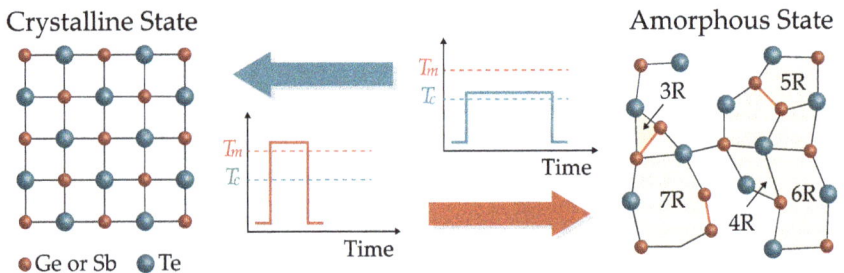

Fig. 4.1.   Reversible switching of phase change material using an electrical pulse, in non-volatile memory applications [19].

provides sufficient energy to melt the material to disorder the atoms, following a rapid quenching to freeze the atoms, thus transforming the material from a crystalline state to an amorphous state. The RESET pulse is illustrated by the red arrow in Fig. 4.1.

4.2.1. *Phase change material: Germanium Telluride (GeTe)*

GeTe is a chemical compound of germanium and tellurium and is a component of chalcogenide glasses. It shows semi-metallic conduction and ferroelectric behavior [21]. GeTe exists in three major crystalline forms, room-temperature α (rhombohedral) and γ (orthorhombic) structures and high-temperature β (cubic rocksalt-type) phase; α phase being the most phase of pure GeTe below the ferroelectric Curie temperature of approximately 670 K [22]. Doped GeTe is a low-temperature superconductor [23]. The ON (crystalline) state resistivity is 2e-4 Ω cm, and the OFF (amorphous) state resistivity is > 1e3 Ω cm. This results in a dynamic range (OFF-state/ON-state resistance ratio) of around 5e6 times. Such values have been the driver of using GeTe for RF applications [20, 24].

Phase change (chalcogenide) material is defined as alloys containing group VI elements such as sulfur (S), selenium (Se), and telluride (Te). Alloys containing germanium (Ge), antimony (Sb), and Te are most common, with the germanium-antimony-telluride ($Ge_2Sb_2Te_5$) alloy being the most thoroughly researched material. Figure 4.2 shows the ternary phase diagram for this system where single-phase alloys that lie on the pseudo-binary line of germanium telluride (GeTe) and antimony telluride (Sb_2Te_3) are indicated. Allies include $Ge_1Sb_2Te_4$, $Ge_2Sb_2Te_5$, and $Ge_1Sb_4Te_7$. Along the pseudo-binary line from GeTe to Sb_2Te_3, the properties change from high stability and low speed to low stability and high speed [20]. A material composition selected from the pseudo-binary line may typically achieve fast crystallization and reasonable stability. However, for a reconfigurable switch in RF applications, low ON-state resistance and large dynamic range is very crucial.

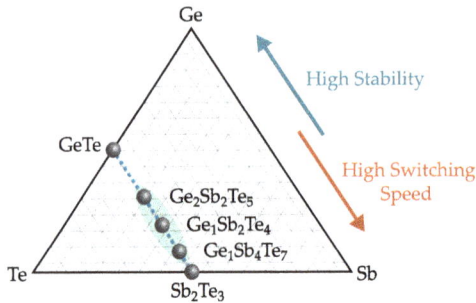

Fig. 4.2.   Ge-Sb-Te ternary phase diagram depicting various phase change alloys along with single phase compositions that reside on the tie-line of GeTe and Sb_2Te_3.

4.2.2.   *Metal-insulator-transition material: Vanadium oxide*

Vanadium (IV) dioxide, or simply VO_2, is an inorganic compound. It is a dark blue solid. VO_2 is amphoteric, dissolving in non-oxidizing acids to give the blue vanadyl ion, $[VO]^{2+}$, and in alkali to give the brown $[V_4O_9]^{2-}$ ion, or at high pH $[VO_4]^{4-}$. VO_2 has a phase transition very close to temperature ($\sim$68°C). Electrical resistivity, opacity, etc., can change up several orders. At the rutile to the monoclinic transition temperature, VO_2 also exhibits a metal-to-semiconductor transition in its electronic structure: the rutile phase is metallic, while the monoclinic phase is semiconducting. The optical band gap of VO_2 in the low-temperature monoclinic phase is about 0.7 eV. Due to these properties, it has been widely used in surface coating, sensors, and imaging. Applications include its use in memory devices and RF systems.

4.2.3.   *GeTe: A clear choice for RF devices*

As can be seen from the above discussion, GeTe is preferred as a perfect material for RF devices due to the latching nature of the material. Just a short duration of the pulse is required to switch the state, while VO_2 requires consistent biasing and thus consumes continuous power. Moreover, deposition of VO_2 is very inconsistent due to the requirement of the reactive sputtering. As depositing thin films of VO_2 requires a precise amount of oxygen in the sputtering chamber, a slight variation of oxygen rate or pressure can produce different undesired oxides like vanadium trioxide (V_2O_3), vanadium pentoxide (V_2O_5) or other oxides, which do not exhibit metal-insulator transition (MIT), or in simple words, do not show resistance change behavior. Moreover, sputtering targets get oxidized fast and need cleaning after 3–4 depositions. On the other hand, GeTe target can be used for DC sputtering. As no reactive component is involved, thus the design to development time is short. These advantages of GeTe make it an ideal PCM for the development of RF devices and systems.

4.3. Characterization of Phase Change Materials

Optimization of GeTe thin films is a crucial step in the development of high-performance RF switches with low ON-state loss and high OFF-state isolation. The performance of the Phase Change Material (PCM) switch depends on the GeTe film quality. Poor-quality films exhibit a lower resistance ratio between crystalline and amorphous states. Various factors, including but not limited to deposition type, base pressure, chamber pressure, material purity, deposition power, DC/RF sputtering, inert gas flow rate, and deposition temperature, can affect the film quality in terms of resistance ratio and surface morphology.

4.3.1. *Deposition conditions*

GeTe thin films are usually deposited using DC magneton sputtering using an ultra-high purity $Ge_{0.5}Te_{0.5}$ target. Deposition parameters are given in [19] and [24]. Deposition pressure is varied from 2 to 10 mTorr [19], and the deposition temperature is varied from 25 to 200°C. GeTe thin films are characterized to get optimal film quality that demonstrate higher resistance ratio between crystalline and amorphous states. The primary issues with GeTe films are cracks and voids formation after annealing. To avoid such problems, films deposited at various deposition conditions are studied by investigating the surface topography using high-resolution AFM,

SEM, and cross-wafer resistance mapping measurements. A study of GeTe thin films was carried out in [19] using two sets with four samples (A–D and E–H).

4.3.2. *Study of surface topography*

Samples ("A", "B", "C", and "D") of 150-nm GeTe are deposited with varying argon flow rates using DC sputtering on a 635-μm Al_2O_3 substrate. The AFM scans of samples "A", "B", and "D" are shown in Fig. 4.3. Films are annealed for 30 mins at 220°C to get a crystalline state of the material [19]. A 3D AFM scan of sample "B" in an amorphous state is shown in Fig. 4.3(a), and the crystalline state of the same sample is shown in Fig. 4.3(b). The average roughness is almost identical in both states except for the larger particle size. The AFM scan area for samples "A" to "D" is 3×3 μm. Surface defects of alumina wafers are prominent in the AFM scans

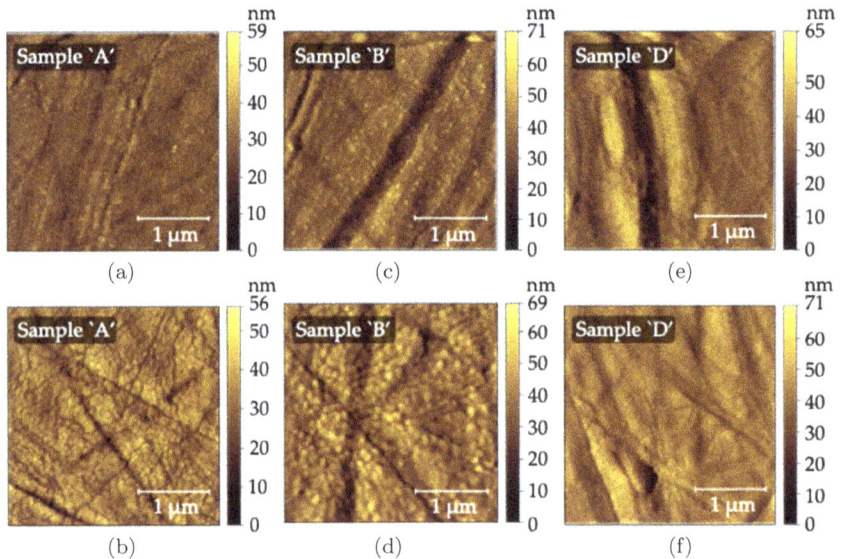

Fig. 4.3.   AFM scans of GeTe thin film deposited on Al_2O_3 substrate. (a) Sample "A" amorphous, (b) Sample "A" crystalline, (c) Sample "B" amorphous, (d) Sample "B" crystalline, (e) Sample "D" amorphous, and (f) Sample "D" annealed with only two orders of resistance change. Streaks are from a substrate with an average roughness of 25 nm [18, 19, 24].

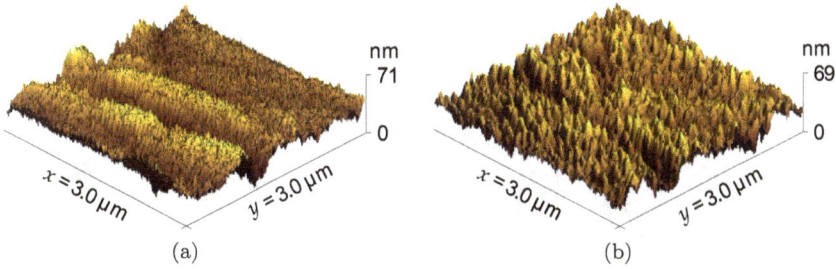

Fig. 4.4.   3D surface scan of GeTe film in (a) amorphous state (Sample B), and (b) crystalline state (Sample B). Total AFM scan area is $3 \times 3\,\mu m$ [18, 19, 24].

of GeTe films, as shown in Fig. 4.3(a–f). Sample "D" in Fig. 4.3(f) shows GeTe film after annealing with only two orders of resistance change, and Fig. 4.3(e) shows sample "D" as-deposited film.

Amorphous films are smoother than crystalline films, as shown in Fig. 4.3. The AFM scan shown in Fig. 4.3(a) is of sample "A", and the image shown in Fig. 4.3(c) is of sample "B". Sample "B" is deposited at 60 W deposition power and 5 mTorr deposition pressure with a higher argon flow of 40 sccm. Although the films are of the same thickness, crystalline films, as shown in Fig. 4.3(b) and (d), show higher roughness. 3D AFM scans are shown in Fig. 4.4.

To study the effects of the deposition temperature, an AFM is used to evaluate the surface topography and film roughness on glass substrate due to its negligible surface roughness [19]. GeTe samples "E", "F", "G", and "H" are deposited at elevated temperatures to study the effect of deposition temperature on GeTe films. In these samples, sputter deposition power is kept at a constant value of 60 W, a deposition pressure of 3 mTorr, and an argon flow of 20 sccm were used while varying the temperature from 25 to 200°C. These samples are also annealed for 30 mins at 220°C after the deposition.

Films deposited at room temperature show a distinct large grain size in annealed GeTe films (crystalline state) and rather smooth surface profile in the as-deposited amorphous state. Voids and cracks formation are more prominent in crystalline films, as shown in Fig. 4.5(b). Amorphous or as-deposited films at different temperatures show uniform roughness, as shown in Fig. 4.5(a), (c), (e),

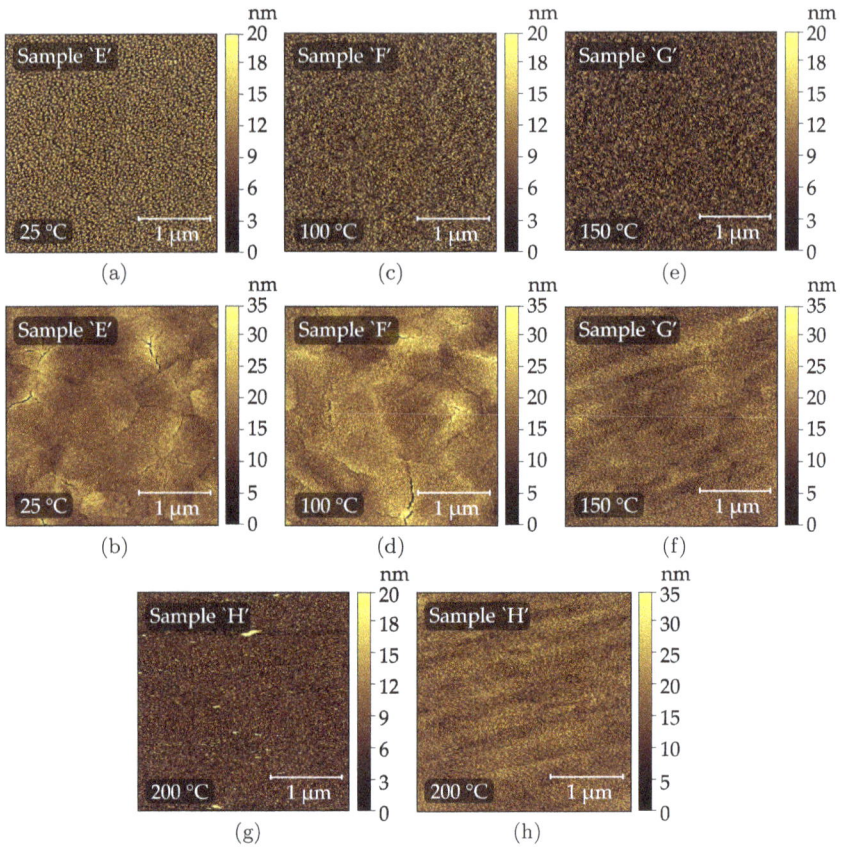

Fig. 4.5.   AFM scan of GeTe thin film deposited on a glass substrate. (a) Sample "E" amorphous, (b) Sample "E" crystalline deposited at room temperature, (c) Sample "F" amorphous, (d) Sample "F" crystalline deposited at 100°C, (e) Sample "G" amorphous, (f) Sample "G" crystalline deposited at 150°C, (g) Sample "H" amorphous, and (h) Sample "H" crystalline deposited at 200°C. Change in grain boundaries is prominent in films deposited at room and elevated temperatures [19, 24].

and (g), while crystalline films show different surface topographies when deposited at elevated temperatures. Figure 4.5(d) shows the crystalline state of film deposited at 100°C.

Fewer cracks are observed in the films deposited at elevated temperatures compared to the ones deposited at room temperature. However, an elevated temperature deposition of GeTe films shows

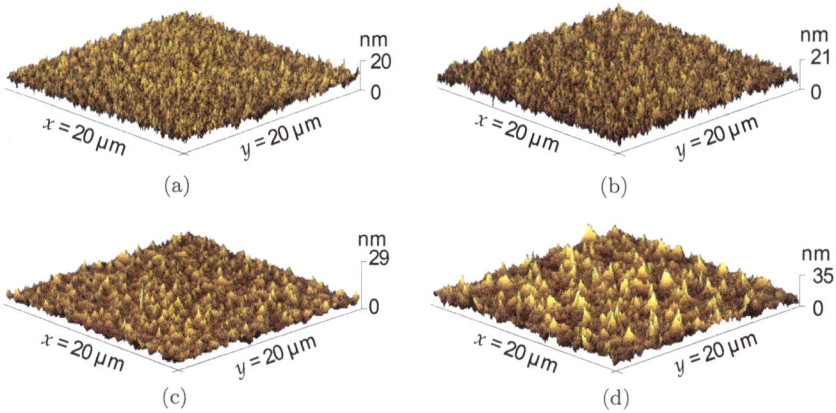

Fig. 4.6.   The 3D surface scan of GeTe films in crystalline states deposited at (a) 25°C (Sample "E"), (b) 100°C (Sample "F"), (c) 150°C (Sample "G"), and (d) 200°C (Sample "H"). Scan area for the samples is 20 × 20 μm [19].

high roughness. Films deposited at 150°C, as shown in Fig. 4.5(f), are almost free from cracks and voids, but roughness is higher than that of samples "E" and "F". Crystalline films deposited at 200°C, as shown in Fig. 4.5(h), do not show any prominent sign of cracks or voids but are not usable for PCM switches. High roughness of PCM thin films is not favorable for achieving narrow PCM channel in series SPST switches [24].

The scan area has been extended from a 3 × 3 μm range to 20 × 20 μm since surface roughness and larger grains become more prominent in a large scan area. 3D surface scans of the same samples are shown in Fig. 4.6(a) for sample "E", (b) for sample "F", (c) for sample "G", and (d) for sample "H".

4.4. Fabrication Process

The fabrication of the PCM-based RF switch involved the use of a high-resistivity substrate and the deposition of GeTe films, high-resistivity heater thin-films, dielectric layers to insulate the heater from the GeTe switching element, and high-conductivity metallization thin-films for input/output ports of the switch. The fabrication process flow is shown in Fig. 4.7 [24–26]. The microfabrication process

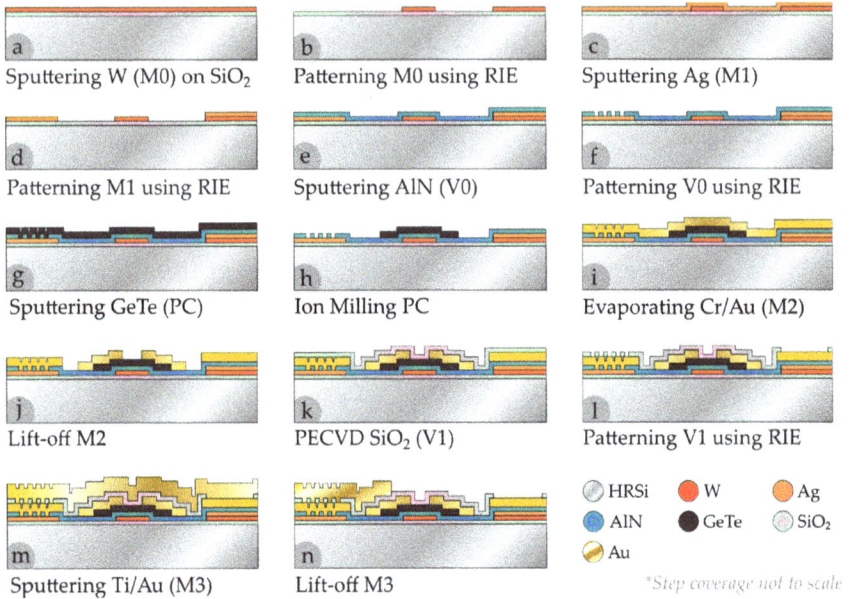

Fig. 4.7.   Cross-section of Gen 3 RF-PCM fabrication process flow using eight layers, with two metal layers for RF signal routing and two metal layers for bias routing and actuation mechanism [24–26].

starts with an RCA-1 and Piranha cleaning of high-resistivity Si wafer (>20 kΩ cm) and 500-μm thickness. The process includes eight layers with four metal layers (M0, M1, M2, and M3), one PCM layer, and three dielectric layers (D0, V0, and V1). D0 is the substrate oxide for thermal isolation. M2 and M3 are used for RF signal routing, and the M0 layer is used for the micro-heaters, while M1 is used for biasing networks. In Fig. 4.7, (a) 50 nm tungsten (W) is sputtered (M0) at an elevated temperature of 850°C deposited on a thin dielectric insulator oxide layer, (b) patterned using RIE, (c) 70 nm silver (Ag) is sputtered (M1), (d) patterned using RIE, followed by (e) deposition of 100 nm AlN layer using RF sputtering, (f) patterned using RIE to act as a barrier layer (V0) between micro-heaters and PCM, (g) 130 nm GeTe thin film is sputtered using $Ge_{0.5}Te_{0.5}$ target (PC), (h) patterned using ion milling, (i) evaporation of 350 nm of Au that serves as M2 with 30 nm of Ti as a seed layer, (j) M2 is patterned using the lift-off technique, (k) 200-nm SiO_2 layer is deposited using low-temperature PECVD, and (l) patterned using RIE that serves as a passivation layer for PCM (V1), (m) sputtering of 450-nm Au layer (M3) with 40-nm Ti layer as a seed layer, and (n) M2 is patterned using the lift-off technique.

4.5. Phase Change RF Switches

The GeTe material exhibits a transition between the crystalline ON-state and amorphous OFF-state that is attained by heating the PCM above its melting temperature (T_m) and followed by quenching the material, which solidifies the atoms in the amorphous state as shown in Fig. 4.1.

The RF PCM switch model in a series single-pole single-throw (SPST) configuration is shown in Fig. 4.8. Terminals 1 and 2 are RF input and output ports, while terminals 3 and 4 are bias pads to provide actuation pulse to the switch. A simplified switch model described in both the ON and OFF state is highlighted with the ON-state represented by a series R_{on} resistor while in the OFF-state, capacitance C_{pc} dominates.

An optical micrograph of a microfabricated RF PCM SPST switch is shown in Fig. 4.9(a), highlighting the RF signal input and output ports and bias ports for the actuation signal. The device is fully passivated and can be used for heterogeneous integration via flip-chip or wire-bonding. A false colored SEM image shows the closeup view of the narrow PCM channel l_s and width of the micro-heater w_h in Fig. 4.9(b). RF electrodes overlap the micro-heater to avoid switch failure due to cracks formation in GeTe near the edges of the micro-heater, as discussed in [19]. Figure 4.9(c) depicts a 3D-rendered view of the switch. The optical micrograph shown in Fig. 4.9(a) shows the natural colors of three different stacked dielectric layers. The overall switch area of the fabricated switch is 400×500 μm.

Switching from crystalline to amorphous state can be achieved by applying short high voltage pulse via the embedded micro-heater. Switching the PCM from an amorphous to a crystalline state is achieved by heating the material beyond its recrystallization

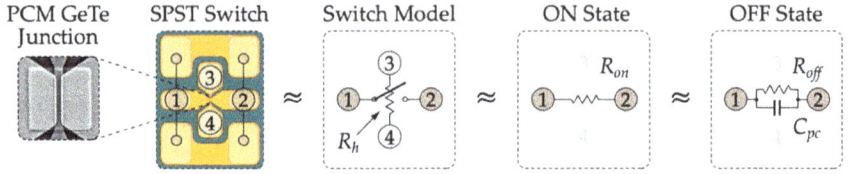

Fig. 4.8.   A four-terminal phase-change RF switch. Terminals 1 and 2 are RF input and output ports, while terminals 3 and 4 are bias pads to provide an actuation pulse to the switch. A simplified switch model in both the ON and OFF states is shown.

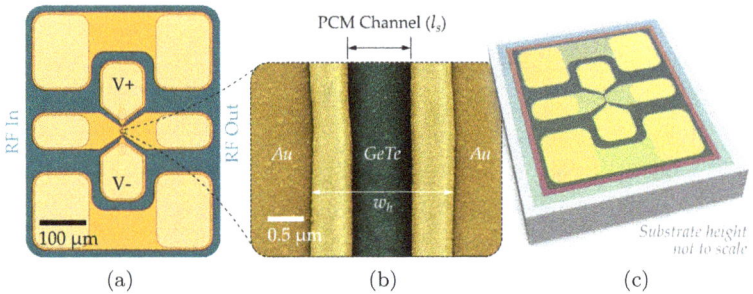

(a)   (b)   (c)

Fig. 4.9.   Fabricated RF PCM SPST switch (Gen 3): (a) Optical micrograph, (b) False colored SEM micrograph of the PCM channel, and (c) 3D rendered view of the PCM SPST switch [24].

temperature (T_c), at which the growth of crystalline grains and nucleation is enabled [18, 27]. High voltage pulses of 12 V and ($\sim$200 ns) are used to amorphize the material, and low voltage pulses of 7.8 V and ($\sim$1.2 μs) are used to crystallize the PCM. Pulse amplitude scales with the micro-fabrication process parameters of the thin-film resistor (TFR) material. The voltage required to change the GeTe states can be downscaled by reducing the resistivity of the micro-heater, either by changing the design dimensions or thickness of the thin film.

The measured and simulated results of the fabricated PCM-based RF SPST switch are shown in Fig. 4.10 [28]. It exhibits an insertion loss of less than 0.4 dB up to 60 GHz. The developed switches not only compete with the state-of-the-art but also outperform in certain aspects. Insertion loss and isolation performance is measured at 20°C

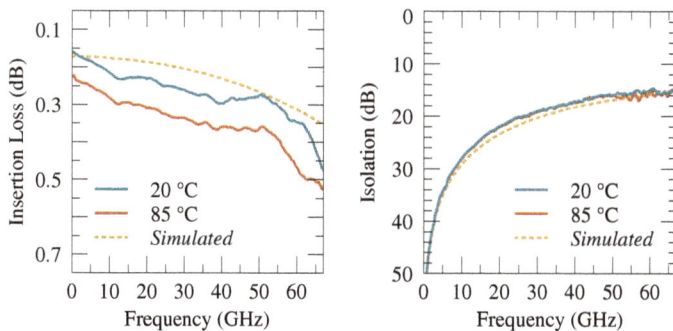

Fig. 4.10. Measured insertion loss and isolation of the PCM RF SPST switch over DC-67 GHz at 20°C and 85°C [28].

and 85°C. An increase in insertion loss is due to the increase in resistance of the material at crystalline state. Various parameters of the switch, including switch cycle testing, RF performance measurements, two-tone linearity, high-power handling, current carrying capacity, self-actuation, and reliability (lifetime cycles) have been experimentally measured and reported in [28].

4.5.1. *Comparison with state-of-the-art*

It is essential to compare the reported PCM-based switches with the current state-of-the-art. Most of the commercial RF switches are available in multi-port configurations. For a fair comparison, the RF PCM GeTe-based multi-port device is compared with the commercially available state-of-the-art multi-port RF switches developed using various technologies. The measured S-parameters of the RF switches used for comparison are taken directly from the manufacturers. The phase change single-pole three-throw (SP3T) switches [24] are compared with various broadband microwave switches developed using commercial RF-MEMS, magnetic relay, GaAs pHEMT, mmWave switches developed in-house using SOI-MEMS [29], and commercial Silicon-CMOS, UltraCMOS SOI, and GaAs MMIC technologies. Part numbers of the compared devices are given in the footnotes of Table 4.1 The RF switches used in

Table 4.1.   Comparison with the commercially available state-of-the-art RF switches [24, 28]

Technology	Device	Frequency Range (GHz)	Control Voltage (V)	Switching Time 10-90%	Insertion Loss (dB) at f_{max}*	Isolation (dB) at f_{max}*	Power Handling (dBm)	Linearity IIP3 (dBm)
PCM GeTe[a]	SP3T	DC-67	12	1.2 μs†	1.2	16	35.5	41
UltraCMOS SOI[b]	SP2T	DC-60	±3	12 ns	2.8	36	27	48
GaAs MMIC[c]	SP2T	DC-50	−5	11 ns	3.0	30	27	40
Silicon SOI[d]	SP4T	DC-44	±3.3	50 ns	3.0	31	27	50
SOI MEMS[e]	SPST	DC-40	8	2 ms	2.8	14	–	–
GaAs pHEMT[f]	SP2T	DC-20	−7	10 ns	1.7	39	25	41
Magnetic Relay[g]	SP2T	DC-18	12	7†	1.1	31	44.7†	–
RF MEMS[h]	SP4T	DC-14	3.6	75 μs	3.0	10	36	69
RF MEMS[i]	SP4T	DC-20	90	15 μs	1.0	18	44	95

[a]This work, developed in-house at the University of Waterloo, Canada

[b]pSemi, Part No. PE42525

[c]Analog Devices, Inc., Part No. HMC986A

[d]Analog Devices, Inc., Part No. ADRF5046

[e]Developed in-house at the University of Waterloo, Canada [29]

[f]Analog Devices, Inc., Part No. HMC347B

[g]Teledyne Relays, Part No. GRF121

[h]Analog Devices, Inc., Part No. ADGM1304

[i]Menlo Micro, Inc., Part No. MM5130

†No steady state power consumption.

*fmax = maximum operation frequency

γPower handling data is of a similar Part No. RF100/RF103.

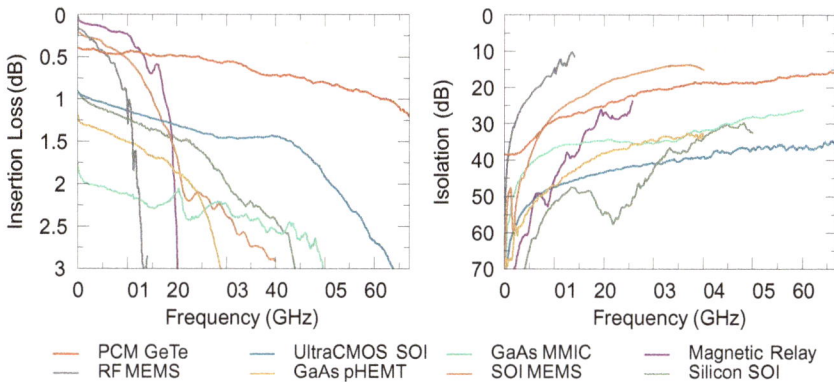

Fig. 4.11.   Measured insertion loss and isolation performance of current state-of-the-art RF switch technologies. Measured S-parameters data of the commercial RF switches are taken directly from the manufacturers [28].

comparison in Table 4.1 have their performance metrics over the others [28].

The RF performance of the compared devices in the ON- and OFF-states is shown in Fig. 4.11, and a comparative summary in tabular form is provided in Table 4.1 The RF PCM GeTe switch developed using the reported microfabrication process [28] outshines other technologies with its exceptional and best-in-class insertion loss performance, especially at the mmWave range. It offers the smallest die size among the state-of-the-art, has non-volatile or latching functionality, and an adequate switching speed. The UltraCMOS SOI technology-based single-port double-throw (SPDT) switch demonstrates excellent isolation, reasonable insertion loss till 60 GHz, and low voltage but has a 15× larger chip area than the PCM GeTe SP3T switch. The IIP3 of the PCM GeTe reported in the comparison is measured for the SPST series switch configuration.

Two commercially available and highly reliable SP4T RF-MEMS switches that guarantee more than one billion switching cycles are also included in the comparison. Currently, available commercial RF-MEMS demonstrate exceptional linearity performance but only work up to 22 GHz. The compared RF-MEMS switch from Analog Devices, Inc. (Part No. ADGM1304) has power handling capability close to the PCM GeTe switches but is much larger. An RF-MEMS switch from Menlo Micro, Inc. (Part No. MM5130) exhibits up to 44 dBm CW RF power handling and exceptional IIP3 of 95 dBm but is large and requires a high 90 V as a bias voltage.

The magnetic relay offers excellent RF performance, latching functionality, and outstanding power handling capability among the competition, but works up to 18 GHz only and is reliable till 3 M cycles with a slow switching time of 7 ms. Magnetic relays also have a relatively much larger size. GaAs MMIC and CMOS are top-of-the-line switches in terms of short switching time and low control voltage requirements but lack RF performance compared to other technologies.

Out of all the RF devices compared, each switch technology is unique in some way or the other and lacks in one or more aspects. Switch technology selection choice all comes down to the applications and frequency range. It depends on the use case scenario or application to decide which performance parameter is the switches' primary selection criteria. The emerging PCM GeTe-based developed switches push the boundaries by offering a technology that leads to devices with broadband operation frequency, low insertion loss, and miniature die size [24, 28].

4.5.2. *Bias signature, resistance change, and thermal crosstalk*

For biasing the monolithic RF PCM switches, the applied actuation voltage pulses are shown in Fig. 4.12(a), and the measured current is shown in Fig. 4.12(b). A high-speed refractory tungsten micro-heater is integrated underneath the PCM GeTe with a sandwiched barrier layer. The refractory micro-heaters utilized in the fabrication process exhibit a measured linear temperature coefficient of resistance, α, of $0.0031°C$ and a DC resistance of 35 Ω at 20°C that is measured using a 1-mV read-out voltage. The measured temperature profile of the micro-heater is shown in Fig. 4.12(c). The measured resistance with varying average pulse power is shown in Fig. 4.12(d). The RF

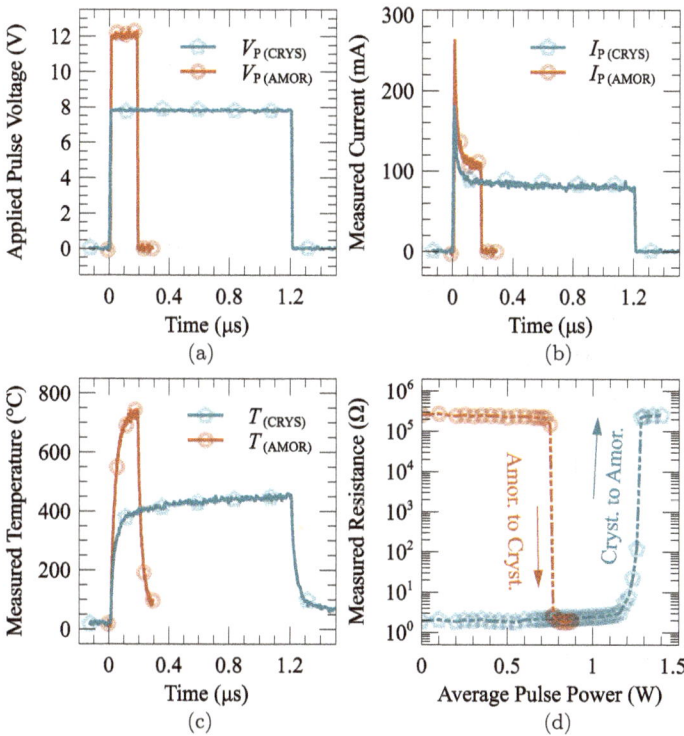

Fig. 4.12. The control signal applied to the RF PCM GeTe-based SPST switches for achieving non-volatile/latching reversible phase change between crystalline and amorphous states. (a) Applied amorphous and crystalline voltage pulse, (b) Measured current, (c) Measured temperature, and (d) Resistance of the device for applied average pulse power highlighting more than five orders of reversible resistance change [28].

PCM switch described above exhibits a resistance change of over five orders of magnitude between the crystalline and amorphous state. Compared to the current state-of-the-art semiconductor-based RF switches, the non-volatile functionality of the PCM GeTe switches does not consume any static DC power to hold the switch state.

It is worth highlighting that despite the high-temperature requirements beyond 750°C, the PCM-based RF switches can be integrated monolithically to develop complex RF devices without any thermal crosstalk concerns [26, 30]. The thermal energy is confined to the PCM junction area for a few nanoseconds only. We have verified the thermal actuation crosstalk and transient heat distribution in PCM switches experimentally utilizing thermoreflectance imaging [28].

Various multi-port miniaturized monolithically integrated complex RF components require several switches to be integrated very close to each other. PCM technology allows very tight integration of switches due to the smaller size of switches compared to other technologies. Experimental investigation of thermal actuation crosstalk using transient thermal imaging proves no actuation crosstalk as detailed in [31]. It also provides safe limits to closely integrate PCM switches monolithically.

To investigate the thermal cross-section in PCM switches, heat profiles are measured across 300 μm A–A' cross-section of the RF PCM switch as shown in Fig. 4.13 using transient thermoreflectance imaging technique. Tc is the recrystallization temperature and T_a is amorphization temperature. With a crystalline actuation pulse (8 V, 1.2 μs), heat distribution across the A–A' cross-section is measured at $t = 1.2$ μm. No thermal crosstalk or any hot spots are observed across 300 μm length. Temperature rises past T_c within 7 μm length and is less than 200°C within 14 μm around the junction. Metal electrodes sink the temperature; thus, heat is extremely localized and confined within the 15-μm junction surface area, as shown in Fig. 4.13(a). Amorphous pulse (12 V, 200 ns) is applied across the micro-heater terminal and across the A–A' cross-section, and no thermal hotspots or crosstalk is observed. Across 300 μm, the temperature above T_a is confined to a 5-μm junction area and is below 200°C within a 16-μm region. With the amorphous and crystalline pulses, the temperature is localized to a 20-μm area of the PCM junction and is below the safe limits of 50°C, as shown in Fig. 4.13(b).

For the rated bias signature, RF PCM switches can be integrated as close as 20 μm. If miniaturization is not the primary design criteria, switches should be placed at least 100 μm apart as ultimate safe limits. Further details on the analysis are described in [24, 31].

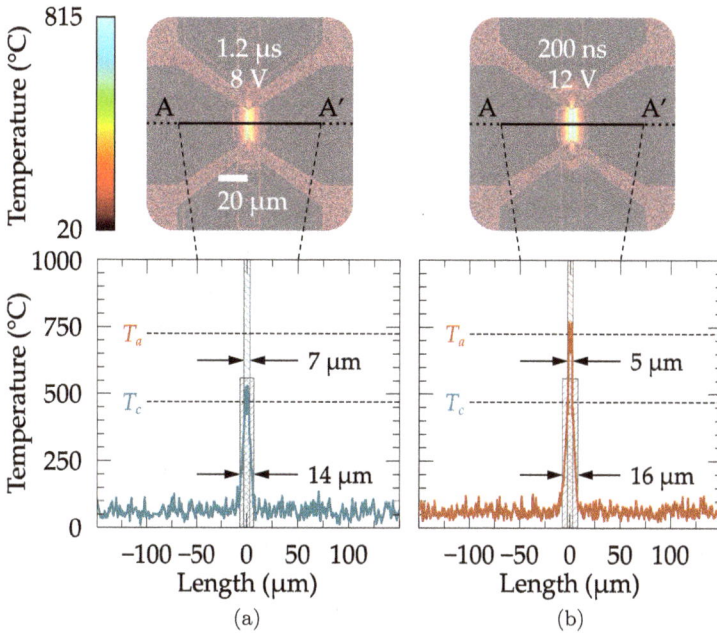

Fig. 4.13. Heat distribution across 300 μm A–A' cross-section of the RF PCM switches to study thermal crosstalk: (a) A crystalline pulse of amplitude 8 V and 1.2 μs width generates sufficient heat (Tc) to crystalline the PCM switch. (b) An amorphous pulse of amplitude 12 V and 200 ns width generates temperature (Ta) to melt the PCM GeTe. No thermal crosstalk is seen in (a) and (b). (c) A 15 V amplitude and 2 μs wide pulse are supplied to the heater, which is beyond its rated actuation limits. A thermal image is captured just before the switch breakdown showing crosstalk across an 80-μm width [31].

4.6. Potential Applications of PCM Switches in 5G Wireless Communication Systems

Miniature, reliable, and high-performance radio frequency (RF) switches are needed to configure antennas, matching networks, phase shifters, filters, and multiband amplifiers in RF front-end modules. RF switches are, therefore, the fundamental building blocks for realizing reconfigurable front ends in wireless and satellite communication systems. The ability to integrate PCM-based switches monolithically/heterogeneously with various other technologies without any need for special packaging makes the PCM technology affordable and attractive for the implementation of reconfigurable RF components for wireless communication systems. Table 4.2 shows a comparison of Figure of Merit (FOM) between semiconductor, MEMS, and PCM switch technologies. R_{ON} is the ON-state resistance that represents the insertion loss of the switch, whereas C_{OFF} is the OFF-state capacitor that represents the switch isolation. The FOM of PCM switches far exceeds that of semiconductor switches; additionally, PCM switches offer a superior linearity performance when compared with semiconductor switches. Research efforts are underway to improve the FOM factor of PCM switches to reach that of contact-type MEMS switches.

PCM switches offer unique advantages over any other known switching technology, particularly when operating at millimeter-wave and sub-terahertz frequencies. A paper was published in [32] on a Vanadium Oxide (VO_2) switch that is capable of operating from DC up to 220 GHz, promising the use of PCM technology in emerging sixth generations (6G) wireless systems. 5G communication networks also employ a wide range of millimeter-wave reconfigurable devices that require the use of switches or switched capacitor banks. Such networks promise a revolution in wireless communications with a wide range of applications in transportation, smart cities, healthcare, smart manufacturing, and wireless home entertainment. The technology will allow data transmission with unprecedented data rates and significantly reduced latency. In contrast to 4G networks that operate at low frequencies (majority below 3 GHz), 5G networks operate at higher frequencies to have enough bandwidth to achieve high data rates. 5G NR (New Radio) has been developed to operate in

Table 4.2.   A comparison of switch technologies. Figure of Merit (FOM) = $1/(2\pi^*\text{Ron}^*\text{Coff})$

Technology	FOM
Semiconductor switches	<1 THz
Capacitive MEMS switches	10 THz
Contact MEMS switches	20 THz
PCM switches	14 THz (at present)

two distinct bands — Sub-6 GHz Frequency Range 1 (FR1: 410 MHz to 7.125 GHz) and mmWave Frequency Range II (FR2: 24.25 GHz to 52.6 GHz). Despite FR1 running into 7 GHz, FR1 continues to be commonly referred to as the "Sub-6 GHz" [33].

Figure 4.14 illustrates the most important elements of RF technology that are essential to enable 5G: Massive MIMO, millimeter wave, beamforming, full-duplex, and machine-to-machine (M2M). These four key elements need to work together to achieve the low latency and high-speed data rate of the 5G technology. Below is a brief description of these key elements:

Massive multiple-input multiple-output (MIMO): The typical 4G-LTE base stations have 4–8 antennas, while 5G base stations are expected to have between 64 and 256 antennas to allow the sending and receiving of signals from many more users/devices at once, increasing the capacity of the communication network.

Millimeter wave: In 5G, more users and devices will be consuming more data than ever before. Therefore, a large bandwidth is needed to accommodate such large numbers of users/devices and to increase throughput to 10 Gbps and reduce latency to 1 ms. The solution

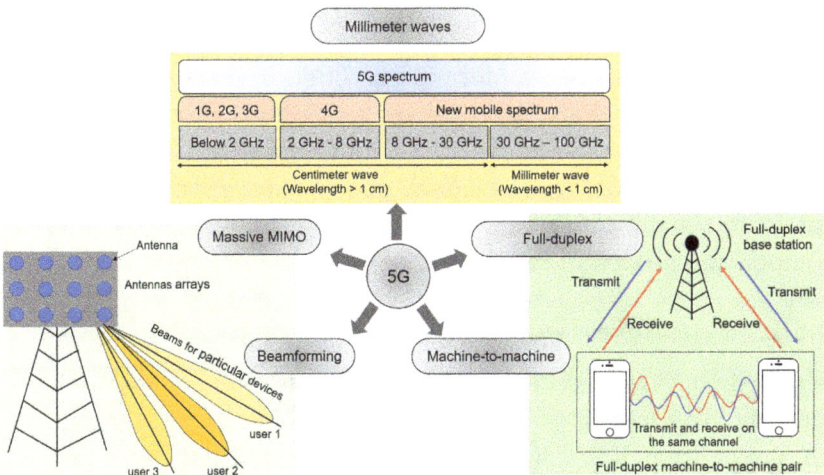

Fig. 4.14. Key technology elements of 5G systems [34].

is to use millimeter-wave frequencies. The millimeter-wave frequency bands that are currently considered for 5G are: (US, 27.5–28.35 GHz, 37–38.6 GHz, 38.6–40 GHz, 64–71 GHz), (EU, 24.25–27.5 GHz, 31.8–33.4 GHz, 40.5–43.5 GHz) and (Asia, 24.25–27.5 GHz, 26.5–29.5 GHz, 27.5–29.5 GHz, 37–42.5 GHz) [33].

Beamforming: For the massive MIMO array to operate efficiently, it needs to focus signals on concentrated beams that point only in the direction of the intended users/devices. This is essential to strengthen the signal's chances of arriving intact and to reduce interference to other users. Beamforming networks are needed to realize this function [33].

Full duplex: Conventional BS transceivers use different frequencies for receiving and transmitting signals. Having transceivers that can transmit and receive data at the same time, on the same frequency, doubles the wireless network capacity and helps to free bandwidth for other users [33–34].

Machine-to-Machine (M2M): 5G wireless networks are expected to drive the evolution of Machine-to-Machine (M2M) communication. New system architectures and technologies will need to be developed to enable 5G wireless networks to deal with M2M communication, while offering high speed and low latency.

Implementation of beamforming and full-duplex systems requires the use of reconfigurable RF devices, which, in turn, employ tunable elements and switches. In particular, full duplex systems require the use of tunable matching networks, phase shifters, tunable attenuators, and tunable delay lines. The following subsection describes the components needed in full-duplex systems.

4.6.1. *Full duplex systems*

Figure 4.15 shows a schematic of a full duplex system that uses one antenna. The receive chain mainly consists of an LNA, a mixer, and an analog-to-digital converter (ADC). The transmit chain consists of a power amplifier and a digital-to-analogue converter (DAC) (only the relevant portion of the system is shown in Fig. 4.15).

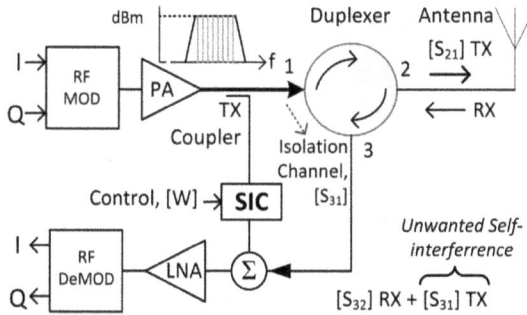

Fig. 4.15.   Basic concept of RF self-interference cancellation (SIC) [35].

The receive and transmit chains are connected to the antenna using a circulator. The interference between the transmit and the receive channels is attributed to three factors: (i) the limited isolation of the circulator, which is typically around 20 dB, (ii) interference that results from having the receive and transmit chains integrated on one board, (iii) interference that results from multipath reflections. Transmit signal reflections from objects near the antenna can lead to received signals of high power levels that can saturate the receiver. A tunable impedance matching network between the circulator and antenna can be employed [36] to circumvent the limited isolation of the circulator, effectively improving its isolation from 20 dB to 40 dB. A tunable matching network is needed since the antenna impedance can dynamically change due to changes in the surrounding environment. Cancellation of the interference due to multipath reflections that dynamically change with time can be implemented using a self-interference cancellation (SIC) network that consists of tunable time delay lines, phase shifters, and tunable attenuators [33]. The RF canceller network couples a portion of the transmit signal and splits it into branches. The signal in each branch acquires a particular delay, phase, and amplitude weighting before being combined into a signal that is used to cancel the self-interference. A total RF cancellation of 70–80 dB can be achieved with the use of the tunable matching network and the RF canceller network. The remaining interference cancellation needed can be realized either in the IF band or digitally in the baseband.

4.6.2. *Beamforming networks*

Beamforming is an essential component of phase array antennas. It is used in steering signals in a specific direction according to the location of connected users/devices. Basically, beamforming is a sum operation of weighted signals (in phase and amplitude) radiated from several antenna elements. There are three main architectures for beamforming: analog, digital, and hybrid. Figure 4.16(a) shows a basic configuration of an analog RF beamforming architecture [34]. The signals from the antenna elements are phase and amplitude reweighted and combined to create a beam in a certain direction. One limitation of this architecture is the difficulty in creating a large number of simultaneous beams in different directions to serve several users. Supporting multiuser transmission necessitates better control of beams direction and interference mitigation, which, in turn, can significantly enhance data rate in wireless networks. This can be realized using the digital beamforming shown in Fig. 4.16(b). While fully digital beamforming is optimum for use in a sub-6 GHz system, it is not practical for deployment in mmWave communications due to the excessive DC power consumption of analog-to-digital conversion (ADC)/digital-to-analog conversion (DAC) when operating at high frequencies. The hybrid beamforming architectures shown in Fig. 4.17 combine the advantages of analog and digital beamforming. It consists of multiple RF chains controlled by digital signal processing (DSP). The RF chains of the hybrid beamforming still need to use phase shifters.

It can be seen that phase shifters are key components in both analog and hybrid beamforming. CMOS-based phase shifters have been employed for a wide range of phase array antenna application due to the ease of integration with LNA/PA on a single chip. Nevertheless, PCM-based phase shifters can offer a superior linearity and insertion loss performance when operating at very high millimeter-wave frequencies. Moreover, they can be heterogeneously integrated with the RF chain or monolithically with Bi-CMOS [37] to build the whole RF chain on a single chip.

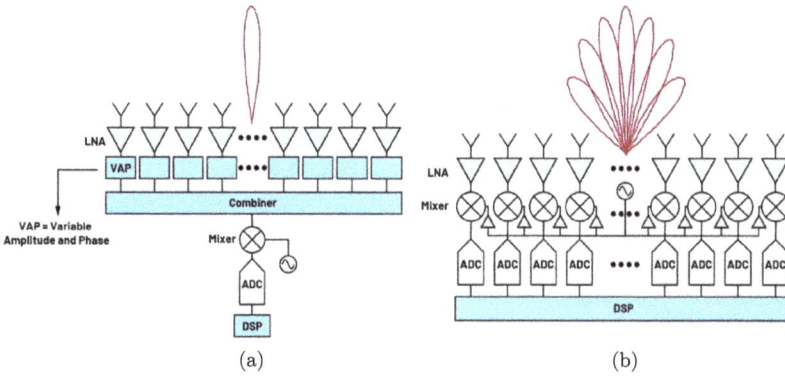

Fig. 4.16. (a) Analog and (b) Digital beamforming architectures [38].

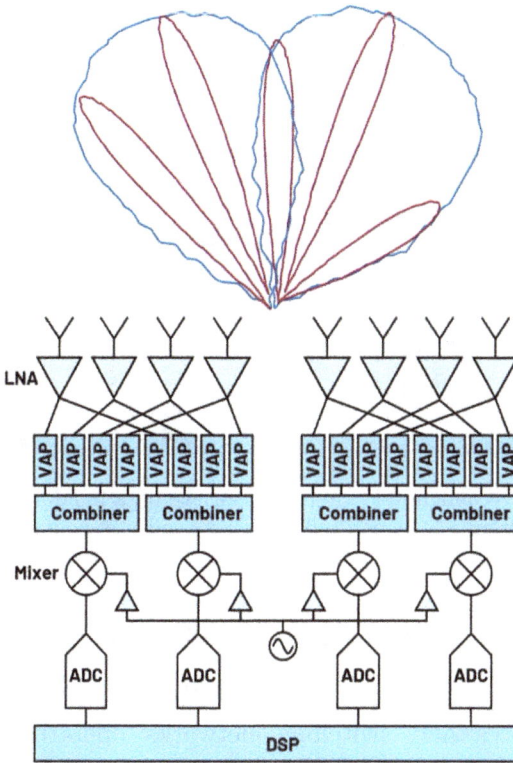

Fig. 4.17. Hybrid beamforming architecture with multiple analog beams [38].

4.6.3. *Reconfigurable intelligent surfaces*

The technology of reconfigurable intelligent surfaces (RISs) is one of the promising technologies for enhancing wireless coverage for 5G and emerging 6G systems [39]. A RIS consists of an array of unit cells whose EM reflection/transmission characteristics can be reconfigured using a tuning element such as semiconductor varactors/switches, liquid crystal, MEMS, or PCM technologies. RISs manipulate the incident electromagnetic (EM) waves, performing functions such as beam steering, anomalous reflection, switching, and beam splitting/collimation, as shown in a concept in Fig. 4.18. A reconfigurable reflect array consisting of unit cells controlled by VO_2 has been presented in [40], enabling beam steering. The PCM technology has good potential to be employed in RIS applications since PCM switches can be easily integrated monolithically with unit cells on a single substrate.

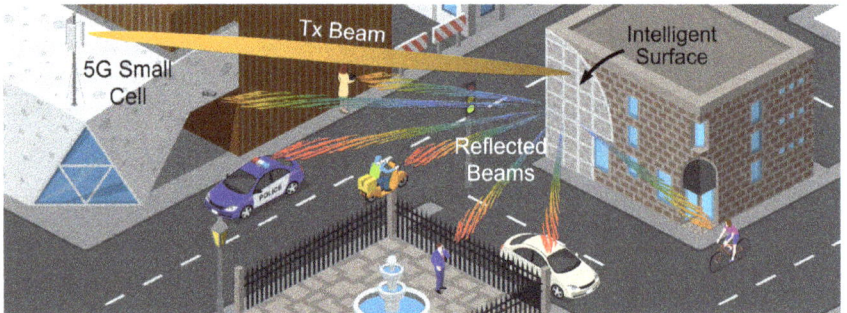

Fig. 4.18.   The reconfigurable intelligent surface on a building helps in steering beams received from a 5G small cell to various end-user equipment (UE).

4.7. PCM-based Reconfigurable mmWave Components

4.7.1. *A scalable crossbar switch matrix*

Switch matrices and switching networks operating at RF and mmWave have tremendous applications in telecommunications, radar systems, and instrumentation for effective bandwidth utilization, providing efficient signal routing, and for enhancing system redundancy. A common approach towards switch matrices is to integrate several single-pole n-throw (SPnT) switching unit cells along with an m×n interconnection network. Crossbar switch matrices offer a highly miniature design and are easily amenable to the realization of large size switch matrices. Crossbar switch matrices typically consist of cascaded cells with signal routing functionality. In general, the crossbar switch unit cells are cascaded to form large matrices.

A scalable crossbar switch matrix architecture shown in Fig. 4.19 consists of an array of switch unit cells (SU$_{x,y}$) and SP2T switches (SD$_{x,y}$) connected in a grid pattern to achieve signal routing. RF signal at any m input ($RFi_1, RFi_2, \ldots RFi_{m-1}, RFi_m$) can be routed to any available n output port (RFo_1, RFo_2, $\ldots$ RFo_{n-1}, RFo_n). Cross-over paths designed in switch unit-cells offer signal routing between two overlapping RF paths. With the switch architecture shown in Fig. 4.19, an $m \times n$ crossbar switch matrix requires a total of $(mn - 1)$ total cells, with $((m-1) \times (n-1))$ switch unit-cells and $((m-1) + (n-1))$ SP2T switches [25, 30].

An optical micrograph of the PCM GeTe-based 4×4 switch matrix is shown in Fig. 4.20. The inset shows magnified optical micrographs of SP2T switch elements, impedance matching, and capacitance compensation of bias network, and a unit-cell implemented in the crossbar switch matrix highlighting turn state. The crossbar switch matrix is highly miniaturized with an overall device size of 0.5×0.75 mm, including RF input and output ports and excluding control pads. With the control pads arrangement depicted in Fig. 4.20, the device periphery is 0.96×0.77 mm.

The RF performance is measured between seven unique routes to get the minimum and maximum performance limits of the 4×4 crossbar switch matrix, as shown in Fig. 4.21. In the best-case scenario (Route 1), the signal passes through two PCM SPST switches, one 90° bend, and one RF crossover junction. While in the worst-case scenario (Route 7), the signal is routed between RFi_4 and RFo_4 ports, which consists of 6 SPDT switches, one 90° bend, and three bias bridges with more than 16 conductive bias wires crossing underneath bias bridges. Despite all the signal degradation elements in the RF path, the switch matrix is designed to exhibit less than 4.2 dB insertion loss in the worst case. Most of the signal routing combinations show an insertion loss of less than 4 dB, measured

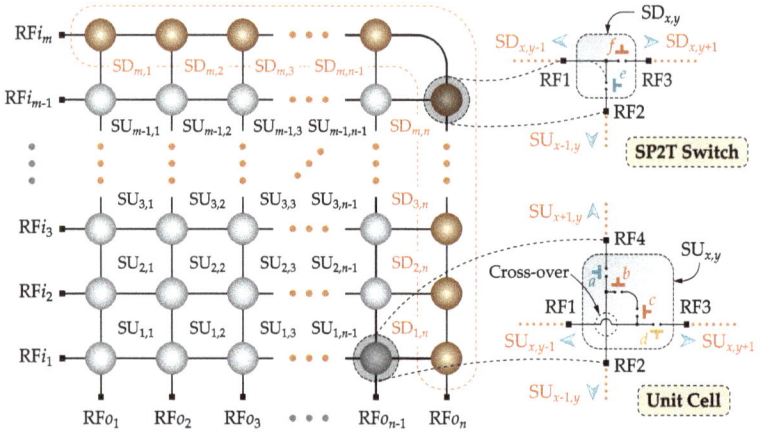

Fig. 4.19.   Scalable m × n crossbar switch matrix architecture utilizing switch unit-cells and SP2T switches arranged in a grid pattern to achieve signal routing [30].

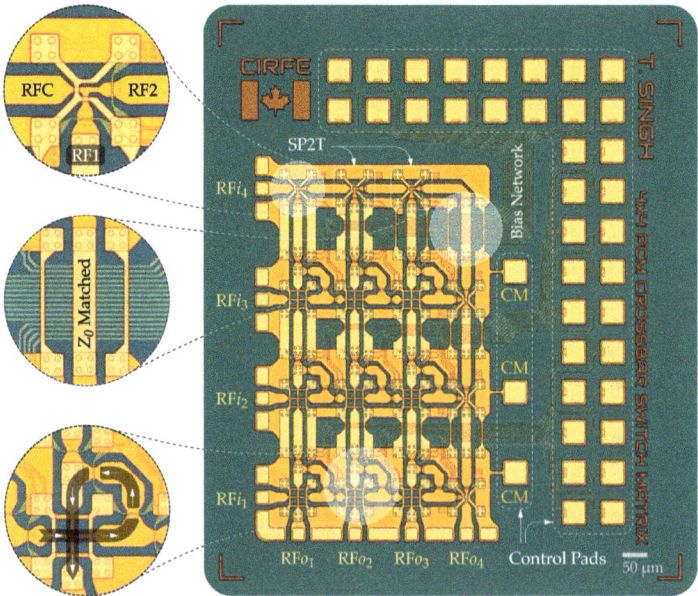

Fig. 4.20.   Optical micrograph of the PCM GeTe-based 4×4 switch matrix. The inset shows magnified optical micrographs of the SP2T switch elements, impedance matching and capacitance compensation of bias network, and a unit-cell implemented in a switch matrix highlighting turn state [30].

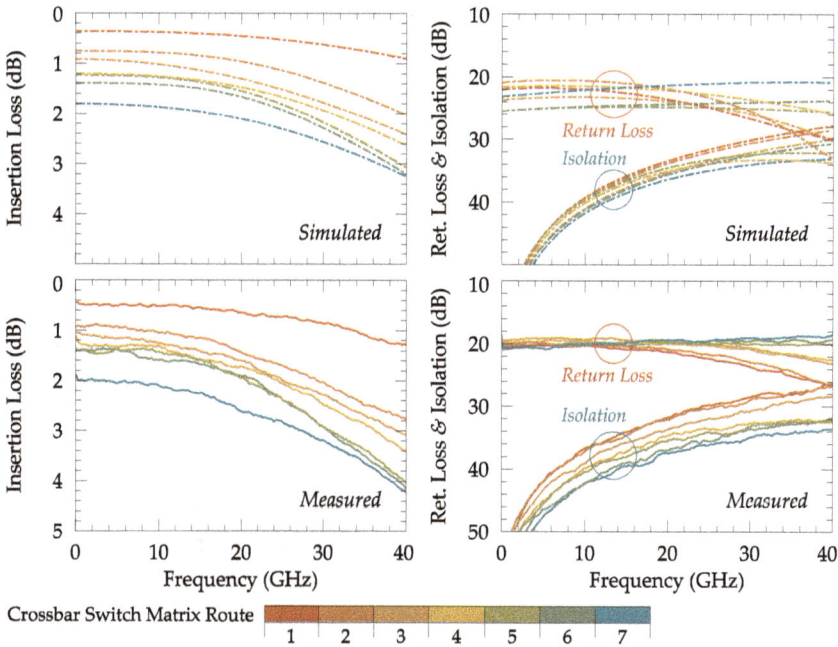

Fig. 4.21. Measured and simulated RF performance of a PCM GeTe-based 4×4 switch matrix over DC to 40 GHz. The RF performance is shown for seven possible routes, with route 7 exhibiting the worst-case scenario [30].

return loss better than 18 dB, and isolation higher than 26 dB over DC to 40 dB. The EM-simulated performance shows a close match with the measured results, with the worst-case loss lower than 3.2 dB, return loss better than 20 dB, and isolation higher than 26 dB. The difference between the measured and simulation insertion loss is due to the fabrication tolerances and deposited material properties being different from simulation models.

4.7.2. *mmWave phase shifters*

Phase shifters are crucial components for electronic beam steering in phased-array systems. Phase shifters are widely used to change the excitation phase of an individual antenna element in phased-array antennas. mmWave phased-array antennas provide the capability of real-time beam steering with high efficiency in a miniaturized package for applications, including but not limited to high-speed 5G cellular communication, automotive radar, and satellite communication.

A 3-bit mmWave switched true-time-delay (TTD) phase shifter based on PCM GeTe is discussed. The phase shifter is designed using four monolithically integrated PCM SP3T switches to route the signal through delay lines. The insertion loss variation between various states is minimized by integrating two fixed PCM GeTe elements maintained in the crystalline state, along with an optimized width of the delay lines. The SP3T switches are connected back-to-back in two stages to provide a 3-bit phase shift with 20° precision.

The monolithically integrated PCM GeTe-based 3-bit switched TTD phase shifter is designed by utilizing SP3T switches to route the RF signal through a combination of delay/transmission line sections. Four delay line sections ($t_1 - t_4$) are cascaded in two stages ("A" and "B") to form a 3-bit device in reference to a delay line (t_0), as shown in Fig. 4.22. Integrated PCM-based SP3T switches utilize series SPST switches in an extremely compact area for routing the signal between three available paths in Fig. 4.22. Delay lines provide phase shift as $t_0 = 0°$ (Reference), $t_1 = 20°, t_2 = 40°, t_3 = 60°$, and $t_4 = 120°$. A short-line segment (t_{0c}) connects two-phase shifter stages ("A" and "B"), as shown in Fig. 4.22. With an adequate selection of the SP3T switching state, the phase shifter would exhibit 20° phase precision in nine discrete states.

Individual delay line sections are designed and optimized to minimize the insertion loss variation by varying the width of the

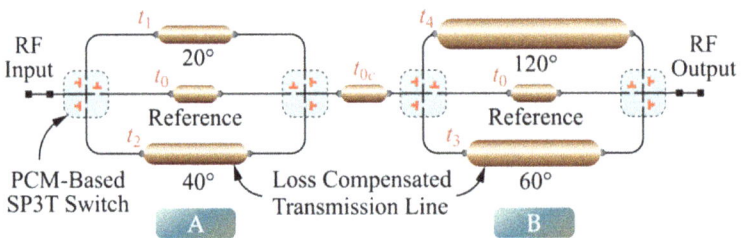

Fig. 4.22.   Schematic of a loss-compensated 3-bit switched TTD phase shifter consisting of SP3T switches and delay lines (t_0 to t_4) cascaded in two stages [26].

lines while keeping impedance matched. A large co-planar waveguide (CPW) signal linewidth is selected for longer phase delay lines, while smaller CPW signal line widths are chosen for shorter delay lines to minimize the variation of the insertion loss. The width of the phase delay line t_4 is larger than subsequent delay lines to match the insertion loss with the lowest bit t_0.

The RF signal passes through t_{0c} in all possible states. In addition to t_{0c}, in state 1, the signal passes through two t_0 sections, while the signal routes through $t_1 + t_0$, $t_2 + t_0$, $t_0 + t_3$, or $t_0 + t_4$ in states 2, 3, 4, and 7, respectively, utilizing one delay line segment in conjunction with t_0. In states 5, 6, 8, and 9, the signal passes through two delay line segments $t_1 + t_3$, $t_2 + t_3$, $t_1 + t_4$, or $t_2 + t_4$, bypassing both t_0 segments, respectively. To compensate for the high insertion loss exhibited in the states when a combination of two delay line segments are selected, two fixed PCM GeTe elements (S_0) are designed without a micro-heater. They are monolithically integrated in two t_0 sections to minimize the insertion loss deviation and to match the loss in states 1, 2, 3, 4, and 7 with the remaining states as highlighted in the schematic and in the optical micrograph shown in Fig. 4.23. PCM elements (S_0) are kept in crystalline state to compensate for the loss in lower order states of the phase shifter.

PCM-based technology provides flexibility to highly miniaturize the integrated phase shifter. The overall device core area is under 0.42 mm^2, making it a highly miniaturized phase shifter, as shown in the layout of the fully integrated phase shifter in Fig. 4.24. The

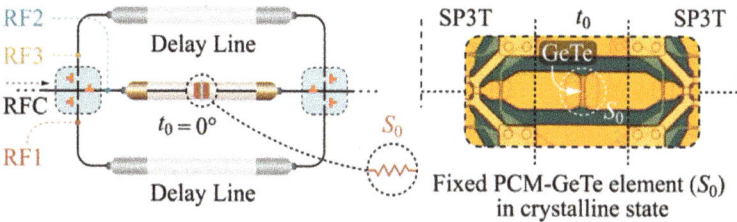

Fig. 4.23.   Loss compensation in t_0 sections using fixed PCM GeTe element (S_0) in each stage of a phase shifter. Both PCM elements S_0 are latched in crystalline state and do not require any actuation. The addition of S_0 minimizes insertion loss variation between phase shifter states [26].

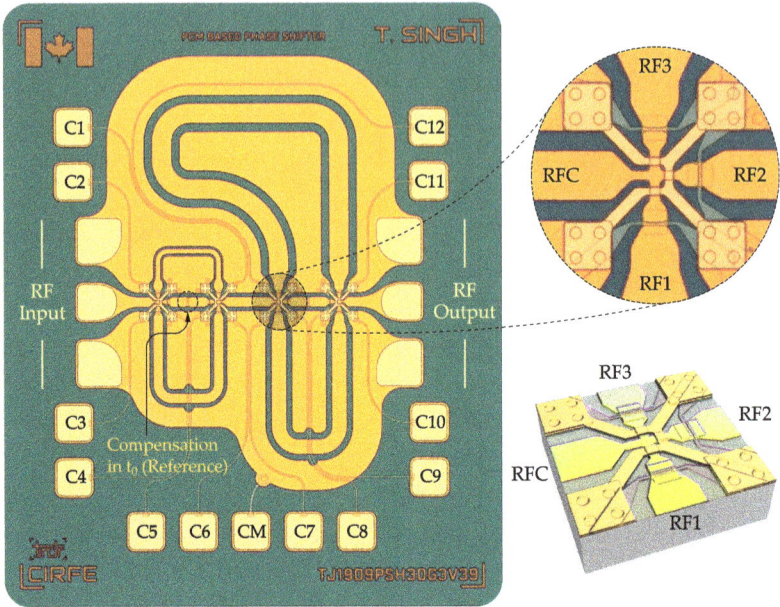

Fig. 4.24.　Optical micrograph of the monolithically integrated PCM-based 3-bit mmWave switched TTD phase shifter. The inset shows an optical micrograph of the zoomed-in view of a PCM-based SP3T switch. The 3D-rendered view of the PCM SP3T switch core is highlighted [26].

PCM SP3T switch has one common input port, "RFC", and three output ports, "RF1, RF2, and RF3", as shown in Fig. 4.23. The desired phase-shift level is achieved by routing the RF signal through delay lines and actuating the cascaded PCM-based switches with the application of an OFF or ON pulse. The control pulse is provided between respective control pads C1, C2, ..., C12 and a common bias pad CM.

An optical micrograph of the fabricated PCM-based 3-bit switched TTD phase shifter is shown in Fig. 4.24, highlighting the overall device size, control pads, delay lines, and loss compensation fixed PCM elements. The inset of Fig. 4.24 shows the optical micrograph of the close-up view of monolithically integrated PCM SP3T switches. The 3D-rendered SP3T switch is also shown in Fig. 4.24. Control pads provide the desired phase shift by reconfiguring the respective PCM switches in the desired phase shift path. Four RF PCM GeTe-based SP3T switches (12 RF SPST switch elements), along with two fixed PCM GeTe elements S_0 in t_0 sections (these elements are always in the ON-state to compensate loss), are monolithically integrated to form the phase shifter, as shown in Fig. 4.24.

The phase shifter is designed to operate over an 8-GHz wide frequency band with a center frequency of 30 GHz. The presented phase shifter is highly miniaturized with an overall device area of 0.42 mm^2. The measured and simulated relative phase shift is depicted in Fig. 4.25. It exhibits a measured average loss of 4.3 dB with a variation of only ±0.3 dB and a return loss better than 20 dB, demonstrating a highly linear 180° phase shift with a 20° step precision over the operational bandwidth with a FOM of 42°/dB.

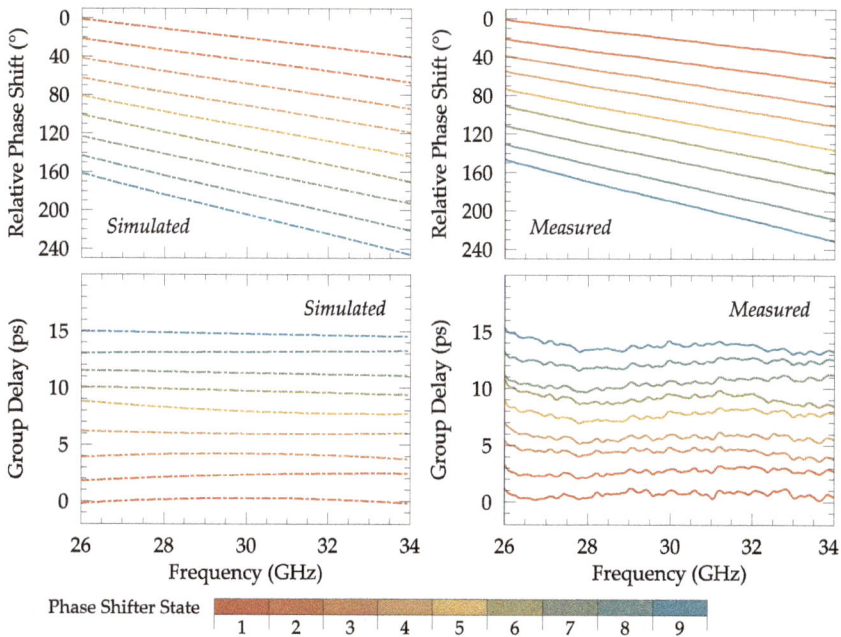

Fig. 4.25.   Measured and simulated phase shift and group delay of the PCM-based 3-bit TTD phase shifter over a 26–34 GHz band. The response of all possible states is shown [26].

The presented phase shifter exhibits less than 15 ps of measured group delay in 2 ps per state precision, as shown in Fig. 4.25. The demonstrated group delay is the lowest compared to the current state-of-the-art [26].

4.7.3. *Wideband mmWave digital attenuator*

Variable attenuators are commonly used for adjusting signal levels in various radio frequency (RF) circuits, such as full duplex wireless systems, radar systems, automatic gain control amplifiers, and vector modulators, to name a few [41]. Miniaturized attenuators with high linearity and precision are highly in demand, particularly for millimeter wave (mmWave) applications [42]. Ka-band, especially at 28 GHz, have tremendous applications for 5G wireless communication systems, especially beamforming networks [43] and in full duplex systems [36].

The monolithically integrated PCM-based 4-bit attenuator is designed by utilizing SPDT switches to route the RF signal through a section of transmission lines or through an attenuator section (bit). Four attenuator bits (A–D) are cascaded to form a 4-bit device. Individual attenuator bits are designed using high-frequency wide-band integrated passive bridged-T resistor networks to provide discrete attenuation levels. Bits (A–D) provide attenuation levels of 3 dB, 6 dB, 9 dB, and 18 dB, respectively. Combining four bits provides 16 discrete attenuation levels [43].

An optical micrograph of the fabricated reconfigurable PCM-based 4-bit attenuator is shown in Fig. 4.26, highlighting the overall device size, control pads, and individual passive bridged-T attenuators. Embedded high-frequency wide-band resistors are optimized for the desired 8 GHz frequency band. Eight SPDT switches are monolithically integrated to load/unload the desired passive bridged-T resistor network, which provides certain attenuation in the RF path.

The measured and simulated RF response of the proposed PCM-based 4-bit variable attenuator is shown in Fig. 4.27, from 24 to

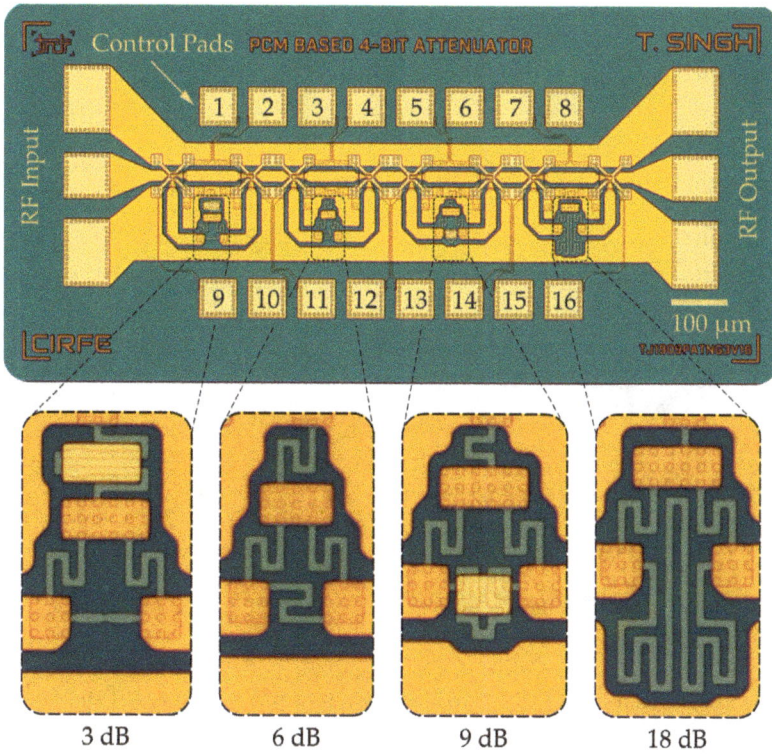

Fig. 4.26. Optical micrograph of the monolithically integrated PCM-based 4-bit variable attenuator. Control pads 1–16 are used to tune the attenuation levels. Four passive bridged-T attenuator sections are shown in a zoomed-in view [43].

32 GHz. The response of all 16 states (4-bit) is demonstrated in Fig. 4.27. The EM-simulated response shows a minimum attenuation of 3.6 dB and maximum attenuation of 38 dB at 28 GHz. All measurements were done on a wafer at room temperature. The measured RF response exhibits a minimum attenuation of 4.7 dB and maximum attenuation of 37 dB at the center frequency. Simulated and measured return loss is better than 20 dB over the bandwidth. Attenuator bits (A–D) are designed to provide fixed attenuation relative to the reset state, where all attenuator bits are unloaded.

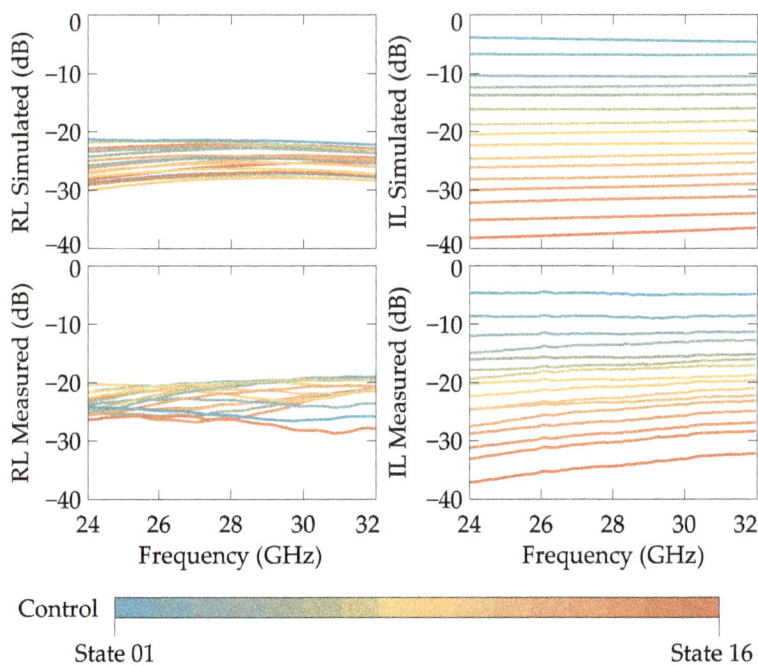

Fig. 4.27.   Measured and simulated insertion loss and return loss of the proposed PCM-based 4-bit variable attenuator over a 24–32-GHz band. The response of all 16 states (4-bit) is shown [43].

4.8. Conclusion

RF switches are the fundamental building blocks for realizing reconfigurable devices in communication systems. While semiconductor RF switches are widely used in the majority of commercial applications, they perform adequately up to a few GHz. The PCM technology has the potential to realize RF switches with superior performance of up to mmWave frequencies. The PCM GeTe-based switch offers a unique latching capability, reducing the DC power consumption. The key advantage of the PCM technology is that it is easily amenable to monolithic integration with RF circuits such as phase shifters, tunable filters, antenna impedance tuners, switched attenuators, and tunable delay lines. While a wide range of PCM-based RF devices has been successfully demonstrated, the RF PCM technology is still considered to be at its infancy. More research efforts are certainly needed to improve the switch reliability and its power handling capability.

References

[1] S. R. Ovshinsky, "Reversible electrical switching phenomena in disordered structures," *Phys. Rev. Lett.*, vol. 21, no. 20, pp. 1450–1453, 1968.

[2] A. T. Waterman, "On the positive ionization from certain hot salts, together with some observations on the electrical properties of molybdenite at high temperatures," *Philo. Mag. J. Science*, vol. 33, no. 195, pp. 225–247, 1971.

[3] S. Raoux and M. E. Wuttig, *Phase Change Materials: Science and Applications*. Springer Science & Business Media, 2010.

[4] R. Neale, D. Nelson and G. E. Moore, "Nonvolatile and reprogrammable, the read-mostly memory is here," *Electronics*, vol. 43, no. 20, pp. 56–60, 1970.

[5] R. Shanks and C. Davis, "A 1024-bit nonvolatile 15ns bipolar read-write memory," *IEEE Int. Solid-State Circuits Conf. Digest*, San Francisco, CA, USA, pp. 112–113, 1978.

[6] A. V. Kolobov, P. Fons, A. I. Frenkel, A. L. Ankudinov, J. Tominaga and T. Uruga, "Understanding the phase-change mechanism of rewritable optical media," *Nat. Mater.*, vol. 3, no. 10, pp. 703–708, 2004.

[7] T. Ohta, "Phase-change optical memory promotes the DVD optical disk," *J. Optoelectron. Adv. Mater.*, vol. 3, no. 3, pp. 609–626, 2001.

[8] T. Ohta and S. Ovshinsky, *Photo-Induced Metastability in Amorphous Semiconductors*. Wiley-VCH Verlag GmbH & Co., Weinheim, Germany, 2003.

[9] E. K. Chua, L. P. Shi, R. Zhao, K. G. Lim, T. C. Chong, T. E. Schlesinger and J. A. Bain, "Low resistance, high dynamic range reconfigurable phase change switch for radio frequency applications," *Appl. Phys. Lett.*, vol. 97, no. 18, pp. 95–98, 2010.

[10] A. Crunteanu, J. Givernaud, J. Leroy, D. Mardivirin, C. Champeaux, J.-C. Orlianges, A. Catherinot and P. Blondy, "Voltage-and current-activated metal–insulator transition in VO2-based electrical switches: A lifetime operation analysis," *Sci. Technol. Adv. Mater.*, vol. 11, no. 6, p. 065002, 2010.

[11] S. D. Ha, Y. Zhou, C. J. Fisher, S. Ramanathan and J. P. Treadway, "Electrical switching dynamics and broadband microwave characteristics of VO2 radio frequency devices," *J. Appl. Phys.*, vol. 113, no. 18, p. 184501, 2013.

[12] M. Field, C. Hillman, P. Stupar, J. Hacker, Z. Griffith and K.-J. Lee, "Vanadium dioxide phase change switches," *Proc. SPIE Open Architecture/Open Business Model Net-Centric Systems and Defense Transformation*, vol. 9479, no. 947908, pp. 1–8.

[13] M. Wang and M. Wais-Zadeh, "Directly heated four-terminal phase change switches," *Proc. IEEE MTT-S Int. Microw. Symp. Digest (IMS)*, Tampa, FL, USA, pp. 1–4, 2014.

[14] N. El-Hinnawy, P. Borodulin, E. B. Jones, B. P. Wagner, M. R. King, J. S. Mason, J. Bain, J. Paramesh, T. E. Schlesinger, R. S. Howell, M. J. Lee and R. M. Young, "12.5 THz Fco GeTe inline phase-change switch technology for reconfigurable RF and switching applications," *Proc. IEEE Compound Semiconductor Integrated Circuit Symposium (CSICS)*, La Jolla, CA, USA, pp. 1–3, 2014.

[15] J.-S. Moon, H.-C. Seo, D. Le, H. Fung, A. Schmitz, T. Oh, S. Kim, K.-A. Son and B. Yang, "10.6 THz figure-of-merit phase-change RF switches with embedded micro-heater," *Proc. 15th IEEE Topical Meeting on Silicon Monolithic Integrated Circuits in RF Systems (SiRF)*, San Diego, CA, USA, pp. 73–75, 2015.

[16] A. Mennai, A. Bessaudou, F. Cosset, C. Guines, P. Blondy and A. Crunteanu, "Bistable RF switches using Ge2Sb2Te5 phase change material," *Proc. 2015 Euro. Microw. Conf. (EuMC)*, Paris, France, pp. 945–947, 2015.

[17] J. Jiang, K. W. Wong and R. R. Mansour, "A VO2-based 30 GHz variable attenuator," *Proc. IEEE MTT-S Int. Microw. Symp. Digest (IMS)*, Honolulu, HI, USA, pp. 911–913, Jun. 2017.

[18] T. Singh and R. R. Mansour, "Chalcogenide phase change material GeTe based inline RF SPST series and shunt switches," *Proc. IEEE MTT-S International Microwave Workshop Series on Advanced Materials and Processes for RF and THz Applications (IMWS-AMP)*, Ann Arbor, MI, USA, pp. 1–3, Jul. 2018.

[19] T. Singh and R. R. Mansour, "Characterization, optimization and fabrication of phase change material germanium telluride based miniaturized DC–67 GHz RF switches," *IEEE Trans. Microw. Theor. Tech.*, vol. 67, no. 8, pp. 3237–3250, Aug. 2019.

[20] E. K. Chua, "Development of phase change switches with low resistance in the "ON" state," PhD Thesis, Carnegie Mellon University, Pittsburg, PA, USA, 2011.

[21] A. I. Lebedev, I. A. Sluchinskaya, V. N. Demin and I. H. Munro, "Influence of Se, Pb and Mn impurities on the ferroelectric phase transition in GeTe studied by EXAFS," *Ph. Transit.*, vol. 60, no. 2, pp. 67–77, 1997.

[22] E. I. Givargizov, A. Melnikova and D. W. Wester, *Growth of Crystals.* Springer, 1986.

[23] R. A. Hein, J. W. Gibson, R. Mazelsky, R. C. Miller and J. K. Hulm, "Superconductivity in germanium telluride," *Phys. Rev. Lett.*, vol. 12, no. 12, p. 320, 1964.

[24] T. Singh, "Monolithically integrated phase change material GeTe-based RF components for millimeter wave applications," University of Waterloo, Waterloo, ON, Canada, 2020.

[25] T. Singh and R. R. Mansour, "Ultra-compact phase-change GeTe-based scalable mmWave latching crossbar switch matrices," *IEEE Trans. Microw. Theor. Tech.*, vol. 70, no. 1, pp. 938–949, 2021.

[26] T. Singh and R. R. Mansour, "Loss compensated PCM GeTe-based wide-band 3-bit switched True-Time-Delay phase shifters for mmWave phased arrays," *IEEE Trans. Microw. Theor. Tech.*, vol. 68, no. 9, pp. 3745–3755, Sept. 2020.

[27] T. Singh and R. R. Mansour, "Investigation into self actuation limitation and current carrying capacity of chalcogenide phase change GeTe-based RF switches," *IEEE Trans. Electron. Dev.*, vol. 67, no. 12, pp. 5717–5722, Dec. 2020.

[28] T. Singh and R. R. Mansour, "Experimental investigation of performance, reliability, and cycle endurance of nonvolatile DC–67 GHz phase-change RF switches," *IEEE Trans. Microw. Theor. Tech.*, vol. 69, no. 11, pp. 4697–4710, Nov. 2021.

[29] N. K. Khaira, T. Singh and R. R. Mansour, "Monolithically integrated RF MEMS-based variable attenuator for millimeter-wave applications," *IEEE Trans. Microw. Theor. Tech.*, vol. 67, no. 8, pp. 3251–3259, Aug. 2019.

[30] T. Singh and R. R. Mansour, "Ultra-compact phase-change GeTe-based scalable mmWave latching crossbar switch matrices," *IEEE Trans. Microw. Theor. Tech.*, vol. 70, no. 1, pp. 938–949, Jan. 2022.

[31] T. Singh and R. R. Mansour, "Experimental investigation of thermal actuation crosstalk in phase-change RF switches using transient thermoreflectance imaging," *IEEE Trans. Electron. Dev.*, vol. 68, no. 7, pp. 3537–3544, Jul. 2021.

[32] C. Hillman, P. A. Stupar and Z. Griffith, "VO2 Switches for millimeter and submillimeter-wave applications," *IEEE Compound Semiconductor Integrated Circuit Symposium (CSICS)*, Orleans, LA, USA, 2015.

[33] W. Hong, Z. Jiang, C. Yu, *et al.*, "The role of millimeter-wave technologies in 5G/6G wireless communications," *IEEE J. Microw.*, vol. 1, no. 1, pp. 101–122, Jan. 2021.

[34] H. Hao, D. Hui and D. Lau, "Material advancement in technological development for the 5G wireless communications," *Nanotechnol. Rev.*, vol. 9, no. 1, pp. 683–699, 2020.

[35] B. Keogh and A. Zhu, "Wideband self-interference cancellation for 5G full-duplex radio using a near-field sensor array," *2018 IEEE MTT-S International Microwave Workshop Series on 5G Hardware and System Technologies (IMWS-5G)*, Dublin, Ireland, 2018.

[36] C. F. Campbell, J. A. Lovseth, S. Warren, A. Weeks and P. B. Schmid, "A BST varactor based circulator self interference canceller for full duplex transmit receive systems," *2020 IEEE/MTT-S Int. Microw. Symp. (IMS)*, Atlanta, GA, USA, 2020.

[37] N. El-Hinnawy, G. Slovin, J. Rose and D. Howard, "A 25 THz Fco (6.3 fs Ron*Coff) phase-change material RF switch fabricated in a high volume manufacturing environment with demonstrated cycling > 1 billion times," *2020 IEEE/MTT-S Int. Microw. Symp. (IMS)*, Atlanta, GA, USA, 2020.

[38] P. Saha, *A Quantitative Analysis of the Power Advantage of Hybrid Beamforming for Multibeam Phased Array Receivers*. Analog Devices, 2021.

[39] M. S. Ayub, P. Adasme, I. Soto and D. Z. Rodriguez, "Reconfigurable intelligent surfaces enabling future wireless communication," *2021 Third*

South American Colloquium on Visible Light Communications (SACVLC), Toledo, Brazil, 2021.

[40] R. Matos, A.-S. Kaddour, S. V. Georgakopoulos and N. Pala, "VO2 based ultra-reconfigurable Ka-band reflectarrays for next-generation communication and radar systems," *IEEE International Symposium on Antennas and Propagation and USNC-URSI Radio Science Meeting (APS/URSI)*, Singapore, 2021.

[41] Y. Huang, W. Woo, Y. Yoon and C. Lee, "Highly linear RF CMOS variable attenuators with adaptive body biasing," *IEEE J. Solid-State Circuits*, vol. 46, no. 5, pp. 1023–1033, May 2011.

[42] H. Dogan, R. G. Meyer and A. M. Niknejad, "Analysis and design of RF CMOS attenuators," *IEEE J. Solid-State Circuits*, vol. 43, no. 10, pp. 2269–2283, Oct. 2008.

[43] T. Singh and R. R. Mansour, "Miniaturized reconfigurable 28 GHz PCM-based 4-bit latching variable attenuator for 5G mmWave applications," *Proc. IEEE MTT-S Int. Microw. Symp. Digest (IMS)*, Los Angeles, CA, USA, pp. 53–56, Aug. 2020.

CONFORMITY

There is cost to every change, but there is also a cost
if there is no change.

Kiat Seng Yeo

Chapter 5

Standardization of 6G Wireless Communication Systems

Sumei Sun, Yonghong Zeng and Amnart Boonkajay

The fifth generation (5G) network promises to deliver enhanced mobile broadband (eMBB), ultra-reliability low-latency communications (URLLC) for mission-critical Internet of Things (IoT), and massive machine type communications (m-MTC) for massive IoT, aiming to drive digital transformation in all industry sectors. Since the third-generation partnership project (3GPP) published Release 15 in 2019, the first release of the 5G new radio (NR) standard, 5G networks have started being deployed in many places in the world. Meanwhile, academia, industry, and national and international regulators have started developing visions, identifying use cases, and specifying enabling technologies for 6G communications. Many white papers have been published, research programs started, and forums and workshops organized, providing good references for the 6G standard's development by 2030.

In this chapter, we will summarize the 6G vision, enabling technologies, and vertical use cases that are being discussed in white papers, forums, and workshops. We will then provide an overview on 6G standardization. The chapter is organized as follows. In Section 5.1, we summarize the 6G vision. In Section 5.2, the candidate-enabling technologies are described. Section 5.3 presents the research initiatives, and Section 5.4 presents the 6G standardization process, including regulations, standard development organizations (SDOs), and industry forums. Finally, Section 5.5 concludes the Chapter.

5.1. The 6G Vision

A number of 6G white papers have been published by industry companies, industry alliances, academic research centers, and public-private partnered research programs. Please see [1–30]. This list is by no means exclusive. It keeps increasing and getting updated. However, there are good consensus and well-aligned views. In this section, we provide a quick summary of the 6G vision from these white papers.

Moving to 2030, the physical world, digital world, and human world will be even more seamlessly connected and interacted, creating brand new experiences in work, leisure, social activities, learning, and study. Real-time fusion of the virtual and physical worlds, extended human senses, and the human world will also further accelerate the digital transformation in processes and practices in all industry sectors and public services. Digital twin, immersive extended reality (XR), teleoperation, and Industry 5.0 are some leading 6G use cases. Drive for sustainability, represented by the Sustainable Development Goals (SDGs) in the United Nations (UN) Agenda 2030 [31], also calls for 6G's contribution. Demands for communication anywhere, anytime, for anyone and anything, and for intelligent digital transformation, form the core driver for 6G research and innovation.

The 6G research and technological innovations include transmission technologies, network architectures, hardware and device technologies, softwarization and open network, artificial intelligence (AI) integration, security and privacy of the system, and information [32, 33]. 6G standards will ensure interoperability and backward compatibility with legacy and older generation systems such as 5G and 4G, and a sustainable supply chain of 6G equipment, devices, and software.

5.2. The 6G Enabling Technologies

In this section, we describe the 6G-enabling technologies to support the envisioned seamless integration of the physical, digital, and human worlds. We focus on the following aspects: (1) machine as the primary user, (2) AI as a performance driver, (3) agile and software-defined everything (SDX) as a necessity, (4) privacy, security, and resilience as warranty, and (5) sustainable networking and energy efficiency as the baseline.

5.2.1. *Machine as the primary user*

Among the three major applications for 5G, namely, eMBB, mMTC, and URLLC, URLLC is designed to support mission-critical applications that require very low latency and ultra-high reliability. The requirement for latency is down to 1 ms, and the requirement for reliability is 99.999%. 6G will enable even more stringent latency and reliability performance to support more tactile internet (TI) applications [34, 35].

In the highly connected physical, digital, and human worlds supported by 6G, the number of IoT-connected devices will continue to grow, and machines will become dominant users. Figure 5.1 shows the

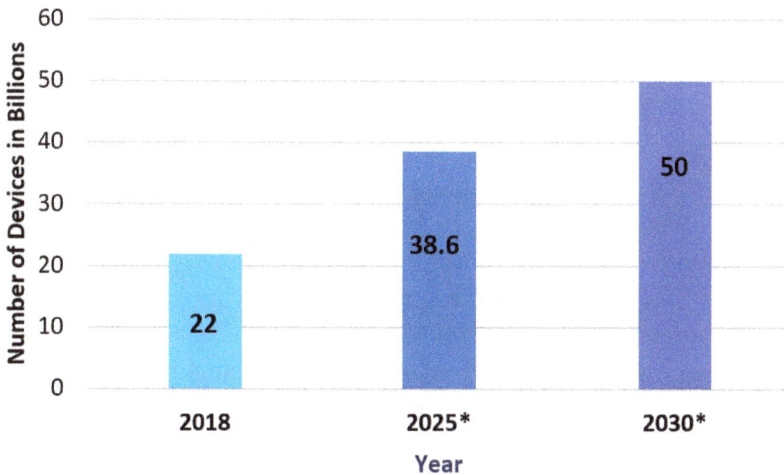

Fig. 5.1. Predicted number of Internet of Things (IoT) connected devices (source: Statista 2022).

predicted number of IoT-connected devices by 2030 (source: Statista 2022, https://www.statista.com/statistics/), where 50 billion of such IoT devices are expected in 2030. The URLLC-type of critical machine-type communications will continue to increase, pushing for support of a higher number of URLLC user volumes and higher area-average URLLC user capacity. We refer to this as *massive URLLC communications*. For example, in the coordinated fleet of robots or unmanned aerial vehicles (UAVs), the number of robots or UAVs can be large, and URLLC communication is generally required. In smart factories supporting flexible and hyper-personalized manufacturing, the number of wireless-connected machines will increase on the manufacturing floor, and wirelessly-connected gadgets will become a necessity for remote supervision, teleoperation, and digital twinning. Massive URLLC communications have to be made available to support these operations.

In 6G, URLLC is not only required in low data rate applications but also in some high data rate scenarios. The boundary between URLLC and eMBB will blur. For example, immersive XR is likely to be applied widely in many applications in the future. XR requires a huge data rate and low latency, which the current 5G cannot support.

To enable massive URLLC communications in 6G, the following technical features are required.

5.2.1.1. *Flexible waveforms*

In 6G networks, the number of connected devices is huge, and the traffic types and data rates of different devices and applications can be quite different. Some of them may generate periodical traffic, but many others may generate sporadic data. It is a big challenge to coordinate such a diversified and heterogeneous network. Especially the waveform and multiple access technologies need to be reinvestigated to meet the new requirements.

Currently, the cyclic prefixed orthogonal frequency division multiplexing (CP-OFDM) is the dominant waveform. In OFDM, coarse time synchronization is required to avoid cross-interferences among users. However, it is difficult and resource-hungry to maintain time synchronization, especially for low-power devices with sporadic traffic. There are a few other waveforms that can relax the synchronization requirements, such as the filter bank multi-carrier (FBMC), universal filtered multicarrier (UFMC), or generalized frequency division multiplexing (GFDM). However, their implementation complexity is higher than the OFDM. Waveforms like the FBMC has also the long tail problem, which is not suitable for short packet transmission [36].

In 6G, we also need to support many applications with high mobility, such as vehicular-to-everything (V2X) and UAV communications. The high mobility induces a high Doppler spread that makes the propagation channel fast-changing in time. If the channel changes significantly within an OFDM symbol, there will be large errors in OFDM demodulation. Recently, the orthogonal time frequency space (OTFS) modulation has been proposed to handle the high Doppler spread or fast time-varying channel [37–39]. In OTFS, the signal is modulated in the delay Doppler domain; that is, it maps the time-varying multipath channel to the time-delay-Doppler domain. In [38], the OTFS is compared with OFDM in high mobility scenarios, which show that OTFS can indeed have much better performance than OFDM at the expense of a more complicated receiver. There are still many unsolved issues in OTFS, which include the pilot design, channel estimation, and complicated equalization. The combination of OTFS with other mature technologies, such as multiple-input multiple-output (MIMO), also needs to be studied further.

One single waveform may not suit the need of all traffic types and applications in 6G while meeting the efficiency and sustainability requirements. But incorporating different waveforms in a harmonized system is a big challenge to address in 6G. Numerology, backward compatibility, hardware, software and network management are some of the very important aspects in the design consideration.

5.2.1.2. *Agile multiple access*

Multiple access techniques manage how the spectrum and time resources are shared among multiple users. In 5G, multiple access is mainly based on orthogonality among users. The orthogonality can be maintained with frequency separation by frequency division multiple access (FDMA), subcarrier separation by OFDM access (OFDMA), time separation by time division multiple access (TDMA), code division multiple access (CDMA), or a combination of them. All these can be called orthogonal multiple access (OMA). The low implementation complexity of OMA is the main reason for its dominance in the past, although it is known to be sub-optimal based on Shannon's information theory [19]. In the coming 6G, non-orthogonal multiple access (NOMA) is likely to be adopted as another multiple access technique in coexistence with OMA. In NOMA, users can share the same spectrum at the same time, leading to higher spectral efficiency. The different channel conditions, quality of service (QoS) requirements, and power levels of users are used to decide on the multiple access scheme and demodulations. In recent years, the rate-splitting multiple access (RSMA) has been studied as a generalization of NOMA [19].

The power-based NOMA and code-based NOMA have been studied extensively [19]. In the 5G standardization process, there were plenty of proposals on NOMA, although it is not adopted in the current 5G standard. Going forward, we have to reconsider the NOMA to cater for the massive connectivity with different QoS in 6G. For this purpose, it is crucial to adopt a general framework for the standardization of NOMA in 6G systems, such that the system maintenance cost is acceptable while keeping the spectral efficiency advantages and reasonable receiver complexity of the NOMA.

It is well known that ultra-low uplink latency is one of the most difficult tasks in URLLC [40–43]. Traditionally, an uplink user needs to request the base station for the grant of transmission before emitting its signal. This "grant-based" uplink, in general, cannot meet the delay requirement due to the time spent in the grant process [40, 41]. In recent years, "grant free" uplink schemes have been proposed to bypass the grant process and allow an uplink user to transmit without a "grant" [40–43]. In 5G, some grant-free uplinks have already been adopted, like the semi-persistent scheduling (SPS) uplink [40–43]. There has been much research on grant free uplink [40–43]. Among them, the contention-based grant-free uplink can overcome the resource wastage problem in SPS uplink [40], which can be treated as a modification and generalization of the SPS uplink to allow multiple users to share the same spectrum and time. In the contention-based grant-free uplink, any user in a group randomly chooses a resource in the pre-scheduled shared resources for uplink. In [40], a scheme called sub-band random sensing (SRS) grant-free uplink is proposed. The SRS grant-free uplink creatively combines the sub-band sensing, user grouping, and random access, which allows a user sensing the unlicensed spectrum to use available subresources in a wideband and reduce collisions with other users. The new scheme overcomes the severe spectrum wastage problem of the SPS uplink in applications with sporadic traffic. Compared with the contention-based grant-free uplink, the new scheme achieves a much lower collision probability.

The combination of the grant-free uplink with NOMA is promising, as it can reduce handshaking time with grant-free transmission, allow more users, and reduce the demodulation error rate [42, 43].

5.2.1.3. *High-precision wireless time-sensitive network*

Timing is critical in many applications like industrial automation, automatic driving, connected drones, teleoperation, etc., which are sometimes called time-sensitive applications. Here, timing has two different meanings:

(1) Clock synchronization: The clocks of the related users and devices must be synchronized to a certain degree of accuracy for them to coordinate in solving a common task. The clock synchronization accuracy is typically required to be below 1 us. In some applications, the requirement even goes to the nanosecond (ns) level. For example, for high precision localization with errors below 30 centimeters (cm), the clock synchronization needs to be below 1 ns if the time difference of arrival (TDoA) based positioning algorithms are used [44–46].

(2) Communication latency: The time taken to transfer a given piece of information from a source to a destination must be below a given threshold. Late arrival of the information will cause the

information to be useless or even a disaster. In 5G, 5 ms end-to-end latency is required in URLLC. This is not enough for many applications like teleoperations, XR, high-speed UAV, automatic driving, etc. [34, 35, 47–49].

Time-sensitive applications also rely heavily on the reliability of communication. Clock synchronization is normally based on exchanging clock information among the users. Information reliability is critical. Latency is even more closely related to reliability. When a transmission has errors, it is normal to retransmit the packet in conventional communication protocols. However, for time-sensitive applications, retransmission is generally not useful, as the information may already be useless after the time limit. In recent years, in addition to time delay, the Age of Information (AoI) has been used as another performance metric for network performance analysis [13–19].

There have been standards, especially for time-sensitive networking (TSN). One such standard is developed by the IEEE 802.1 Time-sensitive Networking Task Group [45, 46]. It is designed to make Ethernet-based networks more deterministic. TSN was developed to provide a way to make sure information travels from point A to point B in a fixed and predictable timeline [45, 46]. There are four components in TSN: (1) clock synchronization, (2) latency, (3) reliability, and (4) management. The latency requirement is dependent on the application. The clock synchronization is based on Precision Time Protocol (PTP), which is included in IEEE 802.1AS and IEEE 802.1ASRev [45, 46]. Though not specifically considered in TSN specification, AoI will be an important metric for the analysis and design of time-sensitive networks.

Another related standard is tactile internet (TI), which is being developed by the IEEE 1918.1 WG [47]. Traditionally, wireless communication was mainly used for content delivery (message, voice, file transfer, video streaming, and so on) and monitoring (information collection) applications [35]. In industrial networks, other information like touch, smell, and taste are also important for real-time steering and controlling [35, 48]. TI, which was named in early 2014, transfers information as well as reproduces (or synthesis) stimuli (sight, hearing, touch, smell, and taste) at the receiver. The IEEE 1918.1 WG has defined TI as follows [47]: "A network (or network of networks) for remotely accessing, perceiving, manipulating, or controlling real or virtual objects or processes in perceived real-time by humans or machines". TI is projected to change the paradigm that enables wireless communication for real-time control and steering.

The goal of TI is that humans should not be able to distinguish between locally executing a manipulative task compared to remotely performing the same task across the TI. The results of machine-in-the-loop physical interactions should ideally be the same as if the machines were interacting with objects directly at, or close to, the locations of those objects. Figure 5.2 shows an example of remote balancing of a basketball through the TI [47]. TI conveys the skills needed for balancing the basketball to the remote machine, and returns feedback of the current balance of the ball. The human must know the situation and react promptly. The reaction must be conveyed timely to the other end of the link. All these must be done before the basketball falls.

Table 5.1 shows the latency and reliability requirements in some typical TI applications [35, 47].

The International Telecommunication Union (ITU) first observed the TI as a new technology with potential impact on future

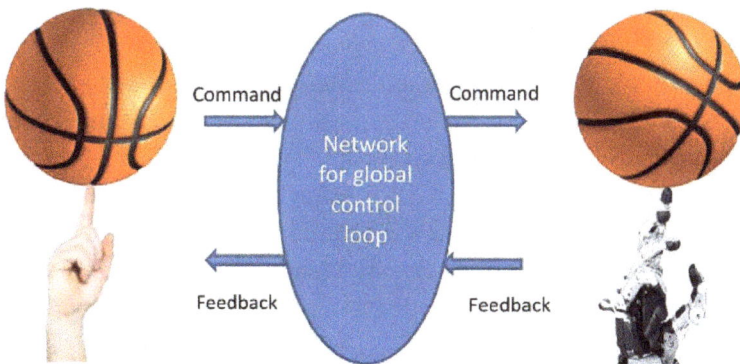

Fig. 5.2.   Remote balancing of a basketball through the TI [47].

Table 5.1.   URLLC requirements in some typical applications.

Use Cases	E2E Latency (ms)	Reliability (%)	Average Data Rate (Mbps)
Teleoperation	1–20	99.999	1–100
Cooperative automated driving	1–50	99.99–99.999	1–40
Immersive virtual reality	1–50	99.99–99.999	1–100
Automotive (self driving, remote driving)	1–100	99.99–99.999	1–10
Internet of drones	2.5–50	99.99–99.999	1–100
Interpersonal communication	10–1,000	99.99–99.999	1–600
Live haptic-enabled broadcast	12–18	TBD	Varying

standardization and applications in ITU's Technology Watch in August 2014. IEEE launched its Digital Senses Initiative (DSI) in June 2015 to advance the technologies. Its purpose is to develop VR and AR standards to harmonize existing standards and develop new ones. It is projected to be finalized before 2022.

5.2.1.4. *Sensing-communication-computing-control convergence*

Integrating sensing into the wireless communication ecosystem has been actively studied recently [14, 18, 50–54]. For example, the IEEE 802.11 standardization committee formed the IEEE 802.11 WLAN Sensing Topic Interest Group as well as the Study Group, respectively, in September and November 2019 [50], which aims to incorporate wireless sensing as a new feature for next-generation Wi-Fi systems. Localization based on cellular networks and signal is an established area. In 5G, there are a few enhancement features for the localization based on time, range, and angles. One of the major enhancements in 5G localization is the newly added dedicated positioning reference signals in NR R-16: NR positioning reference signal (NR PRS) in the downlink and the sounding reference signal in the uplink [2]. The downlink PRS is specifically designed to deliver the highest possible levels of accuracy, coverage, and interference avoidance and suppression. The downlink PRS covers the whole NR bandwidth and over multiple symbols that can be aggregated to accumulate power. Each base station can then transmit the downlink PRS in different sets of subcarriers to avoid interference. Moreover, it is possible to mute the PRS signal from one or more base stations at a given time according to a muting pattern, further lowering the potential interference. For use cases with higher transmission loss, for example, in macro cell deployments, the PRS can be also configured to be repeated to improve hearability.

However, such localization methods are mainly based on the TDoA, signal strength, and angle of arrival (AoA). In recent years, there is a trend to integrate radar sensing and wireless communication as one of the key technologies in future wireless systems [51–54], which is called joint radar and communication (JRC). With radar sensing, range, speed, and angle can be detected at the same time with higher accuracy, which is hardly achievable by another type of sensing. Traditionally, radar and communication are separated by spectrum, hardware, and software. With the increasing usage of software-defined radio and digital signal processing, the hardware and RF front-end for radar and communication tends to be similar. This makes JRC practical. With many emerging applications, an integrated JRC using the same spectrum and sharing many hardware components will help alleviate spectrum congestion and save energy consumption.

Figure 5.3 illustrates two JRC applications [51, 52] — the figure on the left shows a joint V2V communication and localization scenario: a vehicle sends useful information to another vehicle and uses the reflected signal to estimate the range and speed of the vehicle,

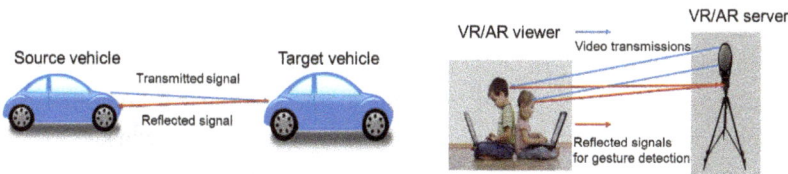

Fig. 5.3.   Example of applications of joint radar and communication systems [51, 52].

while the figure on the right shows a joint video transmission and position/gesture detection scenario in VR/AR: the server transmits video to the persons and obtains the accurate position and gesture of the person using the reflected signal. Both applications need sensing and communication at the same time, and a JRC device can serve the two purposes.

Instead of using the same waveform for simultaneous sensing and communications, a JRC system can also be realized by time-sharing the radar sensing and communication sessions by using dedicated radar and communication waveforms. Resource allocation and anti-jamming performance will be important design considerations [55–57].

In 6G, higher frequencies like the THz band will be used, and massive antenna arrays will be employed. These not only increase the communication throughput and reliability but also create opportunities for very high-accuracy environment sensing, where millimeter-level localization and high-resolution imaging can be achieved. The 6G context-aware intelligent networks will optimize the deployment, operation, and energy usage with information extracted from sensing and communication without human intervention.

However, it is still unclear how to design the joint sensing and communication system and maximize the overall benefit. There are still issues and difficulties to be addressed in the standardization of sensing in 6G networks and infrastructures. These include joint waveform design, spectral sharing among sensing and communication, interference management, system design, hardware and software reuse, cost, and power consumption.

In 6G networks, distributed computation will need to prevail to achieve efficiency, privacy, security, and low latency. The interaction of communication and computation requires an overall joint design for the dual functions. For real-time applications such as factory automation, remote power system control, fleet management of autonomous systems, teleoperation, etc., wireless networked control systems (WNCSs) call for integrated communication-control design and optimization [58, 59]. It is envisioned that integrated sensing and communications (ISAC), computing, and control will be an integral part of 6G, calling for research and technology innovation in network architecture, the various layers in the network, and the interface specification with the application domains.

5.2.2. *Artificial Intelligence (AI) as a performance driver*

There have been active research activities making use of artificial intelligence (AI) to enhance the performance of communication systems and networks. For example, deep learning (DL) algorithms and neural network (NN) models can help mitigate radio frequency circuit impairment and power amplifier (PA) nonlinearity [60–63], machine learning (ML) algorithms can learn wireless channel characteristics for modulation and coding scheme (MCS) selection and link adaptation, and DL models may help enhance the performance in user scheduling, resource allocation, dynamic spectrum management, anomaly detection, etc. [64, 65]. These research activities have also pushed the advancement of AI algorithms development, e.g., reinforcement learning (RL), federated learning (FL), deep Q-learning, etc.

Driven by the great potential of performance improvement by AI, the various standard bodies have started various initiatives and activities to pave the path for standard specification development of incorporating AI in networks. For example, in 2018, the European Telecommunications Standards Institute (ETSI) Experiential Networked Intelligence (ENI) Industry Specification Group (ISG) published a series of Group Specification documents, including the Proof of Concept Framework, terminology for main concepts in ENI, ENI requirements, use cases, etc. [66]. In January 2019, the International Telecommunication Union (ITU) Telecommunication Standardization Sector's (ITU-T) Focus Group on Machine Learning for Future Networks including 5G (FG-ML5G) published "Unified architecture for machine learning in 5G and future networks" [67]. It presents a set of architectural requirements, leading to specific architectural constructs needed to satisfy these requirements. Based on these constructs, a logical ML pipeline is discussed along with the requirements derived and its realizations in various types of architectures. Key architectural issues facing the integration of such an ML pipeline into continuously evolving future networks are also listed. Other standardization development organizations and industry alliances, including 3GPP, Linux Foundation [68], O-RAN Alliance [69], etc., have also been developing a framework to support AI implementation in networks.

The above-listed efforts are mainly focusing on how to enable AI in 5G and future networks after the 5G network standards have already been developed. Moving forward, we will see an even tighter and more native integration of 6G with AI in two aspects: (1) the 6G network is defined to support the AI of everything, and (2) native AI-enabled in the 6G system and network. With this, not only will the 6G system and network performance be significantly enhanced, but so too will the operation efficiency and user experience. For infrastructure, increasingly automated deployment, configuration, and management accelerate time-to-revenue as closed technology stacks give way to open, cloud-native architectures. For end-user applications, the ability to anticipate customer needs and automatically deliver on constantly changing service level agreements simultaneously creates new, differentiated revenue streams while reducing operational costs.

There are remaining technical challenges and research issues to address for the full adoption of AI in future networks, such as explainability, controllability, robustness, training efficiency with small data sets, and privacy preservation. Computation and energy-efficient implementation of AI models at the network edge (mobile edge computing [MEC]), the network server (cloud), and agile edge-cloud interaction for knowledge and data sharing will also remain a challenging research problem in 6G.

5.2.3. *Agile and software-defined everything (SDX) as necessity*

Softwarization and open network will be more widely embraced in 6G, including both core networks (CNs) and radio access networks (RANs). Key initiatives include the Open Radio Access Network (O-RAN) Alliance that aims to provide an open and intelligent RAN and the open network automation platform (ONAP) [70], which develops a platform for network management and its automation through an open-sourced shared architecture. An open network opens more opportunities for new players to enter the market with lower barriers and for network operators more cost-effective options. New challenges in network security need to be addressed.

An open network makes software-defined networking (SDN) at CN and RAN a necessity. Moving to 6G, we will continue to see the usage of heterogeneous spectrums as in 5G. The spectrums in the lower frequency bands that could be deployed for 6G will be highly likely "re-farmed" from the earlier generation of wireless communication systems and other applications, while the higher frequency spectrum that has been less explored will be allocated for 6G. Sharing the license-exempted spectrums with software-defined cognitive radio-based solutions will also be an option. For system design efficiency and flexible spectrum access management, software-defined radio will continue to play an important role in 6G systems.

To overcome the radio wave propagation challenges, especially in the high-band spectrum channels, a large intelligence surface (LIS) [15–18] may make the concept of a software-defined radio wave propagation environment a reality. It is therefore anticipated that software-defined radio (SDR) to support heterogeneous spectrum access, SDN at both CN and RAN to support open networking and heterogeneous applications, and software-defined environment (SDE) by reconfigurable LIS will be essential elements in 6G.

5.2.3.1. *Integrated terrestrial and non-terrestrial network for space-air-ground-sea coverage*

It is generally recognized that pure terrestrial communication networks (TCNs) cannot achieve ubiquitous, high-quality, and high-reliability services anytime and anywhere, especially in coping with the upcoming trillion-level connections of IoT devices in remote areas [15–20]. Thus, it is necessary to integrate terrestrial and non-terrestrial networks (NTNs) to achieve worldwide connectivity and make various applications accessible. The integrated TCN and NTN brings space-air-ground-sea coverage via seamless access to satellite link, UAV communications, terrestrial wireless communications, and fiber-optical backbone networks. In the space-air-ground-sea integrated network, various networking technologies have their pros and cons in providing services in terms of coverage, transmission delay, throughput, reliability, etc. Via effective inter-networking, different network segments can cooperate to support seamless service access and enhance service provision in a resource-efficient and cost-effective manner [19]. Satellite communications can especially complement terrestrial networks for service access in areas with limited or no terrestrial network coverage (e.g., remote areas, disaster scenarios, and open seas). Meanwhile, the complementary properties of satellite link (wide coverage) and fiber-optical backbone (high data rate) can be considered as alternative backbone technologies to wireless backhaul. UAV communications can help to alleviate the terrestrial network burden and to enhance service capacity in congested locations with highly dynamic data traffic load. In addition, satellites/UAVs with remote sensing technologies can support the reliable acquisition of monitoring data and assist terrestrial networks with efficient resource management and planning decisions [19].

However, compared to TCN, NTN will need much further advancement. There are still many issues in integrating the TCN and NTN for 6G; some are listed below [18, 19]:

- *Architecture design for integrated TCN and NTN:* Design of architecture with no distinction between TCN and NTN elements that can therefore be orchestrated to provide cost-efficient network configuration by dynamically moving functionality, creating a flexible network topology. Three main enablers need to be developed, namely, softwarization of the NTN, virtualization, and disaggregation.
- *Satellite/UAV constellation with hierarchical design:* Hierarchical constellations will consist of nodes flying at different altitudes and communicating through horizontal inter-node links, i.e., among nodes at the same altitude, and vertical inter-node links, i.e., among nodes flying at different altitudes, or terrestrial nodes.

- *Smart NTN with computing and storage in the sky:* Flying nodes will become smart edge nodes of the 6G network. Processing, storage, and communication in the sky needs to be enabled in support of the realization of non-terrestrial clouds and space information networks (SINs).

- *Resource optimization with infrastructure as a resource:* Beyond the bandwidth, time, power, and space dimensions, resource optimization needs to address the infrastructure itself as a resource to be configured according to the service requirements.

- *Dynamic spectrum management, coexistence, and sharing:* Dynamic spectrum coexistence and sharing between the terrestrial and NTN segments, as well as among the different layers of the architecture, e.g., geosynchronous equatorial orbit (GEO) and Non-Geostationary-Satellite Orbit (NGSO) nodes, space- and air-borne nodes, need to be developed.

- *New spectrum for NTN beyond THz:* The use of new spectral bands, up to optical communications, needs to be investigated, in close collaboration with the terrestrial counterpart. This includes the characterization of propagation conditions and the development of channel models encountered by the NTN elements, both in inter-node communications and in non-terrestrial to terrestrial communications.

- *Radio access technologies with flexibility and adaptability:* Waveform design shall jointly address the non-terrestrial and terrestrial channel characteristics, such as Doppler effects introduced by the NGSO nodes, the delay and latency aspects of the higher altitude nodes, and the need for efficient support of channel estimation. The integrated access network also needs to support discontinuous backhauling in scenarios where a flying base station cannot assume a continuous connection to the core network, e.g., in Low Earth orbit (LEO) or Very Low Earth orbit (vLEO) incomplete constellations.

- *NTN components with beyond current technologies:* Software-defined payloads, new antenna design and technologies, and new components at THz and beyond.

- *AI for exploitation of NT dynamics:* Autonomous and intelligent network management system shall be appropriately developed, exploiting the predictable dynamics component of the NT segment.

5.2.3.2. *Large Reconfigurable Intelligent Surface*

Large Reconfigurable Intelligent Surface (RIS) or Intelligent Reflecting Surface (IRS) has been proposed as a promising technology for wireless communication [15–20]. RIS is a two-dimensional surface made by electromagnetic metamaterials that are used as a relay to reflect the RF signal. The metamaterials for RIS have unique electromagnetic properties, such as negative refraction, perfect absorption, and anomalous reflection/scattering. In a large-size RIS, there are many subwavelength-sized structural elements, where each element can be controlled independently. The distinct feature of RIS is its adjustable reflection coefficient of each element: the phase of the reflected signal by an element can be electronically tuned to a given value. By adjusting the reflection coefficients of elements, the reflected signals from different elements will be focused on a given direction or point. This makes the RIS similar to the MIMO but with less power consumption and cost. The RIS can be used for enhancing localization in the 6G, where the fixed RIS reflectors can be treated as additional anchors in the positioning process [19, 20]. A special advantage of RIS in localization is the capability of positioning via trilateration without relying on multiple base stations [19]. There have been demonstrations on the RIS for the enhancement of communication in different environments, which has shown notable performance gains.

However, the number of states of the adjustable phases is currently limited due to constraints on material and technology. In most of the currently used RIS, the number of adjustable states is only four, which reduces the achievable performance in practice. The evolution of material and manufacturing technologies may solve the problem and make the RIS more powerful. Other than hardware performance and cost, there are a few major difficulties for using RIS in a real system: (1) Estimation of the channels from the transmitter to elements of RIS and RIS elements to the receiver, and (2) Added overhead due to additional pilots for channel estimation, signaling, and controlling of the RIS.

5.2.3.3. *Heterogeneous spectrum access and management*

In 6G, there are vast different applications. These different applications need different spectrum properties. Overall, the large number of applications and massive number of users require a huge spectrum. Figure 5.4 shows the possible spectrum for 6G. Other than the spectrum used in 5G, it is estimated that a tremendous spectrum is needed in the terahertz band (0.3 THz to 3 THz, wavelength range of 1 mm to 100 μm) and visible light communication (VLC) band (400–790 terahertz, wavelengths from about 380 to about 750 nanometers) [71, 72].

The newly added frequency for 6G is likely to start from around 90 GHz, which is called 6G frequency [71, 72]. The band classification and rectangular waveguide (RW) sizes of the 6G frequency are shown in Table 5.2 [73]. The FCC has opened up experimental 6G spectrum licenses in 2019. Through its rules, the FCC will give innovators a 10-year license to experiment with new 6G products and services. The spectrum, once thought of as useless, could offer super high-speed

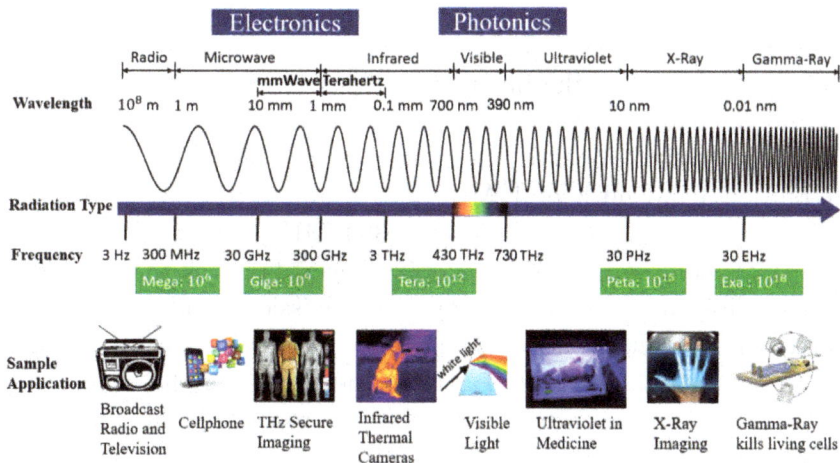

Fig. 5.4.   Potential radio frequencies for 6G [72].

Table 5.2.   The band classification and rectangular waveguide (WR) sizes of 6G frequency.

Band	Frequency	WR-Size
F	90–140 GHz	WR-8
D	110–170 GHz	WR-6
G	140–220 GHz	WR-5
G	170–260 GHz	WR-4
G	220–325 GHz	WR-3
Y	325–500 GHz	WR-2
Y	500–750 GHz	WR-1.5
Y	750–1,100 GHz	WR-1
	up to 3,000 GHz (3 terahertz)	

internet service for data-intensive applications, such as super-high-resolution imaging and sensing applications.

Different spectrums have different regulations [71, 72]. Two requirements are commonly found in regulations: (1) Occupied channel bandwidth (OCB). The OCB requirement is expressed as the bandwidth containing 99% of the power of the signal and shall be between 80 and 100% of the declared Nominal Channel Bandwidth, and (2) Maximum Power Spectral Density (PSD). The regulations are also different in different countries/regions. For example, both these requirements are enforced for 5 GHz carriers according to ETSI 301 893, while only the maximum PSD requirements are enforced in the US for 5 GHz. Maximum PSD requirements exist in many different regions. For most cases, the requirement is stated with a resolution bandwidth of 1 MHz. For example, the ETSI 301 893 requires 10 dBm/MHz for 5,150–5,350 MHz. The implication of the PSD requirement on the physical layer design is that, without proper designs, a signal with small transmission bandwidth will be limited in transmission power. This can negatively affect coverage. That is, the maximum PSD requirement is a binding condition that requires changes to uplink (UL) transmissions in unlicensed spectrums. Learning how to regulate the terahertz band for 6G is still an open question, which must be done before 6G standardization.

Although new spectrum bands such as THz and VLC bands could provide a large bandwidth for 6G for very high data rate transmission, low-frequency bands (for example, a sub-6-GHz band) are essential to provide communication services with low cost, broad coverage, and massive access support. Dynamic spectrum access or cognitive radio is needed to support spectrum sharing and intelligent accessing for both licensed and unlicensed bands. There could be a few spectrum access and management schemes in the 6G [15–20].

- **Centralized spectrum management.** This has been partially used in 5G to adjust the spectrum allocation in different network slices. The scheme needs to set up a database to indicate the spectrum availability and a centralized management unit to dynamically authorize the spectrum use and update the spectrum availability information. The major issues of this scheme are the requirement of real-time information and future traffic prediction of the network. If network dynamics are high, it is very difficult to acquire real-time information and predict future traffic. It is also challenging for the optimization algorithm. As the number of network slices and the range of spectrum bands are large, the optimization algorithm may run a long time and cost a lot of energy.
- **AI for dynamic spectrum access.** Fast capturing/learning the network state and predicting future traffic are the key enablers for efficient dynamic spectrum access. Unsupervised learning, like deep reinforcement learning (DRL), could be used to tackle these challenges. In recent years, we have seen a boom in AI research in wireless networks. It is expected to see a wide usage of AI in various aspects of 6G. However, it is still not crystal clear how important the role of AI is, as the reliability and interpretability of AI are still not well understood.
- **Spectrum sharing in unlicensed bands.** The use of unlicensed bands is already implemented in 5G [15–20]. For example, 5G has defined the standard NR unlicensed to use the unlicensed spectrum, including the industrial, scientific, and medical (ISM)

bands in 2.4 GHz and 5.9 GHz. In 6G, this trend will continue. Further enhancement will be introduced in 6G for smart access schemes to avoid cross-interferences among different networks that are sharing these license-exempted bands, including Wi-Fi, LTE, 5G NR, and 6G. In addition to ISM bands, there may be other traditionally licensed bands for legacy systems that will be freed for spectrum sharing in the future.

The 6G standards and protocols will need to coordinate the dynamic spectrum access in licensed, unlicensed, license-assisted, or tiered-access bands for heterogeneous usage.

5.2.3.4. *Open network*

In 6G, there will be a vast number of use cases with different requirements and network structures, which provide both public and private smart networks and services with a higher degree of flexibility and functionality. To efficiently run such a complicated system, there are a few criteria on network management: (1) The overall architecture complexity of network layers and system functions should not be increased with increased flexibility and functionality, and (2) The visible complexity to the user/developer/app needs to be reduced such that the final user, the developer, and the app can use simple primitives with clear features and error conditions without constraining flexibility and features. Towards this end, the network needs to be softwarized with the full support of cloud-native architecture and service-based architecture (SBA), as well as software and hardware disaggregation.

In recent years, open network design is becoming more and more popular to enhance user selection and interoperability of multiple vendors' equipment and protocol stacks. In an open network, commodity hardware and open standards are adopted, which enables network compatibility in both hardware and software, expandability, and extendibility. With open network design, network designers and administrators can benefit from flexible hardware and software components without having to worry about relying on a single vendor, which helps improve cost efficiency and network configurability. An open network will also facilitate more flexibility and open innovation and allow more industry players to support the expansion of network infrastructure in future networks.

There have been a number of initiatives to develop more open and interoperable open networks, both in RAN and CN. There are a number of challenging issues to address to further develop the open network in 6G.

A software-defined network (SDN) is one aspect of open networks that enables more flexible and agile network control. SDN enables enterprises and service providers to respond quickly and adapt to changing business requirements. Taking advantage of SDN principles, an open network adds the use of open-source platforms and defined standards in the production of networks. In 6G, open network design will need to be considered in different aspects. Firstly, the network architecture should be open so that with network virtualization, the required serving network or network slicing for any vertical application can be generated on top of the shared physical infrastructure. Meanwhile, the interface should be open to supporting the interoperability of different participants and the flexible schedule of virtualized network elements.

It is important to develop the 6G open network platform and open interface to support the interconnection and interoperability of different vendors and share the physical infrastructure and construction of the supplier ecosystem of mobile networks [15–20]. Based on an open network platform/interface, 6G will support service-driven network slicing management as needed, which enables service providers and vertical industries to deploy new services quickly. White box hardware with universal processor architecture will be developed for the open network platform, based on which service providers can customize the required network functions by software definition. SDN and Network Function Virtualization (NFV) will be used to generate the required network slicing flexibly. The decoupling of hardware and software and SDN/NFV provide a flexible and fast way to support various service innovations in 6G. Moreover, since 6G will converge with more and more vertical industries, the ownership of physical infrastructure may belong to not only mobile operators but also various vertical industries. Therefore, open network/interface designs and standardization become urgent in 6G.

With network openness, not only will network management be flexible, but computing and caching resources at different nodes can be fully scheduled and utilized by various mobile edge applications to improve user experience. But it should also be noted that with the increasing openness of the network, security problems will be more prominent in the future network.

5.2.4. *Privacy, security, and resilience as warranty*

As the 6G network is envisioned to seamlessly connect the physical, virtual, and human worlds, a vast amount of information will be collected and traversed in the 6G network. In-network computing, as discussed in earlier sections, for example, at the MEC and other network nodes, will derive insights into industrial and business processes, personal work and lifestyles, and social relations for actionable controls. Networked close loop control will be an essential function of 6G, leading to a significant increase of command and control packets/traffics in 6G networks. The 6G network, therefore, introduces new potential security vulnerabilities and opens to a much wider security threat landscape and design challenges than legacy 4G and 5G networks, calling for new enabling technologies for security protection, privacy preservation, and resilience. The 6G vision would only be successfully realized if security, privacy, and resilience could be designed to support the new requirements arising from the real-time interconnected and interactive physical, virtual, and human worlds.

There are multi-folds of challenges and potential technologies in the security, privacy, and resilience design in 6G networks. While traditional security threats such as denial of service (DoS) attacks, eavesdropping, replay attacks, physical attacks, malware applications, and hardware tampering will remain important in 6G, new challenges and issues will arise. The automatization and softwarization of the 6G network make network security much more complicated than that in the 5G network. The system is not only vulnerable to direct cybersecurity attacks but also easily affected by the misbehaviors of automatized functionalities. The 6G security standardization will need to address these new challenges. We discuss a few in the following.

- Guaranteeing the dependability and trustworthiness of the system will be one of the major challenges. Solutions such as zero-touch micro-segmentation and slicing need to be guaranteed, and distributed AI/ML functions and models need to be hardened to avoid malfunctions and cyberattacks and guarantee their trustworthiness and security, for example, to avoid model poisoning and membership inference attacks in federated-learning, distributed architectures.
- Security quantification in a 6G context should deal with complexity and fragmentation and mandate the exploration of mechanisms to identify, evaluate, certify, and monitor the level of security. Based on such quantification, trust can be given to providers and services, enabling the growth of the sector by answering the diversity of demands from the mass market to specific business-to-business (B2B) verticals.
- Despite commonalities with other digital services, some data-centric or green issues remain a specific challenge for 6G systems. As such, human-centric privacy, relevant trustable AI, orchestrated operations across parties, and impact and sustainability of solutions will be key to shaping an acceptable digital future enabled by 6G.
- The research for security solutions will also need to take into consideration the performance, ease of management, and energy consumption, when targeting to guarantee the security and trustworthiness of 6G telecom and service/application platforms.

New and emerging technologies and innovations will need to be incorporated into 6G to address the above security and privacy challenges. We give a few examples here: (1) Security protocols at different network layers in the architectural framework, which have been decomposed into four segments in [74], namely, platform, functions, orchestration, and specialization. Each segment is presented with its own security and privacy challenges, while their integration and interaction open new threats, (2) security key acquisition using physical layer (PHY) information, referred to as physical layer security (PLS) [68, 75], quantum key distribution (QKD) [76], (3) inter-network security in terrestrial/non-terrestrial integrated space-air-ground-sea communications, (4) privacy-preserving communications and computing, (5) use of distributed ledger technologies (DLT), (6) use of AI for threat analysis and prevention, attack and anomaly prediction, detection, attribution, and mitigation, and (7) security threats and counter-measures in 6G-supported vertical domains. More detailed discussions on 6G security and privacy may be referred to [77, 78] and references therein.

There is a very active standardization and project landscape in the security domain. A number of research projects and standardization efforts have been kicked off by ITU-T FG-ML5G [67], ETSI ISG ENI [66], 3GPP, and the National Institute of Standards and Technology (NIST). These initiatives will impact 6G security standards.

5.2.5. *Sustainable networking and energy efficiency as baseline*

With the emergence of 5G technology and its rapid deployment progress, [79] estimates that about 15% of the total global carbon dioxide (CO_2) emission would be from the information and communication technology (ICT) sector by the end of 2020. Moreover, RAN takes about 57% of the total network energy consumption. With a better energy efficiency of 5G than its predecessors, the RAN energy consumption will reduce to 50.6% by 2025, with 5G to gradually replace 3G and 4G networks [80]. Nevertheless, the ultra-dense deployment, the extremely large data traffic volume, the increased functionalities of sensing and computation, and the tens of billions of devices to be connected by 6G make energy consumption reduction and energy efficiency enhancement a very important area in 6G design. A sustainable network will need to be delivered as a baseline for 6G to support the Sustainable Development Goals (SDGs) [81].

To address energy consumption in 6G, an energy-efficient RAN will be critical. Based on the GSM Association (GSMA) report [82],

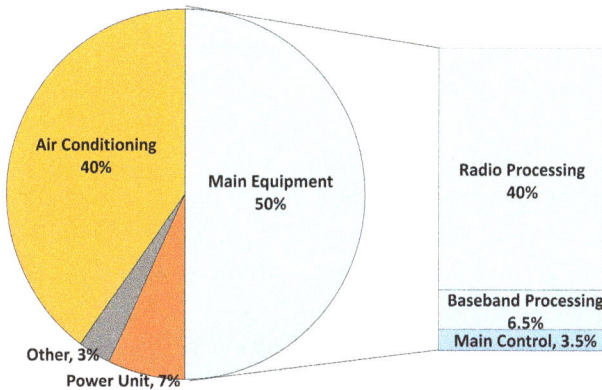

Fig. 5.5.   Energy consumption breakdown at the base station (brought from the Next Generation Mobile Networks report) [84].

RAN accounts for 73% of the total energy consumption of mobile network operators (MNOs), and the energy consumed at base stations in the cellular network can be up to 77% [82]. Vodafone reported in [83] that cellular base station sites consume 73% of the total energy in its cellular network. These numbers clearly indicate that reducing the energy consumption at the cellular base stations will potentially lead to significant improvement in the energy efficiency of the whole system.

Figure 5.5 shows the energy consumption breakdown at a base station site [84]. Although the percentage may vary according to the proprietary settings of MNOs and original equipment manufacturers (OEMs), the main network equipment, consisting of a radio frequency (RF) processing unit, baseband unit (BBU), and main control, requires the largest amount of energy. The radio-processing unit requires 40% of the total energy consumption at the base station, and the high power amplifier (HPA) consumes most of the energy at the radio-processing unit. For a 5G network using massive multiple-input multiple-output (mMIMO), the energy consumption can be broken down into two major parts: the antenna unit and BBU, in which the antenna unit requires about 90% of the total energy consumption. The HPA efficiency has been indicated as a key component of network energy efficiency during the 3G/4G era [85, 86]. However, HPA efficiency has improved over time, in which

Doherty PA architecture with envelope tracking and Gallium Nitride (GaN) semiconductor material has been introduced, resulting in an HPA efficiency of more than 50% [87]. On the other hand, it is observed that energy consumption at the base station varies with the traffic load and signal transmission regularity, i.e., the base station has fewer signals to transmit during a low-load period and, accordingly, consumes less energy than a high-load period. For each base station, a full load has been shown to achieve optimal energy efficiency [88].

A 10% energy consumption reduction target had been set from the 4G system when the concept of 5G system was introduced. For 6G systems, although the exact target number has not been determined yet, a more ambitious target is expected following the SDG. According to the ITU, the ICT sector has committed to reducing greenhouse gas emissions by 45% from 2020 to 2030 [89], while the North American ICT sector is encouraged to achieve net-zero greenhouse gas emission by 2040 [90]. The efforts to reduce network energy consumption in 6G networks can be discussed from an end-to-end perspective, including: (1) specifications and designs from data transmission, signaling, and network architecture, to AI-driven methods, (2) power efficiency improvement at device-level and site-level, e.g., air conditioning and cooling system, (3) energy harvesting and circular energy, and (4) ecosystem and regulatory aspects. As this chapter is on the standardization of 6G communication systems and networks, we will discuss the potential enabling technologies related to system and network specifications, designs, and AI-driven solutions [65, 84, 91, 92].

5.2.5.1. *Specifications and designs*

In aspects of signal transmission and network operation, 3GPP has introduced the following energy-saving techniques in its 5G NR interface.

- **Massive MIMO transmission** — Massive MIMO is introduced at the base station by equipping the base station with a large number of antenna elements [93]. With well-optimized signal transmission at each antenna element, massive MIMO improves spectral efficiency by allowing simultaneous transmission across multiple user equipment (UE). At the same time, energy efficiency is also improved via beamforming, in which the wireless signal from the base station is delivered by a narrow beam directly to a designated UE and, accordingly, improves the received signal quality and extends the network coverage.

As an evolution path to a 6G system, introducing more antenna elements at the base station will enable more radio processing chains and hence increase the total energy consumption. It is therefore necessary to analyze the trade-off between performance gain and energy consumption to develop the massive MIMO specification in 6G systems.

In addition to the single and centralized adaptive antenna array architecture, a distributed antenna system (DAS) can be deployed by placing the antenna elements at different locations in the network coverage area [94, 95]. The antenna elements are connected via optical fiber cable as a fronthaul link, as well as with the central control unit (CCU) and central BBU. Large-scale DAS has been shown to deliver better energy efficiency than centralized massive MIMO [96] due to the advantage of being able to more effectively overcome the radio wave propagation loss and, at the same time, enjoy the beamforming gain. 3GPP and a few other standardization bodies and alliances, e.g., O-RAN, have been developing specifications for DAS implementation in 5G and beyond networks.

- **Lean carrier design** — Lean carrier design refers to a design to minimize the transmission of broadcasting and reference signals when they are not necessary, e.g., when the traffic load is low and/or when the users have low mobility, instead of the "always on" transmission in 4G systems [97]. This can be done in various approaches, for example, increasing the signal cycle, introducing flexibility in reference signal setting, turning the broadcast channel off, etc. In 5G systems, the increasing of the signal cycle, in particular, the cell synchronization signal (SS), which is broadcasted from the base station to provide cell information to new users, can be up to 160 ms, i.e., eight times the original cycle. The use of the lean carrier design saves energy from both the signal transmission and the hardware operation, as some hardware can be switched off during the transmission cycle, as shown in Figure 5.6 [98]. It is expected that the same technique will still be available in 6G systems, with the addition of flexibility to various use cases, e.g., different user mobility and network density.

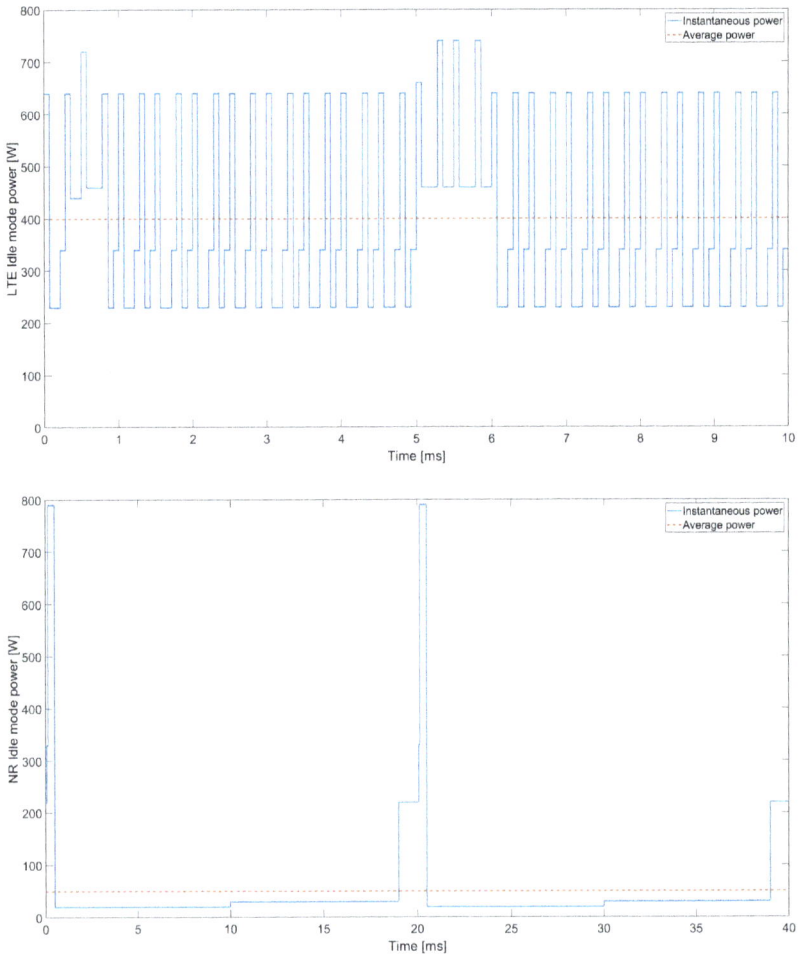

Fig. 5.6.   Power consumption comparison due to signal transmission for 4G (top) and 5G with lean carrier design and reduced SS transmission (bottom) (re-produced from [98]).

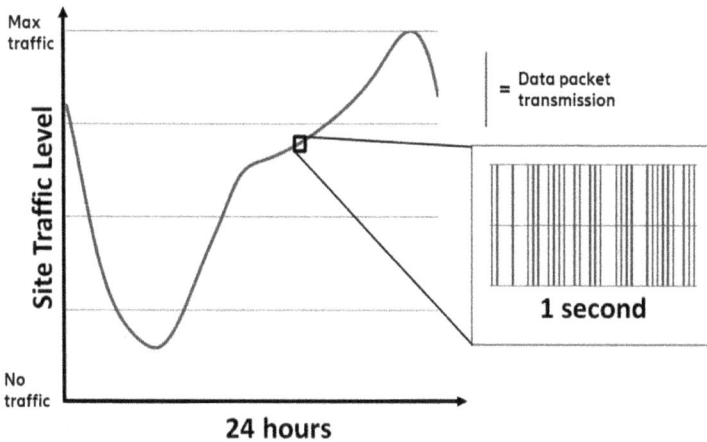

Fig. 5.7. Gaps in data packet transmission (re-produced from [98]).

- **Improved sleep mode** — Sleep mode and its operations are designed under a similar concept to a lean carrier, i.e., switching off some of the network equipment when the traffic load is low or when there is no data to transmit [98]. With the recent development in network equipment, most of the hardware can be switched on/off in the sub-second granularity level without requiring transition time. Therefore, the improved sleep mode is proposed in order to reduce energy waste during the low-load and no-transmission period, even when the gap is very short at the sub-second level, as shown in Figure 5.7.

  The improved sleep mode is now being studied and will be introduced in the next release of 3GPP standards (expected to be Release 18). Four levels of sleep modes, with different minimum sleep durations, are being investigated as summarized in Table 5.3 [99].

  (a) 1st level sleep mode — This level refers to the deactivation of HPA for the duration of at least 71 μs, similar to micro discontinuous transmission (μDTX) in a 4G network.
  (b) 2nd level sleep mode — This level refers to the deactivation of some of the base station components, mostly in the RF units, for at least 1 ms, i.e., one transmission time interval (TTI).

Table 5.3. Sleep mode level comparison [99].

Level of Sleep	Level 1 (4G like)	Level 2	Level 3	Level 4
Sleep time (minimum duration)	71 μs	1 ms	10 ms	Approx. 1 s
Sleep depth (components to switch off)	PA	PA & some RF components	PA & some RF, BB components	PA & some RF, BB components

(c) 3rd level sleep mode — This level deactivates more components in RF and some of the digital baseband units, in which the deactivation duration is at least 10 ms, i.e., one frame.

(d) 4th level sleep mode — This level targets the deactivation duration of approximately 1 s. Most of the components at the base station can be switched off. It is worth noting that the 4th level sleep mode has a longer duration than the maximum SS broadcasting cycle; hence, implementation of this sleep mode should be supported with alternative offloading solutions, e.g., microcell switch off with traffic offloading to a macrocell, in order to prevent any connectivity loss.

The above sleep mode levels are used as a design principle to offer several approaches to switch off or shut down the following network and radio resources:

(a) Symbol shutdown — This is similar to the lean carrier design, which is to minimize any unnecessary signal transmission, especially the cell synchronization signal and beam sweeping. This can reduce energy consumption at the RF unit, especially the HPA.

(b) Channel shutdown — This refers to silencing the base station with multiple channels by switching off the RF components with respect to channels having low traffic, expecting 10–20% of energy saving.

(c) Carrier shutdown — 5G offers flexible radio resource management, in which the whole signal bandwidth can be separated into bandwidth parts (BWP). With multi-layer heterogeneous network deployment, it is possible to switch off the capacity-layer bands, e.g., high-frequency and large-bandwidth bands, while maintaining the coverage-layer bands, e.g., low-frequency and long-coverage bands.

There are also several techniques similar to the above-mentioned approaches, aiming to reduce energy consumption under low-load conditions, e.g., cell zooming and adaptive transmit power control. In addition, since the massive MIMO has been introduced in 5G networks, several OEMs and MNOs are studying approaches to reduce energy consumption at the antenna arrays, including switching off some of the RF chains at the array or using sparse antenna arrays [100].

It is obvious that the sleep mode and network resource switch-off will continue to be studied and developed for adoption in 6G. However, the switch-off technique also leads to drops in quality of services (QoS), which needs to be carefully managed when there are mission-critical services. It is important to take into account the service quality and energy consumption trade-off and ensure the required QoS for all users while driving for better energy efficiency. In addition, other emerging technologies in the antenna, circuit, network architecture, etc., need to be considered holistically for an energy-efficient and sustainable 6G.

5.2.5.2. *AI-driven solution*

With the rapid growth of computational power and computational resource, data collection and storage, and neural network development, AI is being introduced in various fields, including wireless communications. Wireless network design is usually based on optimization, dealing with multiple factors, including the network load, available resources, user priority, required QoS, channel conditions, and the overall network performance metrics, such as reliability, capacity, energy efficiency, etc. It is not possible to solve the wireless network optimization at full-scale. ML and DL have been actively studied to drive the network energy efficiency performance.

Several industry standards are focusing on integrating ML and AI for energy efficiency [99, 101–103]. In particular:

- ETSI Zero-Touch Network and Service Management (ZSM)
- ITU-T Focus Group on AI for Environmental Efficiency
- 3GPP TS 29.520, 5G Systems; Network Data Analytics Service
- O-RAN AI/ML in RAN Intelligent Controller

ML and AI may introduce energy saving but also incur energy consumption due to their computational load. We discuss both aspects in the following.

- **Energy consumption reduction through AI/ML** — Energy consumption reduction across multiple base stations, ranging from hundreds to thousands, is a large-scale optimization problem, which is difficult to solve under time and resource constraints. Within the past recent years, there are studies and experiments of using AI/ML for energy saving, in which energy-saving policies may be based on signal processing, sleep mode setting, or other approaches such as integration with renewable energy resources. Some examples of these works are listed below:

  - AI/ML for optimizing RAN operation and energy flows among power grid and renewable energy resources [104], where the ML can provide 10–40% of energy saving
  - AI/ML for automated network operation, resource management, and QoS assurance [105, 106]
  - Power consumption monitoring, analytics, and predictions for service assurance to optimize the placement of network functions, utilizing 3GPP-defined Network Data Analytics Function (NWDAF), and network slicing [99]
  - UE location tracking and trajectory prediction, then optimizing the sleep mode policy and data traffic offloading across multiple base stations [107].

- **Energy consumption of AI/ML** — Energy consumption of an AI/ML model is not negligible since it requires computational resources to run sophisticated neural networks, especially when some application needs to call the ML function frequently, e.g., every 100 ms. Moreover, energy consumption during model training may need to be taken into account as well, although it is an offline process, due to the following reasons:

  (a) Energy consumption for one-time model training is relatively large, requiring long-time training with a large amount of input data and sometimes requiring a high-performance computer. Furthermore, those processes produce greenhouse gas and contribute to the expenditures of either OEMs or MNOs.

  (b) Some applications may need periodic model re-training, such as once a month, to ensure their functionality and adaptability to changes in the environment.

It is therefore important to take a systematic and holistic evaluation ML/AI-based methods, similar to other energy-efficient approaches. However, there is limited work on the energy consumption assessment of AI/ML models. [84] has proposed a method to account for energy consumption by taking into account AI model parameters and the computational platform. To support the standardization efforts, more research is needed to establish a systematic approach and cover the various key factors.

5.3. Selected 6G Research Initiatives

There has been strong momentum and strong funding support in 6G research since 2019. The European Union (EU), China, Korea, Japan, and the United States (US) are leading the activities.

5.3.1. *European Union*

The European Union (EU) has initiated many beyond 5G and 6G research projects, as summarized in Figure 5.8, courtesy of 6G World. Among them, the European Commission's 6G flagship initiative Hexa-X [30], funded by EU Horizon 2020, and the

Finnish 6G Flagship hosted by the University of Oulu has received much attention.

- Hexa-X (https://hexa-x.eu/)

Kicked off in December 2020, Hexa-X brings stakeholders, including network vendors, communication service providers, verticals, technology providers, and European communications research institutes, aiming to create unique 6G use cases and scenarios, develop fundamental 6G technologies and define a new architecture for an intelligent fabric to integrate the key 6G technologies.

- 6G Flagship (http://6gflagship.com)

6G Flagship is a Finnish initiative under the Academy of Finland, a governmental funding agency. Coordinated by the University of Oulu, it is among the world's first 6G research initiatives. Envisioning a data-driven future society enabled by near-instant and unlimited wireless connectivity, it brings ecosystem players in 6G research, development, and innovation and aims at creating new business opportunities as well as implementing the UN's SDGs and major societal challenges. Its published white papers have been providing useful references in various aspects of 6G.

5.3.2. *The United States*

In October 2020, the Alliance for Telecommunications Industry Solutions (ATIS) announced the launch of the Next G Alliance (NGA), an industry initiative to advance North American mobile technology leadership in 6G and beyond. The NGA will encompass the full lifecycle of research and development, manufacturing, standardization, and market readiness, aiming to establish North American preeminence in the 5G evolutionary path and 6G development. It has three strategic focuses:

○ Develop a 6G national roadmap that addresses the changing competitive landscape and positions North America as the global leader in R&D, standardization, manufacturing, and adoption of Next G technologies.

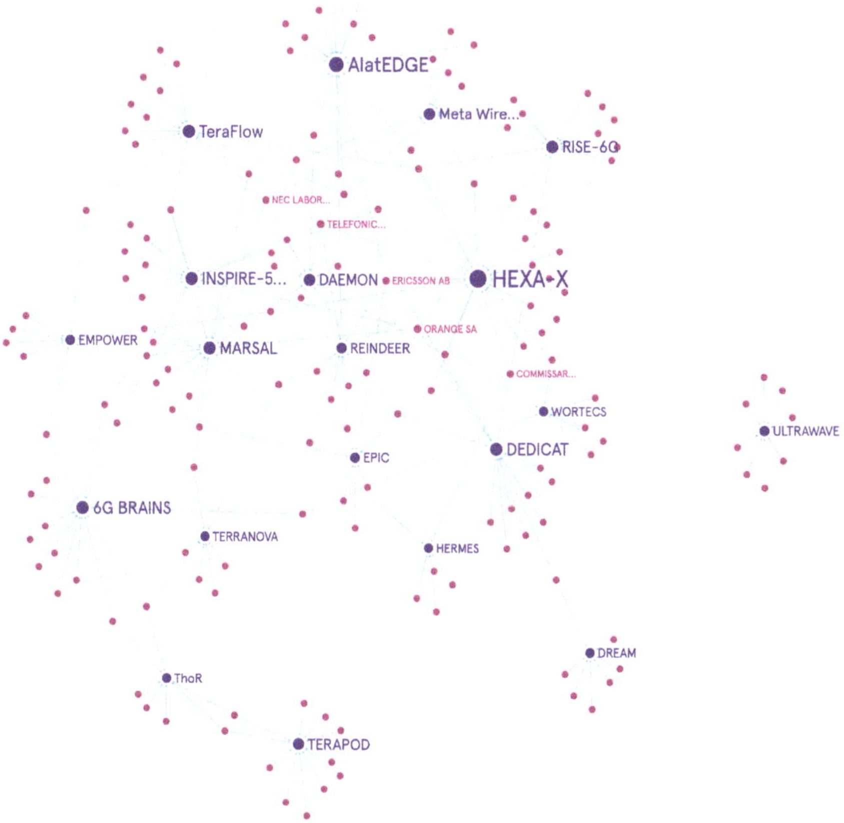

Fig. 5.8. Interactive graph of EU-funded 6G initiatives (source: https://www.6 gworld.com/exclusives/the-ultimate-6g-research-network-graph/).

○ Align the North American technology industry with a core set of priorities that will steer leadership for 6G and beyond to influence government policies and funding.
○ Identify and define the early steps and strategies that will facilitate and lead to rapid commercialization of Next G technologies across new markets and business sectors and promote widescale adoption, both domestically and globally.

Since its launch, the NGA has kicked off a number of working groups to execute its mission and released its white paper recently in February 2022 [90].

5.3.3. *China*

China's government activity on 6G is led by the Ministry of Science and Technology (MOST) and Ministry of Industry and Information Technology (MIIT), and government-industry coordination occurs through the "IMT-2030 (6G) Promotion Group". The "IMT-2030 (6G) Promotion Group" published its white paper on "6G vision and candidate technologies" in June 2021, listing the following use cases:

○ Immersive Cloud XR
○ Holographic Communications
○ Sensory Interconnection
○ Intelligent Interaction
○ Communication for Sensing
○ Proliferation of Intelligence
○ Digital Twins
○ Global Seamless Coverage.

Ten candidate 6G technologies have also been highlighted:

• New network with native AI, including new air interface and new network architecture with native AI
• Enhanced wireless air interface technologies, including basic physical layer technologies, ultra-massive MIMO, and in-band full duplex
• Wireless transmission technologies on new physical dimensions, such as reconfigurable intelligent surface, orbital angular momentum (OAM), and intelligent holographic radio (IHR)
• Terahertz and visible light technologies
• Integrated communications and sensing
• Distributed autonomous network architecture
• Deterministic network
• Computing-aware network
• Integrated terrestrial and non-terrestrial network
• Native network security based on a multi-lateral trust model.

Research and development projects have also been started as part of the national initiative.

5.3.4. *Japan*

Japan established the "Beyond 5G Promotion Consortium" in December 2020 under the Ministry of Internal Affairs and Communications as the hub of 6G activities. The "Beyond 5G Promotion Consortium" aims to achieve an early and smooth introduction of Beyond 5G and to strengthen the international competitiveness of Beyond 5G in order to realize the strong and vibrant society expected in the 2030s. The Consortium consists of the General Assembly and its subordinate committees, the Planning and Strategy Committee and the International Committee, with the Planning and Strategy Committee focusing on the study of comprehensive strategies to promote Beyond 5G and the preparation of the Beyond 5G White Paper, and the International Committee focusing on identifying international trends for promoting Beyond 5G and international dissemination of the status of Japan's efforts.

5.3.5. *Korea*

The Ministry of Science and ICT ("MSIT") of Korea announced its "6G R&D Implementation Plan" in June 2021, which includes the following three goals [108]:

- Secure next-generation key original technologies
- Gain dominance in international standards and patents
- Lay the foundation for 6G research and industry.

5.4. The 6G Standard Development

To create an international 6G standard, a number of organizations will be involved: standard development organizations (SDOs), regulatory bodies and administrations, and industry forums.

5.4.1. *Standard Development Organizations (SDOs)*

3GPP is a global SDO from seven regional and national SDOs in Europe (ETSI), Japan (ARIB and TTC), the United States (ATIS), China (CCSA), Korea (TTA), and India (TSDSI). 3GPP has created the technical specifications of 2G (GSM), 3G (WCDMA/HSPA), 4G

(LTE), and 5G (NR) and is expected to create the 6G technical specifications. These technical specifications are essential for the industry to produce and deploy standardized products and provide interoperability among those products.

- **Release (Rel-) 17** – Rel-17 — the third release of 5G specifications — is expected to be the enhanced version of previous releases, in particular, offering low-latency communication. However, the standard development has been under pressure due to COVID-19 restrictions. The RAN1 physical layer specifications development was initially targeted to freeze by December 2021. Then the Stage 3 freeze (RAN2, RAN3, and RAN4) would follow with the initial targeted timeline of March 2022, and the ASN1 freeze and the completion of performance specifications by September 2022.
- **Rel-18 and the start of work on 5G-Advanced** — During the Rel-18 planning process, some capacity of the working group's effort will be kept in reserve for workload management and to meet late, emerging critical needs arising from commercial deployments. The 3GPP finalized its key topics to be studied and developed in Rel-18 in December 2021, in which the full list can be found in [109]. Below are the topics for further studies:

  - Evolution for downlink MIMO, including an enhancement for channel state information (CSI) and multi-TRP handling
  - Uplink enhancement, which strongly needs to be considered (i.e., the previous releases mainly focused on downlink improvement)
  - Mobility enhancement, especially for high-frequency band (FR2)
  - Reduced Capacity (RedCap) device and specification development for UE, which does not require high network performance but is cost-effective, including power-saving enhancement
  - Non-terrestrial network (NTN), following the first version in Rel-17, including IoT aspects
  - Expanded and improved positioning
  - Evolution of duplex operation, with cross-link interference management
  - AI/ML for air interface and next-generation RAN, starting from use cases and KPI determinations

o Network energy saving

o Additional and open topics, such as carrier aggregation and dual-connectivity enhancement, flexible spectrum integration, RAN support beyond 52.6 GHz, reconfigurable intelligent surfaces (RISs), unmanned aerial vehicles (UAVs), and high altitude platform systems (HAPSs).

The Rel-18 work/study items will be examined over 18 months, starting from the first half of 2022.

5.4.2. *Regulatory bodies and administrations*

The government-led regulatory bodies and administrations are responsible for setting regulatory and legal requirements for selling, deploying, and operating mobile communication systems and other related products. They have two important tasks:

1) Spectrum regulation: Control spectrum use and set licensing conditions for the mobile operators that are awarded licenses to use parts of the *radio frequency (RF)* spectrum. Spectrum regulation is handled at multiple levels: national level, regional level, and global level. There are three regional bodies, namely, the Electronic Communications Committee (ECC) under the European Conference of Postal and Telecommunications Administrations (CEPT) for Europe, the Inter-American Telecommunication Commission (CITEL) for the Americas, and the Asia-Pacific Telecommunity (APT) for Asia. The global regulatory body is the International Telecommunications Union (ITU) [110]. Spectrum regulation specifies the services each spectrum is to be used for and the limits of allowed and unwanted emissions.

2) Regulatory certification and type-approval: This is to ensure devices, base stations, and other equipment are type-approved and meet the relevant regulation.

5.4.3. *ITU*

The ITU is a specialized agency of the United Nations (UN) responsible for all matters related to information and communication technologies [110]. The ITU is preparing high-level performance statements for 6G systems in a form of the International Mobile Telecommunications 2030 (IMT-2030). This is expected to be an iterative and joint effort with SDOs and industry forums. In particular, the following study groups and entities in ITU play active roles in the making of 6G standard.

- The ITU-T Future Networks Study Group SG13:

SG 13 is responsible for developing the requirements and architecture of Future Networks. It considers candidate ideas for IMT-2030 requirements [3], including, but not limited to, network slicing, machine learning, and quantum key distribution. In particular, the Focus Group (FG) on Technologies for Network 2030 (FG-Net2030) in SG 13 is working on a new network vision for 2030 [111–114]. FG-Net2030 published the white paper [10], focusing on the "non-radio-related" aspects of future network capabilities for 2030 and beyond. It has also published [8], specifying the network architecture framework.

- ITU-R:

The ITU-R is the radio communications sector of the ITU. The ITU-R is responsible for ensuring efficient and economical use of the RF spectrum by all radio communication services. The different subgroups and working parties produce reports and recommendations that analyze and define the conditions for using the RF spectrum. The ambitious goal of ITU-R is to "ensure interference-free operations of radio communication systems" by implementing Radio Regulations and regional agreements. The Radio Regulations is an internationally binding treaty for how the RF spectrum is used.

The ITU-R manages global mobile telecommunication standards and provides guidance and a roadmap for next-generation communication R&D by defining the vision for next-generation mobile telecommunication, including 6G. ITU-R Working Party 5D (WP5D) is responsible for the overall radio system aspects of International Mobile Telecommunications (IMT) systems, comprising IMT-2000, IMT-Advanced, IMT-2020, and IMT for 2030 and beyond.

In February 2021, ITU-R WP5D produced the report "Future technology trends towards 2030 and beyond" [4], soliciting inputs from related organizations. WP5D also launched the 6G Vision Group in early 2021 to take charge of establishing the 6G Vision, including defining the key capabilities, working on technology development, and creating timelines for standardization and commercialization of 6G. Starting with completing the 6G Vision by 2023, ITU-R plans to develop technical requirements and recommendations for 6G through industry standards organizations such as 3GPP. Out of those candidate technologies for 6G, the technologies that pass the ITU-R's evaluation will be approved as the global standards for 6G around 2030.

The key objectives of the Vision towards IMT for 2030 and beyond are:

- Focus on the continued need for increased coverage, increased capacity, and extremely high user data rates
- Focus on the continued need for lower latency and both high and low speed of movement of the mobile terminals
- Fully support the development of a Ubiquitous Intelligent Mobile Connected Society
- Focus on tackling societal challenges identified in UN SDGs, in particular, to meet the needs of Industry, Innovation, and Infrastructure
- Consider what the future heterogeneous mobile broadband networks can offer to society and the economy through the applications and services they support
- Target the changing global scenario on how people work and how people stay safe during societal challenges such as the COVID-19 pandemic and global climate changes
- Focus on delivering digital inclusion and connecting rural and remote communities.

There are four key pillars for the vision:

1. Any future technology should help in the development of a Ubiquitous Intelligent Mobile Connected Society.
2. Any future technology should support technologies that can help bridge the digital divide.

3. Any future technology should support technologies that can personalize/localize services.
4. Any future technology should support the connectivity/compute technologies that can address issues of real-world data ownership sensitivities.

It is important to note that the WP5D does not create the actual technical specifications for IMT but defines IMT in cooperation with the regional standardization bodies and maintains a set of recommendations and reports for IMT, including a set of Radio Interface Specifications (RSPCs). These recommendations contain "families" of Radio Interface Technologies (RITs) for each IMT generation — all included on an equal basis. For each radio interface, the RSPC contains an overview of that radio interface, followed by a list of references to the detailed specifications. The actual specifications are maintained by the individual SDO, and the RSPC provides references to the specifications transposed and maintained by each SDO.

- **ITU-R World Radiocommunication Conferences (WRC) [115]:**

A key part of the RF spectrum allocation and coordination in the ITU-R is done through the WRC, held every three to four years. At WRCs, the Radio Regulations are revised and updated, resulting in revised and updated use of the RF spectrum across the world.

The most recent WRC was held in 2019 (WRC-19), and the next one is to be held in 2023 (WRC-23). The Radio Regulations from the WRCs define frequency allocations and technical and regulatory conditions for use of a spectrum by a given service. This is important to ensure compatibility and to prevent and resolve cases of harmful interference. The Radio Regulations are international treaties that are binding to ITU member states. In addition to the radio spectrum regulation, WRC also governs the use of the geostationary and non-geostationary satellite orbits.

ITU-R organizes six study groups to guide the work done in the preparatory process and ready the technical bases for decisions at the WRCs. The six study groups are:

o Study Group 1 (SG1): Spectrum management
o Study Group 3 (SG3): Radiowave propagation
o Study Group 4 (SG4): Satellite services
o Study Group 5 (SG5): Terrestrial services
o Study Group 6 (SG6): Broadcasting services
o Study Group 7 (SG7): Science services.

5.4.4. *Industry forums*

Industry forums are industry-led groups to promote specific technologies or interests. Some examples of industry forums in mobile communications are introduced, including the GSM Association (GSMA) [116], Next Generation Mobile Networks (NGMN) Alliance [117], Next G Alliance [118], and Open Radio Access Network (O-RAN) Alliance [119]. Linux Foundation [120] has been supporting the Open Network Automation Platform (ONAP) [70]. The 6G Smart Networks and Services Industry Association (6G-IA) [121] has also been gaining momentum and attention.

GSMA

The GSMA is a global industry organization representing the interests of mobile operators of more than 750 operators as full members and nearly 400 companies in the broader mobile ecosystem as associate members, including handset and device makers, software companies, equipment providers, and internet companies, as well as organizations in adjacent industry sectors [116]. The GSMA also produces the industry-leading MWC events held annually in Barcelona, Los Angeles, and Shanghai, as well as the Mobile 360 Series of regional conferences [116]. In July 2021, GSMA published its submission to ITU-R titled "Vision 2030 — Insights for mid-band spectrum needs" [25].

NGMN

The NGMN Alliance is a mobile telecommunications association of mobile operators, vendors, manufacturers, and research institutes.

It was founded by major mobile operators in 2006 as an open forum to evaluate candidate technologies to develop a common view of solutions for the next evolution of wireless networks and ensure the successful commercial launch of future mobile broadband networks through a roadmap for technology and friendly user trials.

In its published 6G Vision paper [26], NGMN outlined three drivers for 6G evolution, namely, societal goals at large as well as that expressed in the UN SDGs [31], market expectations of new services and capabilities in a cost-effective manner, and operational necessities of increasingly more efficient planning, deployment, operations, management, and performance of the mobile operator's networks. For 6G standard development, NGMN advocates to broaden, align, and rationalize the scope of standards and technology development process to support a healthier and more vibrant ecosystem, moving from the current approach of piecemeal based on proposals on only specific aspects of the system and ensuring that technologies are developed holistically across different standards organizations.

Next G Alliance

Next G Alliance is an ATIS-led initiative in the North American region to discuss and promote solutions for the 6G era with industry representatives. The US announced an initiative called the Next G Alliance a year ago with the goal of advancing US leadership in 6G. Members of the Next G Alliance include AT&T, Verizon, T-Mobile, Facebook, Qualcomm, Charter Communications, Nokia, Ericsson, Google, and Samsung, plus many others.

O-RAN Alliance

Open-RAN (O-RAN) Alliance [119] was established in 2018 by an initiation of group of MNOs, including AT&T, China Mobile, Deutsche Telekom, NTT DOCOMO, and Orange. Its mission is to re-shape the RAN industry towards more intelligent, open, virtualized, and fully interoperable mobile networks. The O-RAN-developed standards provide a more competitive RAN supplier ecosystem, i.e., supply diversification instead of using the whole solution from one OEM, with faster innovation to improve user experience. O-RAN-based mobile networks also improve the efficiency of RAN deployments as well as operations by the mobile operators. Currently, the alliance has been expanded progressively with a variety of new members from different sectors, including hardware and computation unit producers such as NVIDIA, Intel, and AMD, and universities and research institutions.

It is observed that the standard development in an O-RAN Alliance is, in fact, aligned with 3GPP timeline and technology. While 3GPP focuses on development of new working and study items, O-RAN focuses on system-level realization under its open architecture and enabling open interfaces for various equipment. Further details of O-RAN standard and architecture developments are available in its white paper [69], while its working groups are listed below:

- WG1: Use Cases and Overall Architecture Workgroup
- WG2: The Non-real-time RAN Intelligent Controller (RIC) and A1 Interface Workgroup
- WG3: The Near-real-time RIC and E2 Interface Workgroup
- WG4: The Open Fronthaul Interfaces Workgroup
- WG5: The Open F1/W1/E1/X2/Xn Interface Workgroup
- WG6: The Cloudification and Orchestration Workgroup
- WG7: The White-box Hardware Workgroup
- WG8: Stack Reference Design Workgroup
- WG9: Open X-haul Transport Workgroup
- WG10: Operations, Administration and Maintenance (OAM) Workgroup.

5.4.5. *6G standard timeline*

With the evolution roadmap for the first five generations, 6G is anticipated in the 2030s. While the SDOs and regulators have yet to release the 6G standard development timeline, a few papers have provided a rather aligned roadmap while focusing on different perspectives.

In 2019, the authors of [122] provided a 6G roadmap, which was derived from the strategic plans of various standards bodies. It included the projected timelines for the 6G vision, 6G requirement, and 6G evaluation up to 2030, as reproduced in Fig. 5.9. A similar 6G timeline has been presented in [123], as reproduced in Fig. 5.10.

Fig. 5.9.   Timeline of 6G wireless networks [122].

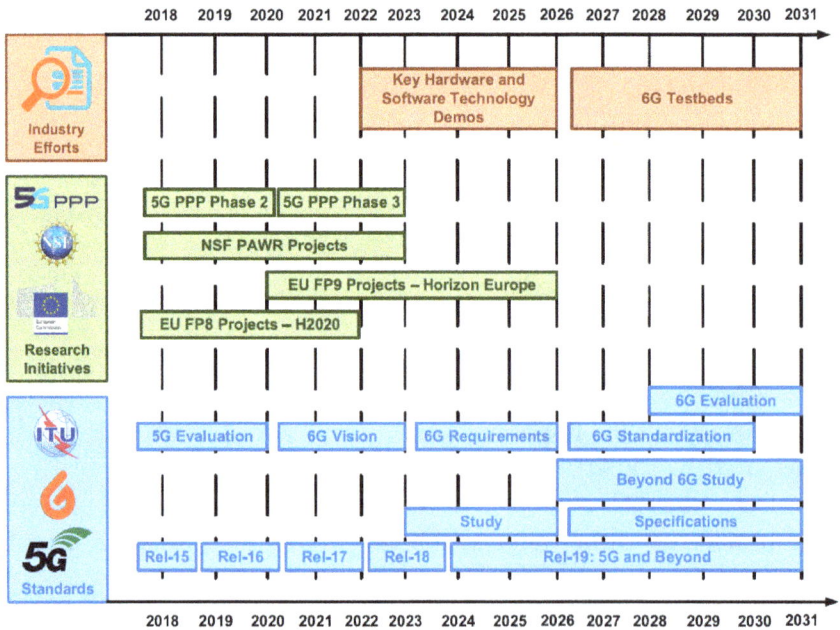

Fig. 5.10.   Projected timeline of 6G and beyond systems [123].

5.5. Summary and Conclusions

This chapter has summarized the 6G vision and briefly discussed the enabling technologies and vertical use cases. The stakeholders in 6G standardization, which include the regulatory organizations, the standard development organizations, and industry forums, are then introduced. The standardization processes are shared, and timelines provided.

References

[1] 6G Flagship, "Key drivers and research challenges for 6G ubiquitous wireless intelligence: 6G Research Visions No. 1," University of Oulu, 2019. http://jultika.oulu.fi/files/isbn9789526223544.pdf

[2] NTT DoCoMo, "White Paper: 5G evolution and 6G," Version 3.0, Feb. 2021 (first version in Jan. 2020).

[3] S. Chen, Y.-C. Liang, S. Sun, S. Kang, W. Cheng and M. Peng, "Vision, requirements, and technology trend of 6G: How to tackle the challenges of system coverage, capacity, user data-rate and movement speed," *IEEE Wirel. Commun.*, pp. 1–11, Feb. 2020.

[4] 6G Flagship, "White paper on 6G drivers and the UN SDGs: 6G research visions no. 2," University of Oulu, Jun. 2020.

[5] 6G Flagship, "White paper on business of 6G: 6G research visions no. 3," University of Oulu, Jun. 2020.

[6] 6G Flagship, "6G white paper on validation and trials for verticals towards 2030's: 6G Research visions no. 4," University of Oulu, Jun. 2020.

[7] 6G Flagship, "6G white paper on connectivity for remote areas: 6G research visions no. 5," University of Oulu, Jun. 2020.

[8] 6G Flagship, "White paper on 6G networking: 6G research visions no. 6," University of Oulu, Jun. 2020.

[9] 6G Flagship, "White paper on machine learning in 6G wireless communication networks: 6G research visions no. 7," University of Oulu, Jun. 2020.

[10] 6G Flagship, "6G white paper on edge intelligence: 6G research visions no. 8," University of Oulu, Jun. 2020.

[11] 6G Flagship, "6G white paper on research challenges for trust, security and privacy: 6G research visions no. 9," University of Oulu, Jun. 2020.

[12] 6G Flagship, "White paper on broadband connectivity in 6G: 6G research visions no. 10," University of Oulu, Jun. 2020.

[13] 6G Flagship, "White paper on critical and massive machine type communication towards 6G: 6G research visions no. 11," University of Oulu, Jun. 2020.

[14] 6G Flagship, "6G white paper on localization and sensing: 6G research visions no. 12," University of Oulu, Jun. 2020.

[15]   6G Flagship, "White paper on RF enabling 6G — Opportunities and challenges from technology to spectrum: 6G research visions no. 13," University of Oulu, Apr. 2021.

[16]   W. Saad, M. Bennis and M. Chen, "A vision of 6G wireless systems: Applications, trends, technologies, and open research problems," *IEEE Netw.*, vol. 34, no. 3, pp. 134–142, May/Jun. 2020. doi: 10.1109/MNET. 001.1900287

[17]   Samsung Research, "6G: The next hyper-connected experience for all," Jul. 2020.

[18]   G. Liu, Y. Hung, N. Li, *et al.*, "Vision, requirements and network architecture of 6G mobile network beyond 2030," *China Communications*, vol. 17, no. 9, pp. 92–104, 2020.

[19]   X. You, *et al.*, "Towards 6G wireless communication networks: Vision, enabling technologies, and new paradigm shifts," *Science China Information Sciences*, Nov. 2020. https://doi.org/10.1007/s11432-020-2955-6

[20]   G. Wikström, P. Persson, S. Parkvall, G. Mildh, E. Dahlman, B. Balakrishnan, P. Öhlén, E. Trojer, G. Rune and J. Arkko, "Ever-present intelligent communication: A research outlook towards 6G," Nov. 2020. https://www.ericsson.com/en/reports-and-papers/white-papers/a-r esearch-outlook-towards-6g

[21]   The 5G Infrastructure Association (5G-IA), "European vision for the 6G network ecosystem," Jun. 2021.

[22]   ITU-T F-NET-2030, "Network 2030: A blueprint of technology, applications and market drivers towards the year 2030 and beyond," 2019.

[23]   ITU, "ITU completes evaluation for global affirmation of IMT-2020 technologies," 26 Nov. 2020. https://www.itu.int/en/mediacentre/Pag es/pr26-2020-evaluation-global-affirmation-imt-2020-5g.aspx

[24]   J. R. Bhat and S. A. Alqahtani, "6G ecosystem: Current status and future," *IEEE Access*, vol. 9, 2021, pp. 43134–43167, Jan. 2021.

[25]   GSMA, "Vision 2030 — Insights for mid-band spectrum needs," Jul. 2021.

[26]   NGMN, "6G drivers and vision," Apr. 2021.

[27]   University of Surrey, "6G wireless: A new strategic vision," 2021.

[28]   5G Americas, "Mobile communications towards 2030," Nov. 2021.

[29]   J. T. J. Penttinen, "On 6G visions and requirements," *J. ICT Stand.*, vol. 9, no. 3, pp. 311–326, Dec. 2021. doi: 10.13052/jicts2245-800X.931

[30]   EU, "Hexa-X vision on 6G and research challenges," Horizon, 2020. https://hexa-x.eu/ [Accessed 14 October 2021].

[31]   UN SDG Report 2019. https://unstats.un.org/sdgs/report/2019/The-Sus tainable-Development-Goals-Report-2019.pdf

[32]   W. Tong and P. Zhu (eds.). *6G: The Next Horizon: From Connected People and Things to Connected Intelligence.* Cambridge University Press, Cambridge, 2021. doi: 10.1017/9781108989817

[33]   Y. Wu, S. Singh, T. Taleb, A. Roy, H. S. Dhillon, M. Raj, Kanagarathinam and A. De, *6G Mobile Wireless Networks.* Springer, Cham, 2021. https:// doi.org/10.1007/978-3-030-72777-2

[34] G. P. Fettweis, "The tactile internet: Applications and challenges," *IEEE Veh. Technol. Mag.*, vol. 9, no. 1, pp. 64–70, Mar. 2014.

[35] A. Aijaz and M. Sooriyabandara, "The tactile internet for industries: A review," *Proc. IEEE*, vol. 107, no. 2, pp. 414–435, Feb. 2019.

[36] B. Khan and F. J. Velez, "Multicarrier waveform candidates for beyond 5G," *12th International Symposium on Communication Systems, Networks and Digital Signal Processing (CSNDSP)*, 2020.

[37] R. Hadani, S. Rakib, M. Tsatsanis, A. Monk, A. J. Goldsmith, A. F. Molisch and R. Calderbank, "Orthogonal time frequency space modulation," *2017 IEEE Wireless Communications and Networking Conference (WCNC)*, pp. 1–6.

[38] B. Parka and H.-G. Ryua, "Performance evaluation of OTFS communication system in doubly selective channel," *15th International Conference on Future Networks and Communications (FNC)*, Leuven, Belgium, 9–12 Aug. 2020.

[39] P. Raviteja, K. T. Phany, Y. Hong and E. Viterbo, "Orthogonal Time Frequency Space (OTFS) modulation based radar system," *IEEE Radar Conference (RadarConf)*, Apr. 2019.

[40] Y. Zeng, Y. Wang, S. Sun and Y. Ma, "Subband random sensing grant free uplink for URLLC in unlicensed spectrum," *IEEE VTC-Fall*, 2021.

[41] T. Jacobsen, R. B. Abreu, G. Berardinelli, K. I. Pedersen, P. E. Mogensen, I. Kovacs and T. K. Madsen, "System level analysis of uplink grant-free transmission for URLLC," *IEEE Globecom Workshops*, Dec. 2017.

[42] Y. Wu, C. Wang, Y. Chen and A. Bayesteh, "Sparse code multiple access for 5G radio transmission," *IEEE VTC-Fall*, 2017.

[43] F. Wang, Y. Zhang, H. Zhao, H. Huang and J. Li, "Active user detection of uplink grant-free SCMA in frequency selective channel," arXiv:1805.03973v1, May 2018.

[44] Y. Wang, Y. Zeng, S. Sun and Y. Nan, "High accuracy uplink timing synchronization for 5G NR in unlicensed spectrum," *IEEE Wireless Comm. Lett.*, Nov. 2020.

[45] P. Varis and T. Leyrer, "Time-sensitive networking for industrial automation," Texas Instruments, 2018.

[46] W. Steiner, "Time-sensitive networking," *16th International Workshop on Real-time Networks (RTN 2018)*, Barcelona, Spain, Jul. 2018.

[47] O. Holland, E. Steinbach, R. V. Prasad, Q. Liu, Z. Dawy, A. Aijaz, N. Pappas, K. Chandra, V. S. Rao, S. Oteafy, M. Eid, M. Luden, A. Bhardwaj, X. Liu, J. Sachs and J. Araújo, "The IEEE 1918.1 'tactile internet' standards working group and its standards," *Proc. IEEE*, vol. 107, no. 2, pp. 25–279, Feb. 2019.

[48] E. Steinbach, M. Strese, M. Eid, X. Liu, A. Bhardwaj, Q. Liu, M. Al-Ja'afreh, T. Mahmoodi, R. Hassen, A. M. El Saddik and O. Holland, "Haptic codecs for the tactile internet," *Proc. IEEE*, vol. 107, no. 2, pp. 447–470, Feb. 2019.

[49] S. Haddadin, L. Johannsmeier and F. D. Ledezma, "Tactile robots as a central embodiment of the tactile internet," *Proc. IEEE*, vol. 107, no. 2, pp. 471–487, Feb. 2019.

[50] IEEE 802.11 WLAN Sensing Topic Interest Group and Study Group reports, 2020. http://www.ieee802.org/11/Reports/senstig_update.htm

[51] Y. Zeng, Y. Ma and S. Sun, "Joint Radar-comm.: Low complexity algorithm and self-interference cancellation," *IEEE Globecom*, Dec. 2018.

[52] Y. Zeng, Y. Ma and S. Sun, "Joint Radar-communication with cyclic prefixed single carrier waveforms," *IEEE Trans. Veh. Technol.*, vol. 69, no. 4, pp. 4069–4079, 2020.

[53] F. Liu, Y. Cui, C. Masouros, J. Xu, T. X. Han, Y. C. Eldar and S. Buzzi, "Integrated sensing and communications: Towards dual-functional wireless networks for 6G and beyond," 2021, arXiv preprint arXiv:2108.07165

[54] J. A. Zhang, *et al.*, "An overview of signal processing techniques for joint communication and radar sensing," *IEEE J. Select. Top. Signal Process.*, vol. 15, no. 6, pp. 1295–1315, Nov. 2021.

[55] I. Lotfi, D. Niyato, S. Sun, H. T. Dinh, Y. Li and D. I. Kim, "Protecting multi-function wireless systems from jammers with backscatter assistance: An intelligent strategy," *IEEE Trans. Veh. Technol.*, vol. 70, issue 11, pp. 11812–11826, Nov. 2021.

[56] I. Lotfi, D. Niyato, S. Sun, H. T. Dinh, D. I Kim and Y.-C. Liang, "Jamming mitigation in JRC systems via deep reinforcement learning and backscatter-supported intelligent deception strategy," *IEEE 6th International Conference on Computer and Communication Systems (ICCCS)*, 2021.

[57] I. Lotfi, D. Niyato, S. Sun and D. I Kim, "Social welfare maximization auction in joint radar communication systems for autonomous vehicles," *IEEE Globecom*, 2021.

[58] W. Liu, P. Popovski, Y. Li and B. Vucetic, "Wireless networked control systems with coding-free data transmission for industrial IoT," *IEEE IoT J.*, vol. 7, issue 3, pp. 1788–1801, Mar. 2020.

[59] J. Cao, X. Zhu, S. Sun, S. Feng, Y. Jiang and P. Popovski, "AoI-oriented wireless networked control system: Communication and control co-design in the finite block length regime," *IEEE Infocomm 2022 Workshop*.

[60] P. Jaraut, M. Helaoui, W. Chen, M. Rawat, N, Boulejfen and F. M. Ghannouchi, "Review of the neural network based digital predistortion linearization of multi-band/MIMO transmitters," *IEEE MTT-S International Wireless Symposium (IWS)*, 2021.

[61] A. Fawzy, S. Sun, T. J. Lim, Y. X. Guo and P. H. Tan, "Iterative learning control for pre-distortion design in wideband direct-conversion transmitters," *IEEE Globecom*, 2020.

[62] F. Daylak, E. O. Gunes, O. Bayat and S. Ozoguz, "Deep neural network based digital predistorter of power amplifiers," *13th International Conference on Electrical and Electronics Engineering (ELECO)*, 2021.

[63] A. Fawzy, S. Sun, T. J. Lim and Y. X. Guo, "An efficient deep neural network structure for RF power amplifier linearization," *IEEE Globecom*, 2021.

[64] E. Kurniawan, Z. Lin and S. Sun, "Machine learning-based channel classification and its application to IEEE 802.11ad communications," *IEEE Globecom*, 2017.

[65] D. López-Pérez, A. De Domenico, N. Piovesan, X. Geng, H. Bao, Q. Song and M. Debbah, "A survey on 5G radio access network energy efficiency: Massive MIMO, lean carrier design, sleep modes, and machine learning," *IEEE Comm. Surv. Tutor.*, vol. 24, issue 1, pp. 653–697, 2022.

[66] ETSI ENI, "Experiential Networked Intelligence (ENI)," https://www.etsi.org/technologies/experiential-networked-intelligence

[67] ITU-T, "Focus Group on Machine Learning for Future Networks including 5G (FG-ML5G)," https://www.itu.int/en/ITU-T/focusgroups/ml5g/Pages/default.aspx

[68] IEEE Communications Society, "Best readings in physical layer security," https://www.comsoc.org/publications/best-readings/physical-layer-security

[69] O-RAN Alliance, "O-RAN: Towards an open and smart RAN," White Paper, Oct. 2018.

[70] The Open Network Automation Platform (ONAP), https://www.onap.org/

[71] ITU, "Radio spectrum for IMT-2020 and beyond: Fostering commercial and innovative use," ITU. https://www.itu.int/en/ITU-R/study-groups/Pages/Commercial-and-Innovative-Use-e-workshop.aspx

[72] T. S. Rappaport, Y. Xing, O. Kanhere, *et al.*, "Wireless communications and applications above 100 GHz: Opportunities and challenges for 6G and beyond," *IEEE Access*, vol. 7, pp. 7872978757, 2019.

[73] International Commission on Non-Ionizing Radiation Committee (ICNIRP), https://www.icnirp.org/en/frequencies/index.html. https://www.miwv.com/what-is-6g/

[74] V. Ziegler, H. Viswanathan, H. Flinck, M. Hoffmann, V. Räisänen and K. Hätönen, "6G architecture to connect the worlds," *IEEE Access*, vol. 8, pp. 173508–173520, 2020.

[75] S. Sun, Y. Wu, B. S. Lim and H. D. Nguyen, "A high bit-rate shared key generator with time-frequency features of wireless channels," *IEEE Globecom*, 2017.

[76] Y. Cao, Y. Zhao, Q. Wang, J. Zhang, S. X. Ng and L. Hanzo, "The evolution of quantum key distribution networks: On the road to the Qinternet," *IEEE Comm. Surv. Tutor.*, 2022.

[77] P. Porambage, *et al.*, "The roadmap to 6G security and privacy," *IEEE Open J. Commun. Soc.*, vol. 2, 2021.

[78] Y. Siriwardhana, P. Porambagey, M. Liyanagez and M. Ylianttila, "AI and 6G security: Opportunities and challenges," *2021 Joint European Conference on Networks and Communications & 6G Summit (EuCNC/6G Summit)*.

[79] J. Banerjee and C. Bueti, "Sustainable ICT in corporate organizations," ITU Technical Report, 2012. https://www.itu.int/dmspub/itu-t/opb/tut/T-TUT-ICT-2012-10-PDF-E.pdf

[80] J. Lorincz, A. Capone and J. Wu, "Greener, energy-efficient and sustainable networks: State-of-the-art and new trends," Sensors (Basel), vol. 19, no. 22, Nov. 2019.

[81] GSMA, "2019 mobile industry SDG impact report," Sep. 2019. https://www.gsma.com/betterfuture/wp-content/uploads/2019/10/2019-09-24-a6 0d6541465e86561f37f0f77ebee0f7-1.pdf

[82] GSMA, "Going green: Benchmarking the energy efficiency of mobile," Jun. 2021. https://data.gsmaintelligence.com/api-web/v2/research-file-d ownload?id=60621137&file=300621-Going-Green-efficiency-mobile.pdf

[83] https://investors.vodafone.com/sites/vodafone-ir/files/2021-05/vodafone-esg-addendum-2021.xlsx

[84] NGMN Alliance, "Green future network; network energy efficiency," Ver. 1.1, Dec. 2021.

[85] J. Joung, C. K. Ho, K. Adachi and S. Sun, "A survey on power-amplifier-centric techniques for spectrum- and energy-efficient wireless communications," *IEEE Comm. Surv. Tutor.*, vol. 17, issue 1, pp. 315–333, 2015.

[86] J. Joung, C. K. Ho and S. Sun, "Spectral efficiency and energy efficiency of OFDM Systems: Impact of power amplifiers and countermeasures," *IEEE J. Select. Areas Commun.*, vol. 32, issue 2, pp. 208–220, Feb. 2014.

[87] Z. Wang, D. Holmes, M. Acar, R. Wesson, M. P. van der Heijden and R. Pengelly, "Modern high efficiency amplifier design: Envelope tracking, doherty and outphasing techniques," *Microwav. J.*, Apr. 2014.

[88] C. K. Ho, D. Yuan, L. Lei and S. Sun, "Power and load coupling in cellular networks for energy optimization," *IEEE Trans. Wireless Commun.*, volume 14, issue 1, pp. 509–519, Jan. 2015.

[89] ITU, "ICT industry to reduce greenhouse gas emissions by 45 per cent by 2030," 27 Feb. 2020. https://www.itu.int/en/mediacentre/Pages/PR04-2020-ICT-industry-to-reduce-greenhouse-gas-emissions-by-45-perce nt-by-2030.aspx

[90] Next G Alliance, "Next G alliance report: Roadmap to 6G," Feb. 2022.

[91] 3GPP TR 32.972 version 16.1.0 Release 16, "5G; Telecommunication management; Study on system and functional aspects of energy efficiency in 5G networks," Nov. 2020.

[92] 3GPP TS 38.840 version 16.0.0 Release 16, "Technical specification group radio access network; NR; Study on User Equipment (UE) power saving in NR," Jun. 2019.

[93] T. L. Marzetta, "Noncooperative cellular wireless with unlimited numbers of base station antennas," *IEEE Trans. Wireless Commun.*, vol. 9, no. 11, pp. 3590–3600, 2010.

[94] E. Björnson and L. Sanguinetti, "Scalable cell-free massive MIMO systems," *IEEE Trans. Commun.*, vol. 68, no. 7, pp. 4247–4261, July 2020. doi: 10.1109/TCOMM.2020.2987311

[95]   D. Castanheira and A. Gameiro, "Distributed antenna system capacity scaling [Coordinated and Distributed MIMO]," *IEEE Wireless Commun.*, vol. 17, no. 3, pp. 68–75, Jun. 2010. doi: 10.1109/MWC.2010.5490981

[96]   J. Joung, Y. K. Chia and S. Sun, "Energy-efficient, Large-scale Distributed-Antenna System (L-DAS) for multiple users," *IEEE J. Select. Areas Commun.*, vol. 8, issue 5, pp. 954–965, Oct. 2014.

[97]   A. Prasad, A. Bhamri and P. Lundén, "Enhanced mobility and energy efficiency in 5G ultra-dense networks with lean carrier design," *2016 IEEE Conference on Standards for Communications and Networking (CSCN)*, pp. 1–5. doi: 10.1109/CSCN.2016.7784884

[98]   P. Frenger, Y. Jading and J. Bengtsson, "5G energy consumption: What's the impact of 5G NR in real networks?" Ericsson.com. https://www.erics son.com/en/blog/2021/10/5g-energy-consumption-impact-5g-nr

[99]   "3GPP portal," https://portal.3gpp.org/desktopmodules/Specifications/ SpecificationDetails.aspx?specificationId=3355

[100] M. Lou, J. Jin, H. Wang, L. Xia, Q. Wang and Y. Yuan, "Applying sparse array in massive MIMO via convex optimization," *2020 IEEE Asia-Pacific Microwave Conference (APMC)*, 2020, pp. 721–723. doi: 10.1109/APMC47863.2020.9331675

[101] ETSI, "Zero touch network & Service Management (ZSM)," https:// www.etsi.org/technologies/zero-touch-network-service-management

[102] ITU, "Focus Group on Environmental Efficiency for Artificial Intelligence and other Emerging Technologies (FG-AI4EE)," ITU. www.itu.int/en/IT U-T/focusgroups/ai4ee

[103] O-Ran Alliance, "O-RAN specifications," https://www.o-ran.org/specific ations

[104] G. Vallero, D. Renga, M. Meo and M. A. Marsan, "Greener RANoperation through machine learning," *IEEE Trans. Netw. Service Manag.*, vol. 16, no. 3, 2019.

[105] Safeguard, "Lean, green telco machine: How AI is greening mobile networks — Telenor Group," https://www.telenor.com/stories/safeguar d/lean-green-telco-machine-how-ai-is-greening-mobile-networks/

[106] GSMA, "Turkcell and PI Works AI use cases in service assurance," 6 Jan. 2020. https://www.gsma.com/futurenetworks/wiki/case-study-ai-use-cas es-in-service-assurance

[107] K. Pedersen, D. Chandramouli, M. Baker, A. Toskala, S. Nielsen and J. Moilanen, "5G-advanced: Expanding 5G for the connected world," Nokia White Paper, 2022.

[108] MSIT, "6G, Korea takes the lead once again: 6G R&D implementation plan established," Ministry of Science and ICT of Korea. https://www. msit.go.kr/eng/bbs/view.do?sCode=eng&mId=4&mPid=2&pageIndex= 9&bbsSeqNo=42&nttSeqNo=517&searchOpt=ALL&searchTxt)

[109] X. Lin, "An overview of 5G advanced evolution in 3GPP release 18," arXiv preprint, 2022. arXiv:2201.01358

[110] ITU, "World radiocommunication conferences," https://www.itu.int/pub/ R-ACT-WRC/en

[111] ITU, "Focus group on technologies for network 2030," https://www.
itu.int/en/ITU-T/focusgroups/net2030/Pages/default.aspx [Accessed 14
October 2021].

[112] ITU, "Beyond 5G: What's next for IMT?" 2 Feb. 2021. https://www.
itu.int/en/myitu/News/2021/02/02/09/20/Beyond-5G-IMT-2020-update
-new-Recommendation [Accessed 14 October 2021].

[113] FG-NET2030, "Technical specification: Network 2030 architecture framework," ITU-T, 2020.

[114] ITU-T, "Network 2030: A blueprint of technology, applications and market
drivers towards the year 2030 and beyond," ITU T NET-2030, 2020.

[115] ITU, "World Radiocommunication Conferences (WRC)," https://www.
itu.int/en/ITU-R/conferences/wrc/Pages/default.aspx#:~:text=instruct
%20the%20Radio%20Regulations%20Board,preparation%20for%20future
%20Radiocommunication%20Conferences

[116] GSMA, www.gsma.com

[117] The Next Generation Mobile Networks (NGMN), https://www.ngmn. org/

[118] Next G Alliance, "Next G Alliance FAQ," ATIS. https://nextgalliance.
org/about/ [Accessed 14 October 2021].

[119] O-Ran Alliance, "About us," https://www.o-ran.org/about

[120] The Linux Foundation, https://www.linuxfoundation.org/category/ai-ml
-data-analytics/

[121] The 6G Smart Networks and Services Industry Association (6G-IA),
https://6g-ia.eu/

[122] K. B. Letaief, W. Chen, Y. Shi, J. Zhang and Y.-J. A. Zhang, "The
roadmap to 6G: AI empowered wireless networks," *IEEE Commun. Mag.*,
pp. 84–90, vol. 57, issue 8, 2019.

[123] I. Akyildiz, A. Kak and S. Nie, "6G and beyond: The future of wireless
communications systems," *IEEE Access*, 2020.

CIRCUIT

Property is about location; management is about leadership; talent is about people; collaboration is about trust; and success is about will and determination.

Kiat Seng Yeo

Chapter 6

Generalized Multiple Tanks Based RF/mm-wave IC for Future Communications

Kaixue Ma, Zhen Yang, Chenyang Meng, Xinbo Zhang
and Kiat Seng Yeo

6.1. Substrate Loss

Substrate loss has always been a major problem in commercial silicon processes, especially when the operating frequency rises to the mm-wave bands, where more energy leaks into the low resistance substrate and generates additional losses and cross talk due to leakage enhancement.

The substrate loss consists of two parts: finite resistance due to electrically induced conductive and displacement currents and magnetically induced eddy current resistance. These losses are known as capacitive and magnetic, respectively [1]. Figure 6.1 shows the losses associated with a silicon-based inductor, which is the most used passive device in radio frequency integrated circuits (RFICs) and mm-wave IC. When the frequency rises, the ohmic loss deteriorates sharply due to the skin effect, while the magnetic coupling strength is also enhanced, which, in turn, leads to the enhancement of displacement and eddy currents. Typically, two strategies are used to reduce substrate loss: one is to try to prevent the signal from entering the substrate, such as Silicon-On-Isolator (SOI), Silicon-On-Sapphire (SOS), High-Resistive Silicon (HRS) [2], and the other is to try to

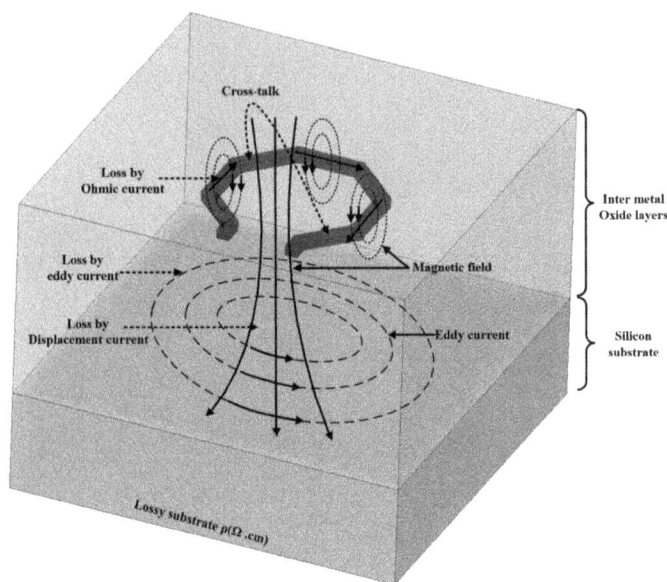

Fig. 6.1.　Integrated coil related losses.

form a shield between the signal and the substrate such as Deep Trench Isolation (DTI) [3] or using a metal shield. The former adds additional cost, while the latter has little effect in the millimeter band.

6.2.　Generalized Multiple Tanks

As said before, magnetic coupling also increases with frequency, and according to Maxwell's equations, the binding capacity of the magnetic field is stronger at high dielectric constants. This motivated us to employ multiple resonators to capture more of the magnetic field in the oxide layer as opposed to the substrate (current dissipation due to the low resistance of the silicon-based substrate). Then, to decrease commercial silicon-based IC substrate loss, we presented a multiple tank approach in a US patent [4].

The approach under a single port scenario is shown in Fig. 6.2. Under a magnetic coupling state, the approach uses numerous coupled coils as a transformer to enhance the equivalent Q factor

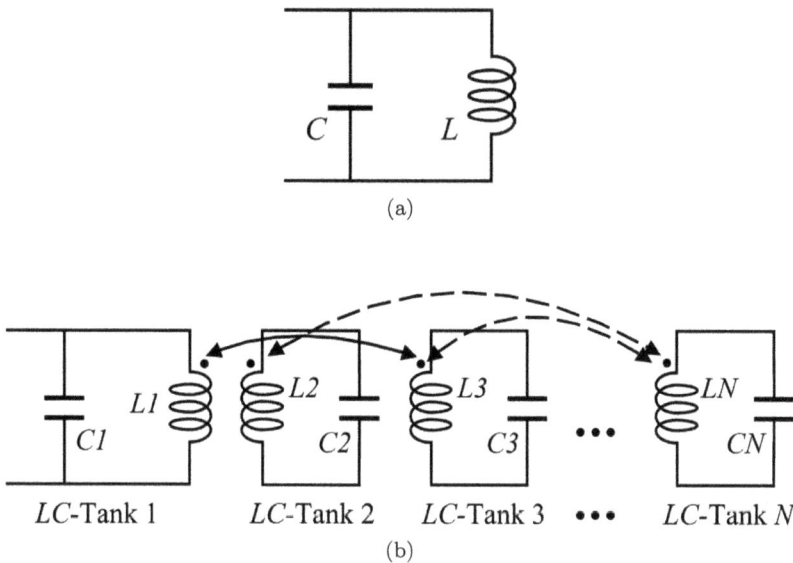

(a)

(b)

Fig. 6.2. Topology of (a) single LC tank and (b) multiple-coupled LC tank.

of the tanks, hence enhancing the performance of the circuit. The improved Q factor of the tanks minimizes matching loss, hence increasing the efficiency of PAs [5–7]. A multiple tanks-based high Q-factor resonator might minimize phase noise and expand the tuning range of voltage-controlled oscillators (VCOs) [8–12], and the technology could reduce die sizes, as shown by a number of RFIC designs [13–15]. In this chapter, the approach's broad idea and analysis are provided. Several different tanks-based circuits and SOC transceivers are examined, and the approach's possible applications are outlined.

6.2.1. *Theory of multiple tanks*

The architecture of a single LC tank and a multiple-coupled LC tank is seen in Fig. 6.2. The multiple-connected LC tank is composed of many single LC tanks that are linked to one another. Multi-coil transformers that comprise the inductors of individual LC tanks provide the connections between these LC tanks. N is intended to indicate the tank order, which is the number of single LC tanks in

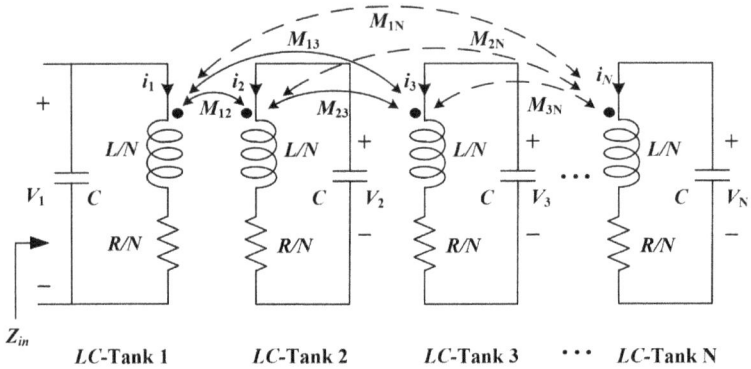

Fig. 6.3.   The circuit schematic of single-port multiple-tank topology.

the multiple-coupled LC tank topology, for purposes of deduction. From the impedance looking into the main LC tank, the effective Q of the multiple-coupled LC tank may be computed (i.e., the LC tank directly connecting to a cross-coupled pair).

The analogous model for a multiple-coupled LC tank is shown in Fig. 6.3. To simplify the calculation and facilitate the comparison with a single LC tank, we assume that each transformer coil in the multiple-coupled LC tank has identical self-inductance equal to L/N, i.e., $L1 = L2 = \cdots = LN = L/N$, and that the capacitors in the multiple-coupled LC tank are also identical and equal to C. Each transformer coil's parasitic resistance is represented by a resistor, and the resistance is considered to be proportional to the self-inductance, which is near to its actual condition [16].

If the coupling coefficients of any two individual tanks are identical, the mutual inductance of the transformer may be expressed as

$$M_{ij} = k\sqrt{L_i L_j} = \frac{kL}{N},$$

(6.1)

where k is the coefficient of coupling. As shown in Fig. 6.3, $i_{1,2,\ldots,N}$ represents the currents on each transformer coil and $V_{1,2,\ldots,N}$ represents the voltage drop on each tank. On the basis of transformer V-I equations, the relationship between $i_{1,2,\ldots,N}$ and $V_{1,2,\ldots,N}$ may be expressed as

$$
\begin{bmatrix}
\dfrac{R}{N}+\dfrac{Ls}{N} & \dfrac{kLs}{N} & \cdots & \dfrac{kLs}{N} \\
\dfrac{kLs}{N} & \dfrac{R}{N}+\dfrac{Ls}{N} & \cdots & \dfrac{kLs}{N} \\
\cdots & \cdots & \cdots & \dfrac{kLs}{N} \\
\dfrac{kLs}{N} & \dfrac{kLs}{N} & \cdots & \dfrac{R}{N}+\dfrac{Ls}{N}
\end{bmatrix}
\begin{bmatrix} i_1 \\ i_2 \\ \cdots \\ i_N \end{bmatrix}
=
\begin{bmatrix} V_1 \\ V_2 \\ \cdots \\ V_N \end{bmatrix}.
\tag{6.2}
$$

Also, $V_1, V, \ldots, V_N$ can be expressed as the voltage drop on the corresponding capacitor, which is given by

$$
V_n = -i_n \frac{1}{sC}, n = 1, 2, 3 \ldots N.
\tag{6.3}
$$

Substituting Eq. (6.3) into Eq. (6.2), the equation systems concerned with $i_2, i_3, \ldots, i_N$ can be obtained as

$$
\begin{bmatrix}
\dfrac{R}{N}+\dfrac{Ls}{N}+\dfrac{1}{sC} & \dfrac{kLs}{N} & \cdots & \dfrac{kLs}{N} \\
\dfrac{kLs}{N} & \dfrac{R}{N}+\dfrac{Ls}{N}+\dfrac{1}{sC} & \cdots & \dfrac{kLs}{N} \\
\cdots & \cdots & \cdots & \dfrac{kLs}{N} \\
\dfrac{kLs}{N} & \dfrac{kLs}{N} & \cdots & \dfrac{R}{N}+\dfrac{Ls}{N}+\dfrac{1}{sC}
\end{bmatrix}
\begin{bmatrix} i_2 \\ i_3 \\ \cdots \\ i_N \end{bmatrix}
$$

$$
=
\begin{bmatrix}
-\dfrac{i_1 kLs}{N} \\
-\dfrac{i_1 kLs}{N} \\
\cdots \\
-\dfrac{i_1 kLs}{N}
\end{bmatrix}.
\tag{6.4}
$$

By solving the linear Eq. (6.4) using Cramer's rule, the expressions of $i_2, i_3, \ldots, i_n$ can be obtained as

$$
i_2 = i_3 = i_4 \ldots i_N = \frac{-i_1 kLCs^2}{N + RCs + [1 + (N-2)k] LCs^2}.
\tag{6.5}
$$

Then, V_1 can be expressed as

$$V_1 = i_1 \left(\frac{Ls}{N} + \frac{R}{N} \right) + (i_2 + i_3 + \cdots + i_N) \frac{kLs}{N}$$

$$= i_1 \left(\frac{R + Ls}{N} + \sum_{n=2}^{N} \frac{i_n}{i_1} \frac{kLs}{N} \right),$$ (6.6)

where

$$\sum_{n=2}^{N} \frac{i_n}{i_1} = \frac{-(N-1)kLCs^2}{N + RCs + [1 + (N-2)k]LCs^2}.$$ (6.7)

Therefore, the input impedance of the multiple-coupled LC tank Z_{in} can be expressed as

$$Z_{in} = \frac{V_1}{i_1 + sCV_1} = \frac{R + Ls + \left(\sum\limits_{n=2}^{N} \frac{i_n}{i_1} \right) kLs}{N + RCs + LCs^2 + \left(\sum\limits_{n=2}^{N} \frac{i_n}{i_1} \right) kLCs^2}.$$ (6.8)

To calculate resonant frequencies of tank, we can make R zero. Then the resonant frequencies can be obtained as

$$\omega_1 = \sqrt{\frac{N}{LC[1 + (N-1)k]}},$$ (6.9)

$$\omega_2 = \sqrt{\frac{N}{LC[1-k]}}.$$ (6.10)

Even though the multiple-coupled LC tank is a high-order network with a high degree of complexity, there are only two resonant frequencies since our assumptions (i.e., identical individual tanks and the same coupling coefficient k) have greatly simplified the computation.

Figure 6.4 plots the resonant frequency, ω_1, ω_2, with the changing of coupling coefficient when choosing a set of typical tank component values $L = 600$ pH and $C = 0.8$ pF. When k is raised, the lower

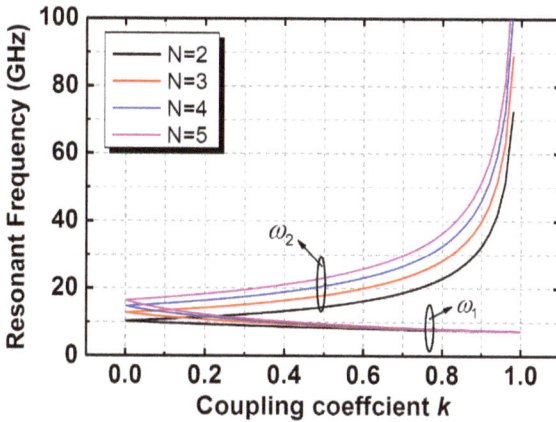

Fig. 6.4.   Resonant frequencies ω_1 and ω_2 versus coupling coefficient with different values of N.

resonant ω_1 decreases and the higher resonant ω_2 increases, resulting in a large separation between the two resonant frequencies.

The effective Q of the multiple-coupled LC tank at ω_1 or ω_2 can be estimated using the well-known formula in [17]

$$Q_{eff} = -\frac{\omega_{1,2}}{2} \frac{\partial \psi}{\partial \omega} |\omega = \omega_{1,2}, \qquad (6.11)$$

where ψ is the phase response of Z_{in} [18]. For a special case when $k = 1, \omega_1 = 1/\sqrt{LC}$, and ω_2 approaches to infinity, the expression of Q_{eff} at ω_1 can be given by

$$Q_{\text{eff}} = \left(\frac{N^3 + 1.5N\frac{CR^2}{L} + 0.5\left(\frac{CR^2}{L}\right)^2}{N^3 + N(1+N)\frac{CR^2}{L} + \left(\frac{CR^2}{L}\right)^2} \right) \frac{N\omega L}{R}. \qquad (6.12)$$

If $R^2 C/L \ll 1$, which is the case in most RFIC oscillators [18], the expression of effective Q can be simplified as

$$Q_{\text{eff}} = \frac{N\omega L}{R}. \qquad (6.13)$$

Since tank Q of a single LC tank with inductance L and capacitance $C = L/R$, tank Q of a multiple-coupled LC tank may be increased by a factor of N when compared to a single LC tank. According to

Leeson's formula [19], the phase noise of a VCO with a multiple-coupled LC tank may be reduced by $20\log(N)$ decibels when Q is increased by N.

When k is smaller than 1, the effective Q of multiple-coupled LC tank Q_{eff} at ω_1 and ω_2 may be computed and plotted against k, as seen in Fig. 6.5(a) and 6.5(b). During the computation, the normal tank component values of $L = 600$ pH, $C = 0.8$ pF, and $R = 3.6$ are used. It is evident that when k increases, Q_1 increases and Q_2 decreases. Though there are two resonant frequencies, we may choose one as the VCO's oscillation frequency by configuring the loop gain at the resonant frequency of interest. It is observed that the multiple-coupled LC tank has a higher Q at ω_1 than at ω_2; hence, for better phase noise performance, we would use ω_1 as the oscillation frequency of the VCO.

In addition to the coupling coefficient, the tank order N is a crucial design parameter for multiple-coupled LC tanks. For instance, in Fig. 6.5(a), for $N = 2, N = 3, N = 4$, and $N = 5$, Q_1 is 13, 19, 25, and 31 when $k = 0.6$. Compared to a single LC tank with a Q value of 7 ($L = 600$ pH, $C = 0.8$ pF, $R = 3.6$), Q_1 has been significantly enhanced, and this enhancement increases with tank order N. Consequently, both a big tank order and a high coupling coefficient are necessary for the Q improvement of a multiple-coupled LC tank to be realized.

The loss of capacitors has been disregarded in the model (Fig. 6.3) in order to simplify the derivation. Considering the capacitor's loss, we may use an additional resistor R_c in series with the capacitor to represent its loss and get tank Q.

Similarly, tank Q at a lower resonant frequency Q_1 may be plotted against a set of tank parameters, as seen in Fig. 6.6. $L = 200$ pH, $C = 100$ fF, $R = 1.46$, and $RC = 1.51$ are selected, and it is possible to compute that the Q of a single LC tank with L and C is 15. Despite the loss of the capacitor, it can be demonstrated that the multiple-coupled LC tank can still enhance tank Q. The distinction is that the increase of Q by a connected LC tank must compensate for the loss caused by capacitors; therefore, the improvement of Q is less than in the absence of capacitor loss. Fortunately, when tank order N and

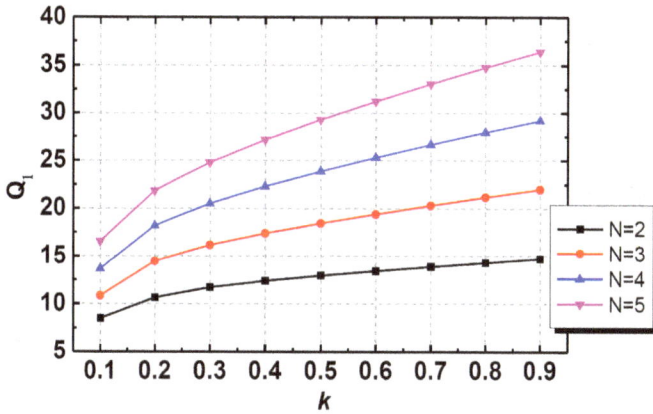

(a)

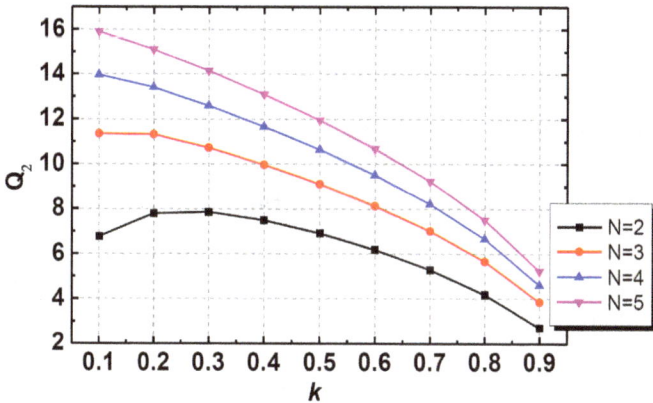

(b)

Fig. 6.5. Effective Q of multiple-coupled LC tanks at the two resonant frequencies (a) $\omega 1$ and (b) $\omega 2$.

coupling coefficient improves, so does the augmentation of tank Q. Therefore, if the transformer coupling is sufficiently strong and the tank order is big, tank Q may still profit from the multiple-coupled LC tank.

In conclusion, multiple-coupled LC tanks may increase tank Q in comparison to a single LC tank. Transformer coupling coefficients and a tank order are associated with the enhancement of tank Q. Transformers with higher coupling coefficients and a greater number

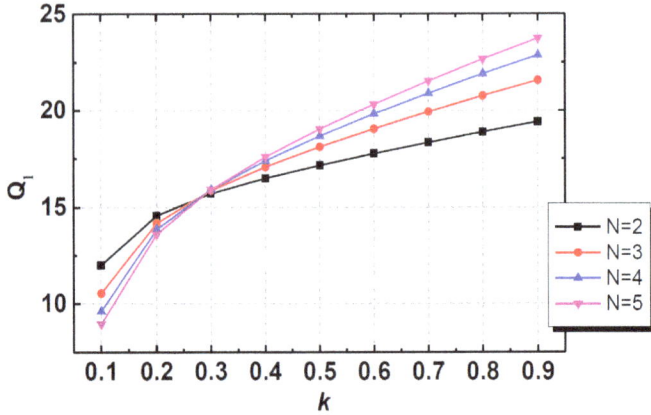

Fig. 6.6.   Effective Q of multiple-coupled LC tanks at ω_1 considering the loss of capacitor in the tanks.

of tank orders are required to increase tank Q. It should be observed that when the tank order grows, the coupling between the distant windings weakens, resulting in a decrease in coupling coefficient and, consequently, tank Q. To get a large tank order and maintain high coupling coefficients among transformer windings, the transformer structure must be constructed according to the target process's available metal layers.

6.3. Multiple Tanks Technique for VCO and Dividers

The VCO and frequency divider are very important modules in microwave systems, and their performance directly affects system performance. The traditional resonant tank is a parallel LC second-order network. After the negative resistance components are connected in parallel, the functions such as frequency selection and filtering of the current provided by the negative resistance components can be realized. The resonant network based on multiple LC tanks not only occupies almost the same area in layout and single inductor but also provides the following additional characteristics:

(1) *Bimodal characteristic*

The multi-tank LC resonator can have multiple peaks in the frequency domain response. It also provides a higher degree of design freedom for an increased tuning frequency range and reduced phase noise.

The dual-band design can be realized by taking advantage of the inherent dual resonance point characteristic of the resonator to achieve wider frequency coverage. Secondly, the oscillator can be operated at the fundamental frequency, and high-order harmonics can be extracted so as to improve the Q of the fundamental wave loop and obtain better noise performance. In addition, the multi-peak feature enables multiple tuning and tuning gain linearization to further increase the tuning frequency range.

(2) *Gain enhancement*

Compared with a single LC resonator, the multi-tanks LC resonator can further increase the peak impedance of the resonator and thus increase the gain. This can be explained in three aspects: (a) The parasitic capacitance is separated by multiple LC tanks, and the parasitic capacitance seen by each inductor becomes smaller; (b) the coupling between coils enhances the effective Q value of the inductor; and (c) more inductors are allowed for the same resonant frequency.

(3) *Compact chip area*

A multi-tank LC resonator and a single inductor occupy nearly the same area.

(4) *High quality factor*

According to the Leeson model, the smaller the parallel equivalent resistance of the resonator is, the smaller the phase noise of the VCO is when other parameters are kept unchanged. When the operating frequency is relatively low, the equivalent parallel resistance of the resonance tank is mainly determined by the value of the inductance, and the smaller the inductance value, the smaller the parallel impedance. Therefore, by reducing the inductance value of the inductor and increasing the capacitance value of the capacitor, the equivalent parallel resistance can be reduced while the resonant frequency remains unchanged, thereby optimizing the phase noise.

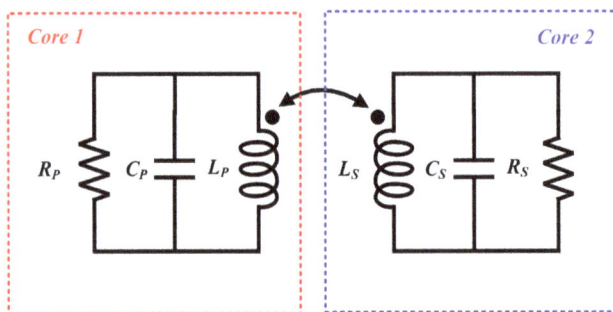

Fig. 6.7. Schematic diagram of dual-core coupling.

However, in the actual design process, when the inductance is reduced to a certain extent, the coupling between the differential inductances is enhanced on the physical layout, and the quality factor deteriorates, thereby deteriorating the phase noise [17].

The multi-core-coupled oscillator can get rid of the influence of the layout on the quality factor of the resonator and realize the optimization of the phase noise. Take dual-core parallel coupling as an example, as shown in Fig. 6.7.

When the two networks are connected in parallel, the quality factor of the entire resonant tank is $Q_{all} = (R_P/2)/(\omega_1 L_P/2) = R_P/(\omega_1 L_P) = Q_1 = Q_2$, while the resonant frequency remains unchanged, and the parallel resistance is halved. According to Leeson's formula, phase noise can be optimized by 3 dB. Similarly, for the coupling of N oscillators, the phase noise can be optimized in $10 \log N$ dB compared to a single oscillator.

In this section, some design approaches for VCOs and dividers using multiple tanks technique will be analyzed in detail.

6.3.1. *Application of multiple tanks bimodal characteristics*

(1) *Dual-mode wideband Voltage Controlled Oscillator (VCO)*

The dual-mode VCO implemented by a transformer-based coupled resonant tank for increased frequency tuning range [20]. The resonant tank Q increases at the lower resonant frequency, but decreases for tank Q with the higher resonant frequency. This results in an improvement in PN for modes operating at the lower resonant frequency and a reduction in PN for modes operating at the higher resonant frequency.

Taking advantage of the bimodal characteristics of multiple tanks, Zou *et al.* [21] propose a novel dual-mode mm-Wave VCO topology with switchable-coupled cores.

In the switching state, using different parasitic capacitances of cross-coupled pairs as the corresponding VCO cores can realize dual-mode operation at millimeter wave frequency, and generate two modes of vibration at the lower resonant frequency of the LC tank, so as to improve the PN performance of the LC tank Q and the VCO in both modes.

Figure 6.8(a) shows the schematic diagram of the proposed dual-mode VCO. Core1 and Core2 are two switchable-coupled VCO cores. Bipolar transistors $Q_1 \sim Q_2$ and $Q_3 \sim Q_4$ are two cross-coupled pairs, C_{v1} and C_{v2} are varactors, and I_{B1} and I_{B2} are the bias currents for the two VCO cores. By controlling I_{B1} and I_{B2}, the VCO has two working modes (Mode A and Mode B), as shown in Fig. 6.8(b).

The cross-coupled pairs have different input capacitances C_{on} or C_{off} at different bias currents I_{on} or I_{off}. The capacitance of the load on the primary and secondary windings of the transformer has different expressions in different modes. For A mode, $C_1 = C_{v1} + C_{on1}$, $C_2 = C_{off2} + C_{v2}$; for B mode, $C_1 = C_{v1} + C_{off1}$, $C_2 = C_{on2} + C_{v2}$. Therefore, the lower resonant frequency ω_L of mode A and mode B have the same expression (6.14), but the ω_L value of mode A and mode B are different due to the change of C_1 and C_2.

$$\omega_{H/L} = \sqrt{\frac{L_1 C_1 + L_2 C_2 \pm \sqrt{(L_1 C_1 - L_2 C_2)^2 + 4k^2 L_1 C_1 L_2 C_2}}{2 L_1 C_1 L_2 C_2 (1 - k^2)}}.$$

(6.14)

Switching between the two modes is manipulated by controlling the bias current of the cross-coupled pair, which avoids the direct connection of the lossy switch to the tank and improves the Q of the resonator. Since the target resonant frequency for both modes

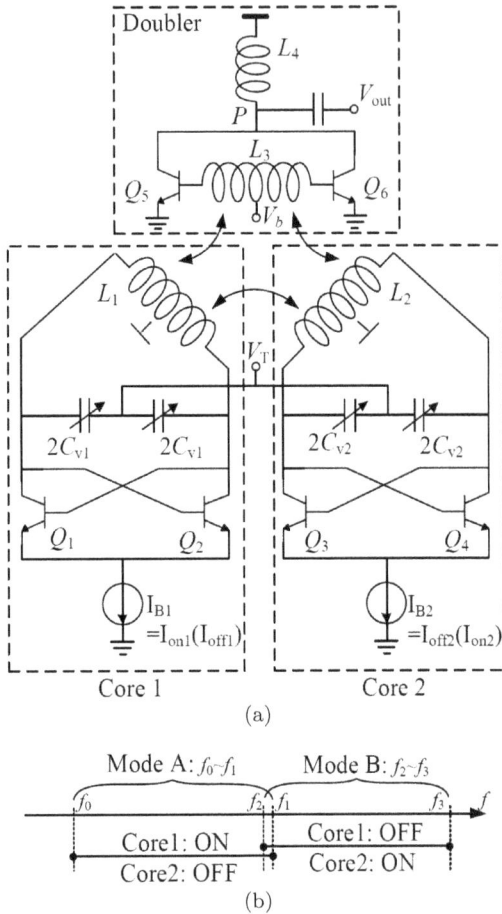

Fig. 6.8.   (a) Topology of the proposed VCO. (b) Working status diagram.

is the lower resonant frequency ω_L, it has a higher LC tank Q and better PN performance. The chip micrograph of the proposed VCO is shown in Fig. 6.9; a strong coupling transformer with a large k value is used to improve the oscillation stability and resonance Q value.

Figure 6.10(a) shows the worst, and Fig. 6.10(b) shows the best PN versus frequency offset over the entire tuning range. The 1 MHz phase noise for Mode A is -91.4 dBc/Hz to -93.45 dBc/Hz, and the phase noise for Mode B is -87.6 dBc/Hz to -89.4 dBc/Hz.

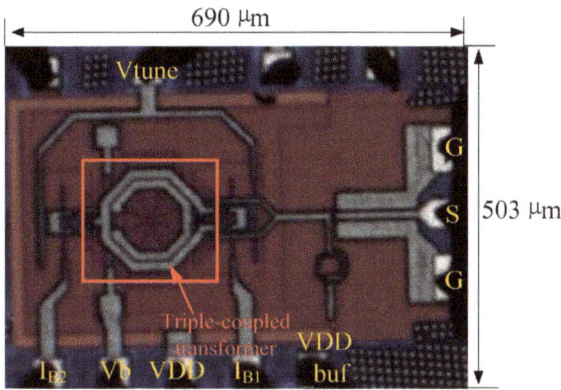

Fig. 6.9.   Chip photomicrograph of the proposed VCO.

The proposed VCO in the switchable-coupled VCO core has a wide tuning range of 17.2% from 55.7 GHz to 66 GHz.

(2) *Harmonic-extraction technique*

In the millimeter wave and sub-terahertz frequency bands, there are many problems with a high oscillation frequency of the VCO, such as large parasitics of transistors, high switching losses at high frequencies, and low-quality factor (Q) of varactors. Harmonic oscillators [22] operate at relatively low frequencies, which can be used in designs to achieve wide tuning ranges and low phase noise.

The bimodal characteristics of multiple tanks provide new ideas for the design of high-performance harmonic oscillators. Liu *et al.* [11] proposed a harmonic extraction technique based on the LC tank of a three-coil transformer to enhance the amplitude and phase noise of the oscillator at the third-harmonic frequency, as shown in Fig. 6.11. Compared with the conventional method, a third coil L_S is added at the source of the cross-coupled pair and forms a positive coupling with the L_D.

Coil L_S provides additional positive coupling, and the total flux of L_D is boosted, resulting in enhanced tank resistance $R_{p3,\text{eff}}$ at the third harmonic. L_S can also be used as a source degeneration, increasing the equivalent output resistance $R_{o,\text{eff}}$. Therefore the

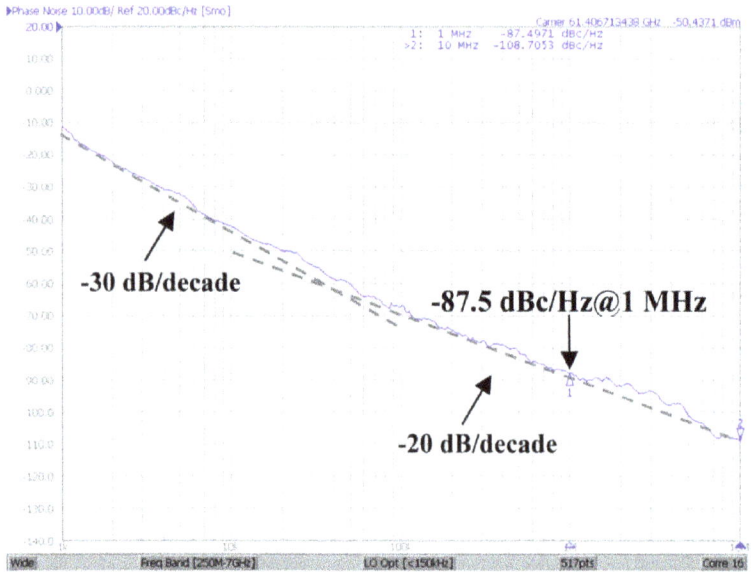

(a)

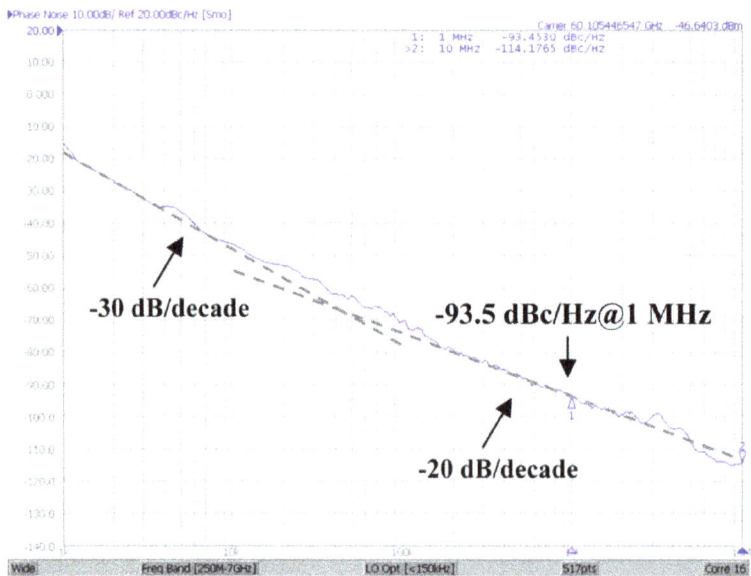

(b)

Fig. 6.10.   (a) Worst and (b) best PN versus frequency offset over the entire tuning range.

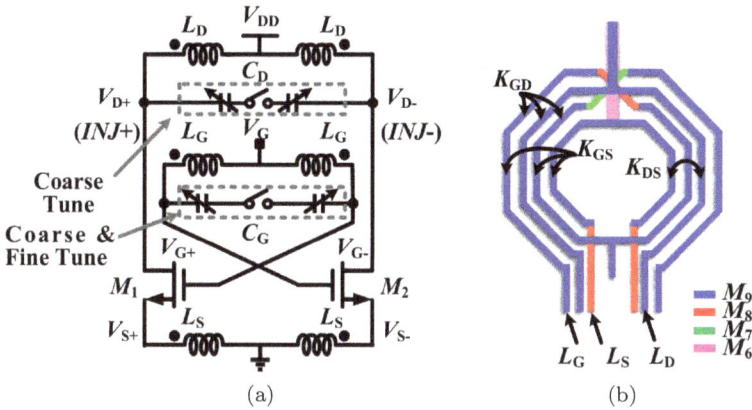

(a)   (b)

Fig. 6.11.   (a) Third-harmonic-enhanced VCO and (b) three-coil transformer layout.

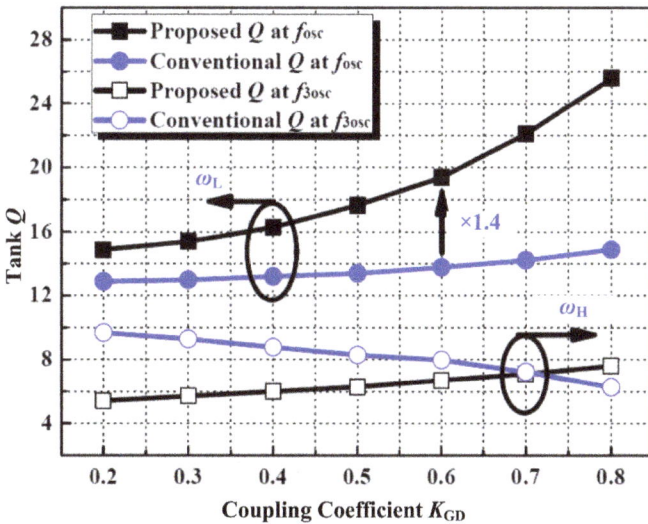

Fig. 6.12.   Calculated resonant tank Q at ω_L and ω_H.

third coil L_S can significantly increase the drain output resistance $R_{p3,eff}//R_{o,eff}$.

As shown in Fig. 6.12, when $K_{GD} = 0.6$, the tank Q at low resonant frequency ω_L is improved by a factor of 1.4, and the tank

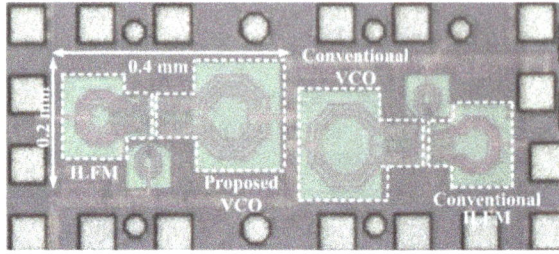

(a)

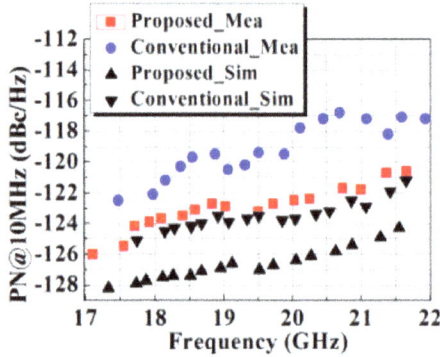

(b)

Fig. 6.13.   (a) Chip micrograph and (b) measured and simulated PN.

Q at high resonant frequency ω_H decreases slightly. This is beneficial to avoid frequency misalignment between ω_H and $3\omega_L$.

Figure 6.11(a) shows the topology of the proposed VCO, which is based on the enhanced third-harmonic extraction technique. The VCO can generate two output signals simultaneously: the fundamental signal (17.1 ∼ 21.7 GHz) and the third-harmonic signal (51.3 ∼ 65.1 GHz). Figure 6.13 shows the chip micrograph and the measured phase noise. The PN at 10 MHz offset is −120.6 ∼ −126 dBc/Hz and −116.3 dBc/Hz, and the output power of the third harmonic is −19.1 dBm.

(3) Multi-tank for dual-tuned VCO

Multiple LC VCOs can be used to cover a larger frequency range, but there are many disadvantages, such as high power consumption, large

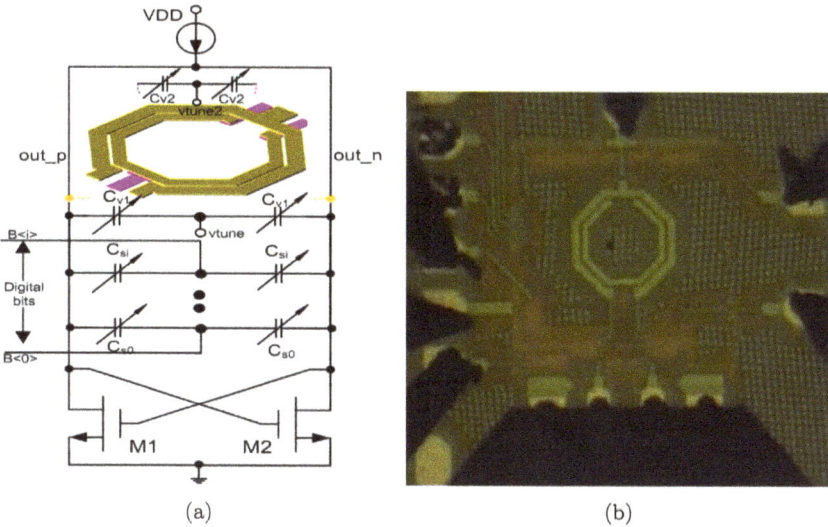

Fig. 6.14. (a) Schematic of the VCO and (b) micrograph of the chip.

chip area, and so on. Mou *et al.* [23] designed a VCO with multiple strongly magnetically coupled LC tanks. The proposed VCO can achieve a larger tuning range and lower PN. Figure 6.14 shows a schematic diagram of the VCO and a micrograph of the chip. The proposed VCO has a secondary coupled LC loop, which can be used to greatly increase the tuning range.

The proposed VCO achieves a tuning range from 10.1 GHz to 14.4 GHz (36%) with a PN of −87.94 dBc/Hz at 1 MHz offset. Figure 6.15 shows the PN.

(4) *Multi-tank for tuning gain linearization*

The VCO should have high tuning linearity, as large changes in the VCO gain will cause significant changes in the PLL loop bandwidth, degrading performance and stability [24].

Chen *et al.* [25] proposed a linear LC-VCO topology based on triple-coupled LC tanks. L_1, L_2, and L_3 are used to couple their respective varactors to compensate for the tuning nonlinearity of the resonant tank.

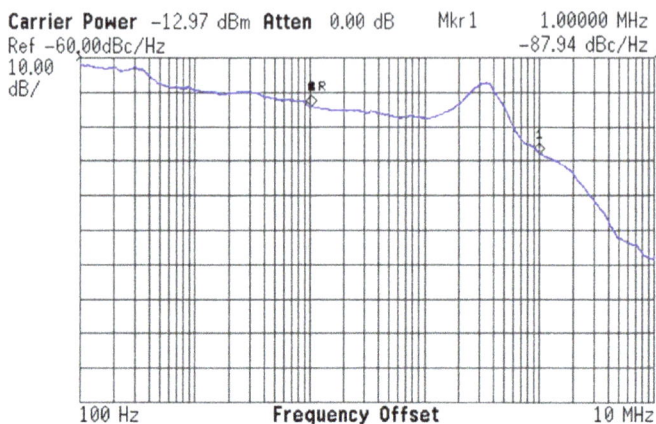

Fig. 6.15.   Measured PN curve.

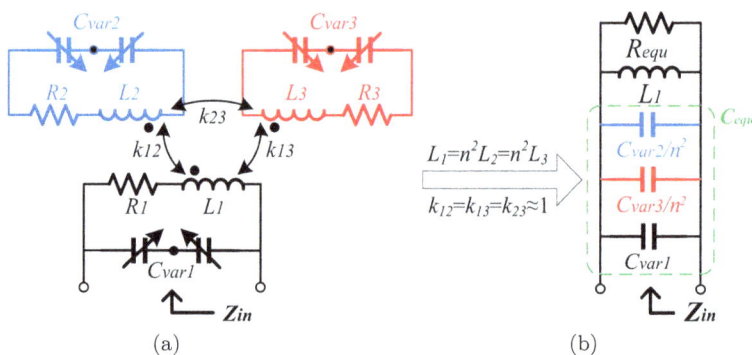

Fig. 6.16.   (a) Schematic diagram of a triple-coupled LC tank. (b) Equivalent parallel LC tank.

This technology can achieve C-V curve linearization without tuning voltage shift circuits or DC blocking capacitors, effectively improving the Q of the resonant tank and achieving high linearity, wide tuning range, and good noise performance in the mm-wave frequency band.

The equivalent circuit of a three-coupled LC tank is shown in Fig. 6.16(a). $L_1 \sim L_3$ are the three inductors, $R_1 \sim R_3$ correspond

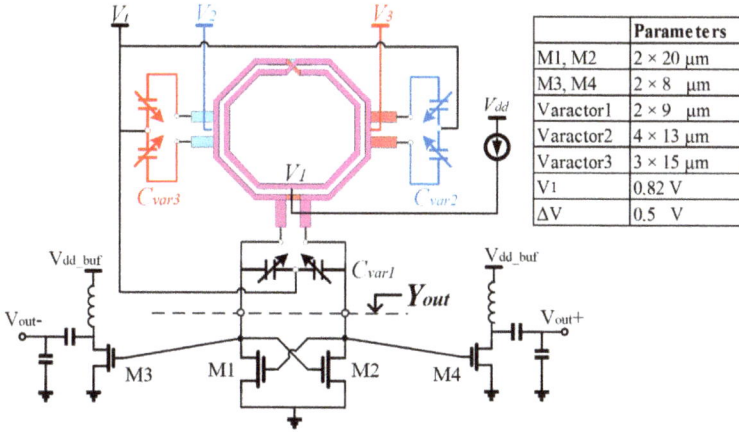

	Parameters
M1, M2	2×20 μm
M3, M4	2×8 μm
Varactor1	2×9 μm
Varactor2	4×13 μm
Varactor3	3×15 μm
V1	0.82 V
ΔV	0.5 V

Fig. 6.17.   Schematic of the linear LC-VCO with a triple-coupled LC tank.

to the loss of each inductor, and $C_{var1} \sim C_{var3}$ are the capacitances of the three pairs of varactors. k_{12}, k_{13}, and k_{23} are coupling coefficients.

Assuming $L_1 = n^2 L_2 = n^2 L_3$ (n is the turns ratio of the coupled inductance), the resonant frequency ω_{RES} of the triple-coupled LC tank can be obtained under strong coupling conditions.

$$\omega_{RES} = \frac{1}{\sqrt{L_1 \left(C_{var1} + C_{var2}/n^2 + C_{var3}/n^3\right)}}. \tag{6.15}$$

The above equation shows that the triple-coupled LC tank is equivalent to a parallel LC resonator with an equivalent inductor $L_{eq} = L_1$ and an equivalent capacitor $C_{eq} = (C_{var1} + C_{var2}/n^2 + C_{var3}/n^2)$, as shown in Fig. 6.16(b). Therefore, the nonlinearity of the CV curve of C_{var1} can be compensated by choosing an appropriate value of n and the varactors $C_{var2} \sim C_{var3}$.

The schematic diagram of the proposed linear LC-VCO with a triple-coupled LC tank is shown in Fig. 6.17. In order to increase the coupling coefficient between the coupled inductances, the stacking inductance method is used, and M9 $\sim$ M7 metal layers are used to design $L1 \sim L3$ with thicknesses of 3.4 μm, 0.85 μm, and 0.31 μm, respectively. The primary inductor $L1$ uses a two-turn coil to reduce the chip area.

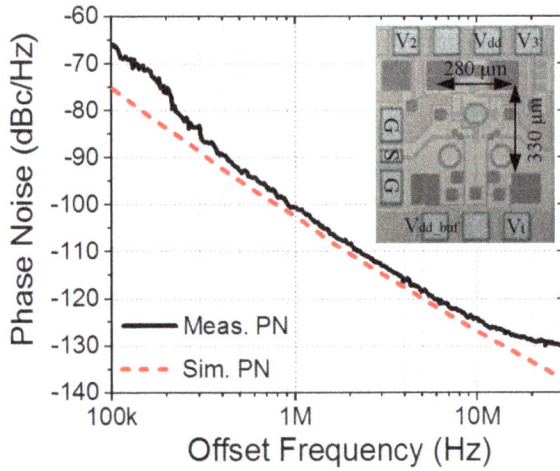

Fig. 6.18.   The proposed linear VCO: (a) simulated and measured PN at 20.299 GHz and (b) the chip photograph.

Figure 6.18 shows the chip photo of the proposed linear VCO and the simulated and measured PN at 20.299 GHz. The measured tuning bandwidth of this linear VCO can reach 15.8%, the PN at 1 MHz offset is −100.7 dBc/Hz, and the figure of merit (FOM$_T$) is −181.8 dBc/Hz.

6.3.2. *Gain enhancement characteristics*

(1) *Gm-boosting technology*

The Gm-boosting technology can significantly enhance the negative transconductance by adding two interstage inductors between the Gm unit and the tank, which can overcome the problems of poor startup conditions and high-power consumption. Kashani *et al.* [26, 27] used this technique to design a low noise, low power differential Colpitts VCO with a 20% tuning range. The complete schematic is shown in Fig. 6.19.

The input admittance of a conventional VCO can be expressed as

$$\text{Re}[Y_{in\_Conv}] = G_{eq}, \tag{6.16}$$

$$\text{Im}[Y_{in\_Conv}] = \omega C_B + \omega(C_P + C_{eq}), \tag{6.17}$$

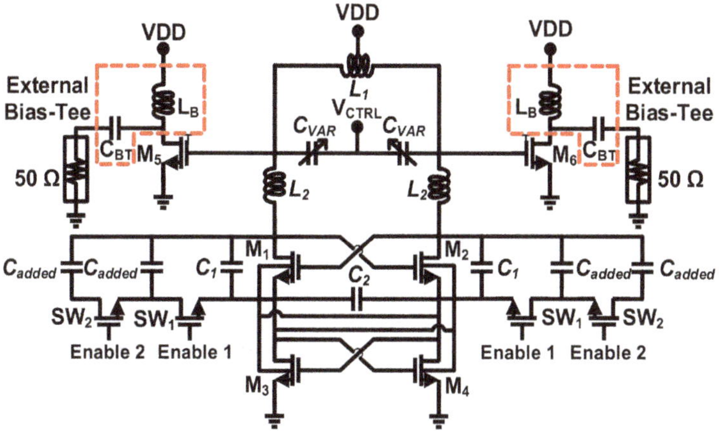

Fig. 6.19. Complete schematic of the proposed VCO.

where C_p is the drain parasitic capacitance of M_1, and G_{eq} and C_{eq} are the equivalent parallel negative transconductance and capacitance of the G_m cell, respectively. The input admittance of the proposed VCO can be expressed as

$$\text{Re}\left[Y_{in\_proposed}\right] = \frac{\frac{G_{eq}}{\omega^2 L_2^2}}{G_{eq}^2 + \left[\omega\left(C_p + C_{eq}\right) - \frac{1}{\omega L_2}\right]^2} \approx \frac{1}{G_{eq}\omega_0^2 L_2^2},$$

(6.18)

$$\text{Im}\left[Y_{in\_proposed}\right] = \approx \omega C_B - \frac{1}{\omega L_2}.$$

(6.19)

By comparing Eq. (6.16) and Eq. (6.18), the proposed VCO has a larger equivalent negative transconductance. The peak value of equivalent transconductance decreases with increasing frequency, indicating that Gm-boosting technology can be used in different operating frequency bands. Equations (6.17) and (6.19) show that the proposed VCO tank has low parasitic capacitance.

Figure 6.20(a) shows another advantage of the Gm-boosting technique, where the interstage inductor acts as an impedance transformer network. The total admittance seen by the tank is much larger than that seen by L_2, which indicates an increased voltage

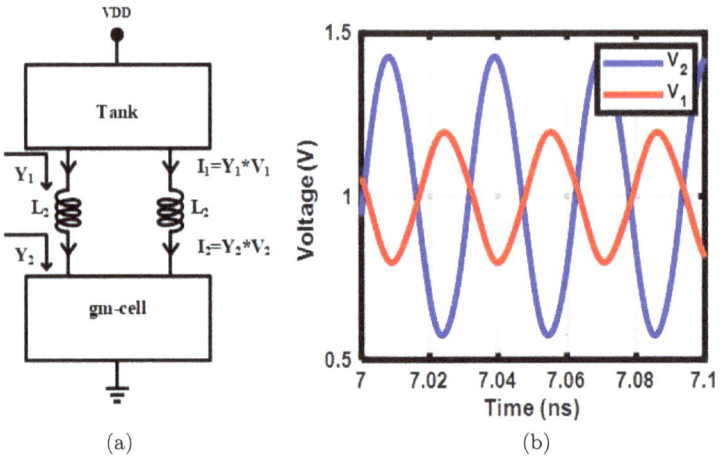

Fig. 6.20. (a) Schematic diagram of the interstage inductor acting as an impedance transformer network. (b) Simulated transient signal across L_2.

swing on the gate of M_1, resulting in good noise performance. The transient signal across L_2 is shown in Fig. 6.20(b).

In summary, the proposed technique extends tuning range, enhances negative transconductance, and reduces power consumption and phase noise.

The differential Colpitts VCO with a center frequency of 26.3 GHz is designed in a 65-nm CMOS process. Its chip micrograph with body bias and the new Gm-boosting technology is shown in Fig. 6.21.

Figure 6.22 shows the phase noise measurement at 26.3 GHz, which shows a PN of -86 at 300 kHz offsets. The VCO has a tuning range of 20% and has a FOM of -187 dBc/Hz at a 10 MHz offset from the center frequency. The VCO with Gm-boosting technology simultaneously achieves a wide-locking range, low-power consumption, and low-phase noise.

(2) *Multi-tank for high Power-Added Efficiency (PAE)*

Although many scholars have reported on VCOs in different aspects, such as phase noise, power consumption, and tuning range, it is still

Fig. 6.21.   Chip micrograph of the proposed VCO.

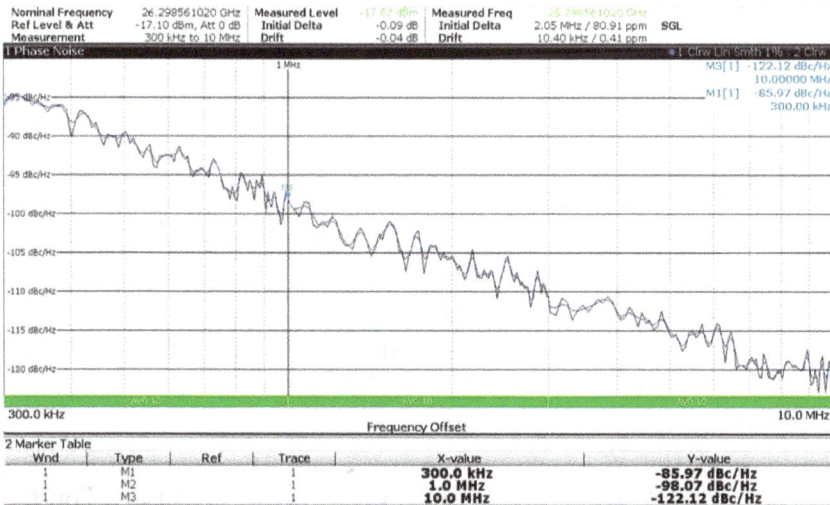

Fig. 6.22.   PN measurement results at 26.3 GHz.

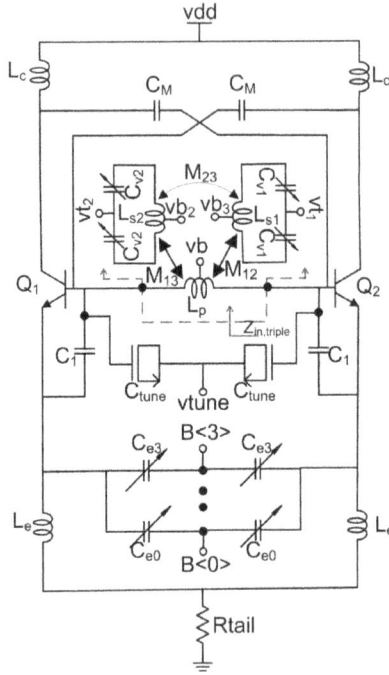

Fig. 6.23.   Topology of the proposed triple-coupled LC tank VCO.

a difficult problem to simultaneously achieve wide tuning range, low PN, and good PAE [28].

As shown in Fig. 6.23, Mahalingam *et al.* [29] proposed a fully differential K-band Colpitts VCO using a triple-coupled LC resonant tank with varactors and capacitor banks while achieving low PN, wide tuning range, and good PAE.

The proposed VCO consists of the degenerate inductor L_e and a collector tuned-circuit consisting of L_c and C_M. The degenerate inductor L_e and collector tuning circuit can improve the signal feedback of the VCO and improve the negative resistance in the frequency band of interest, reducing power consumption. For the Colpitts oscillator, the negative resistance can be expressed as

$$\operatorname{Re}\left(Z_{in}\right) = R_{neg} = -\frac{g_m}{\omega^2 C_{1T} C_{equi}}, \tag{6.20}$$

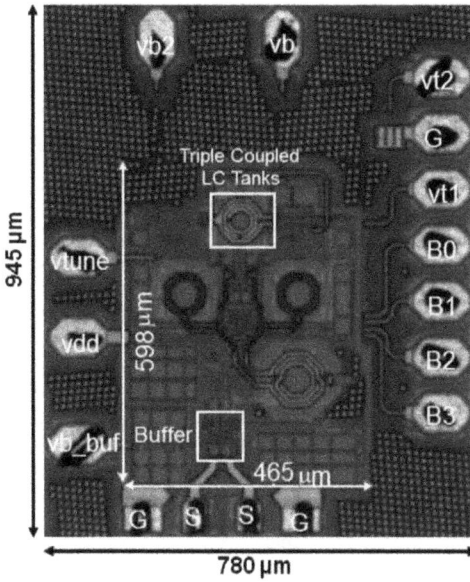

Fig. 6.24. The chip micrograph of the proposed VCO.

where C_{1T} is the sum of C_1 and the base-emitter capacitance $C\pi$, and C_{equi} is the equivalent capacitance connected to the transistor emitter.

For low-power Colpitts VCOs operating in the K-band and above, the collector-base capacitance C_{BC} of the transistor makes the negative resistance in the above formula insufficient. The combination of C_M and L_c can effectively increase the negative resistance and avoid the VCO startup problem.

Furthermore, the strong coupling of the LC resonator improves the slot quality factor and improves the output power and PAE. Figures 6.24 and 6.25 show the chip micrographs of the proposed VCO and the measured VCO phase noise, respectively.

The proposed VCO uses a BiCMOS process with a power consumption of 8.2 mW and a wide oscillation range from 22.50 GHz to 26.23 GHz, i.e., the frequency tuning range reaches 15.3%. Meanwhile, the PAE is 5.97%, and the PN at 1 MHz offset is −107.7 dBc/Hz.

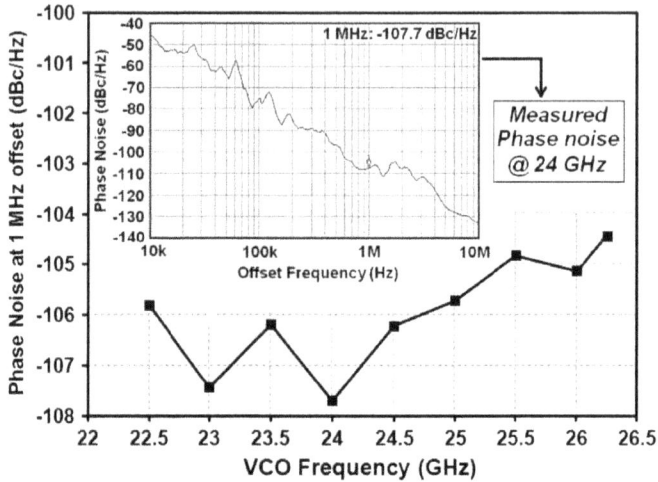

Fig. 6.25.   The measured VCO phase noise.

6.3.3. *Compact chip area applications*

In low power transceivers, the ILFD is required to be able to operate at low injection power and provide higher-power-efficiency (P_{out}/P_{DC}). The locking range of the injection-locked divider can be expressed as

$$\frac{\Delta\omega}{\omega} = \frac{I_{inj}}{2Q \times I_{OSC}}, \tag{6.21}$$

where I_{inj} is the injection current, and I_{osc} is the oscillator core current. Low tank Q or low I_{osc} helps to achieve wide-locking range, but this does not meet the requirements of high-power efficiency.

Mahalingam *et al.* [30] employed coupled dual LC tanks along with differential injection transistor pairs to increase the locking range (LR) of ILFDs with low-injection power, while maintaining low PN and high-power-efficiency. The proposed divide-by-2 injection-locked divider is shown in Fig. 6.26. The LR for this topology is given by Eq. (6.22), where $g_{m,inj}$ and $g_{m,osc}$ are the transconductances of the injection transistor and cross-coupled transistors M_1–M_4, respectively, R_{inj} represents the resistance of the injection transistor,

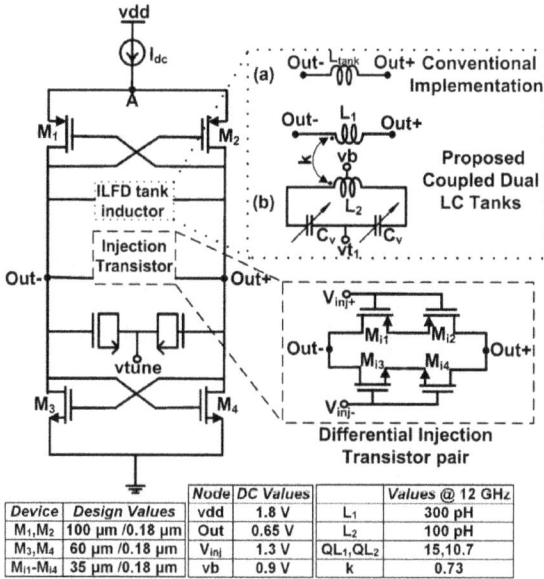

Device	Design Values
M_1, M_2	100 μm /0.18 μm
M_3, M_4	60 μm /0.18 μm
$M_{i1}-M_{i4}$	35 μm /0.18 μm

Node	DC Values
vdd	1.8 V
Out	0.65 V
V_{inj}	1.3 V
vb	0.9 V

	Values @ 12 GHz
L_1	300 pH
L_2	100 pH
QL_1, QL_2	15,10.7
k	0.73

Fig. 6.26. Schematic diagram of the proposed injection-locked divider.

and I_{dc} is the bias current. It can be seen from the above equation that increasing the tank inductance and increasing the size of the differential injection transistor pair can increase the locking range.

$$\frac{\Delta\omega}{\omega} = \frac{g_{m,inj}V_i\omega L_p}{g_{m,osc}I_{dc}R_{inj}^2}. \tag{6.22}$$

Compared with conventional methods, the proposed strongly magnetically coupled dual LC tank ILFD can achieve larger effective inductance in a compact chip area, and the series connection of injected transistor pairs provides higher closed loop gain and wider LR. The chip micrograph of the proposed ILFD is shown in Fig. 6.27.

At 0 dBm and −16 dBm input power, the ILFD achieves locked input frequency ranges of 20.09 to 25.86 GHz and 21.41 to 25.18 GHz, respectively. At 25 GHz, the input-referenced and the PN at 10 kHz offset is −190.7 dBc/Hz and −116 dBc/Hz, respectively.

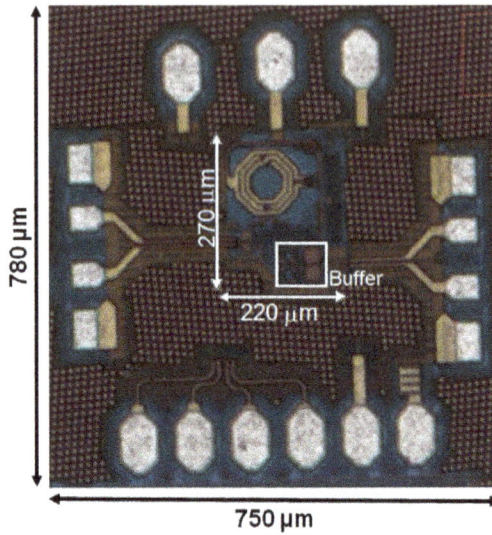

Fig. 6.27.   Die photomicrograph of the proposed strongly magnetically coupled dual LC tank ILFD.

6.3.4. *High quality factor applications*

The PN of an oscillator is related to the quality factor, so as the resonator quality factor Q increases, the noise performance is optimized. For multi-tanks LC resonators, since part of the energy exists between the two coils in the form of magnetic energy, the Q factor of the entire resonant LC tank can be greater than the Q factor of the resonant LC tank formed by the coil and the corresponding coil capacitance. The class-C VCO topology with transformer-enhanced tuning range and tank quality factor proposed by Lightbody *et al.* [31] is shown in Fig. 6.28.

Connecting the secondary loop of the transformer to the varactor reduces phase noise and increases the frequency tuning range, as this method reduces parasitic effects in the secondary tank circuit and improves the quality factor of the resonant LC tank. In addition, varactors allow for more flexible biasing and control of most frequency tuning.

The VCO has a tuning range of 29.8% (20.77 to 28.02 GHz) and consumes 12.65 to 15.12 mW. As shown in Fig. 6.29, a PN of −106.6 dBc/Hz is obtained at the 1 MHz offset.

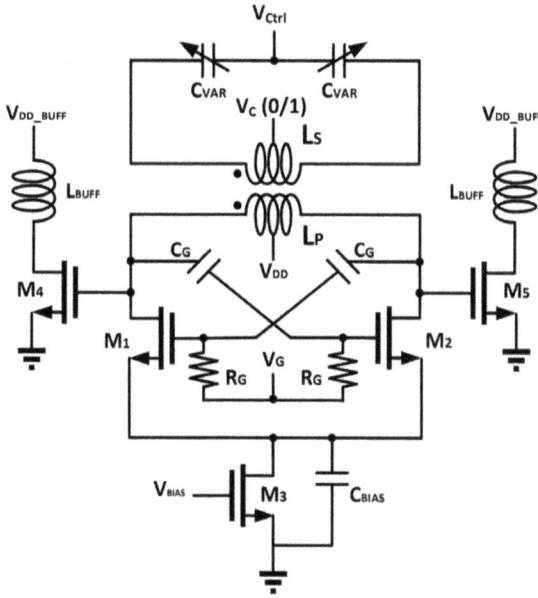

Fig. 6.28. The proposed VCO with wide tuning range and high resonant LC tank quality factor.

Fig. 6.29. Measured PN at 26.45 GHz.

6.4. Multiple Tanks Technique for Power Amplifiers

RF and mm-wave amplifiers are key building blocks in accurate RF microwave sources and exist in almost all RF and microwave transceivers for smartphones, equipment, etc. In RF and mm-wave circuits and systems, the PAs are key blocks in the TX chain and have a great impact on the overall performance of the transmission link. The Q-factor of the output matching network may directly affect the output power and power-added efficiency (PAE) of the PA. Since the multiple tanks technique can effectively improve the Q-factor of LC tanks, the technique is widely used in PAs.

6.4.1. *Reconfigurable power amplifier with high PAE*

As exhibited in Fig. 6.30, a reconfigurable K-/Ka-Band PA using the current reuse topology with magnetically coupled LC tanks is proposed in [6]. The loaded tank consists of a coupled dual tank and a MOS varactor bank as a frequency tunable load. A MOS varactor bank is employed to provide frequency-band switching, and the dual tank achieves not only Q enhancement of the loaded tank circuit, but also high amplifier linearity. The current reuse improves amplifier efficiency by reducing the overall dc power consumption. It achieves a 55.9% PAE and a variable gain from 1.8 dB to 16 dB with a tunable 3-dB bandwidth of 24.3 GHz to 35 GHz. Figure 6.31(a) and 6.31(b) show the microphotograph of the amplifier with the tunable load proposed in [6] and its measured S-parameters, respectively. Figure 6.32(a) and 6.32(b) show the P_{sat} and PAE, respectively, of the amplifier with the tunable load proposed in [6].

6.4.2. *CMOS power amplifier using hybrid coupling combiner*

A CMOS PA using a 4-way hybrid coupling combiner concentric winding topology is proposed in [5], as shown in Fig. 6.33. In the 4-way hybrid power combiner comprising multiple-coupled LC tanks, the blue or green primary loops are traditional series combinations, while another two primary loops parallel combine with concentric winding. This configuration not only reduces the size of the power

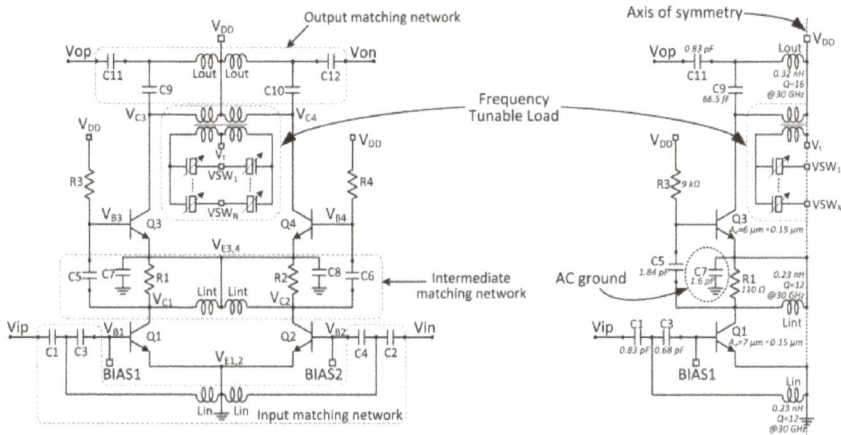

Fig. 6.30.   The schematic of the reconfigurable multi-band amplifier and half circuit equivalent proposed in [6].

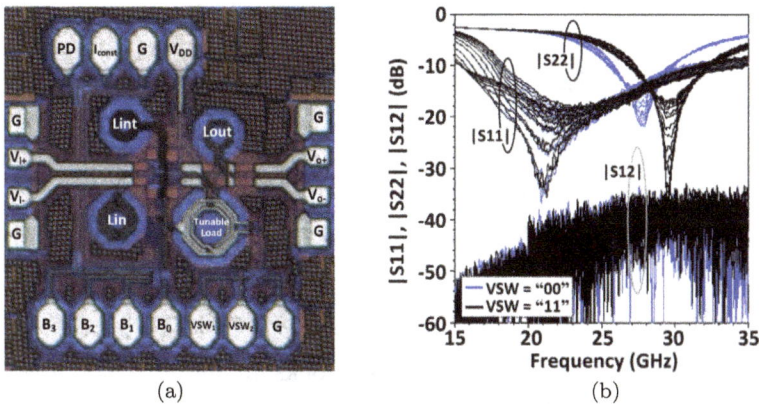

Fig. 6.31.   (a) The microphotograph of the amplifier with the tunable load proposed in [6]. (b) Measured S-parameter.

combiner and overall power dissipation but also increases the efficiency. The proposed PA achieves an output power of 17.5 dBm/19.7 dBm with 10.3%/13.4% PAE at 1.2 V/1.4 V external supply voltage, respectively, and offers a 26.8 dB linear gain with 3 dB bandwidth of 51–67 GHz with the chip area of only 0.32 mm^2.

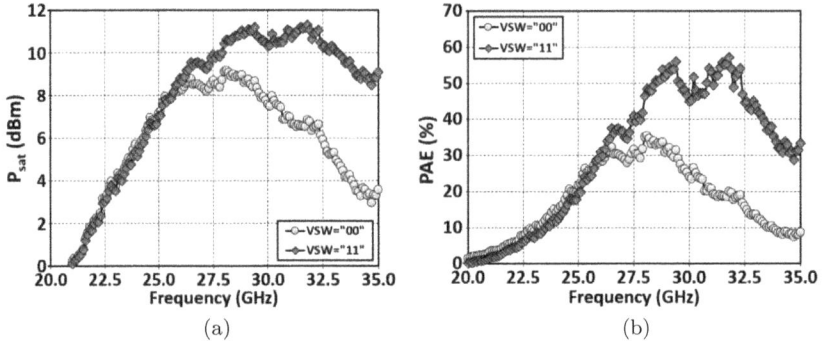

Fig. 6.32.   Measured (a) P_{sat} and (b) PAE of the amplifier with the tunable load proposed in [6].

6.4.3. *Transformer-based power amplifier*

In [7], a PA with a power combiner based on coupled LC tanks is proposed, as shown in Fig. 6.34. The transformers of the LC tanks use current-combining topology to reduce the output power loss and achieve a high PAE. At the same time, the transformers may mitigate the imbalance problem. At 27 GHz, the proposed PA achieves a 1-dB gain compression output power (P_{1dB}) of 14.7 dBm and a peak PAE of 43%. Measured results of the L-tank-based power amplifier are illustrated in Fig. 6.35.

6.5. Multiple Tanks Technique for Switch Applications

Switches are fundamental building blocks in modern wireless systems. Low insertion loss and compact size are the main requirements for RF and mm-wave switches in silicon-based technologies. The multiple tanks technique could effectively improve the Q-factor of the LC tanks keeping the chip size compact; therefore, the multiple tanks have a wide range of applications in switches.

6.5.1. *Switchable artificial resonators introduced for SPDT based on multiple tanks technique*

In [13], a standing-wave single-pole single-throw (SPST) filtering switch using a switchable artificial resonator is proposed. The ON

(a)

(b)

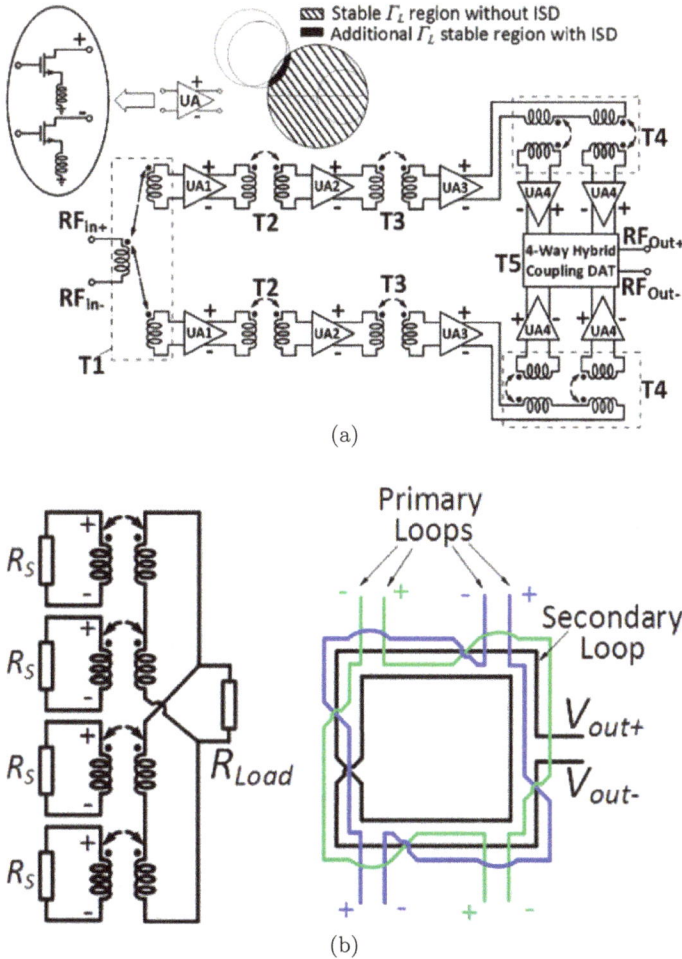

Fig. 6.33. A 4-way hybrid coupling combiner PA proposed in [5]. (a) The architecture of PA. (b) 4-way hybrid coupling combiner topology.

status of the switch can be modeled as three strongly coupled LC tanks, which are equivalent to a single LC tank resonator at the first resonant frequency but are made much smaller due to the strong mutual coupling among the three ground stubs. Meanwhile, the switchable artificial resonators are also introduced for single-pole double-throw (SPDT) filtering switches to achieve low insertion loss, compact size, and high P_{1dB}, as given in Fig. 6.36. The proposed

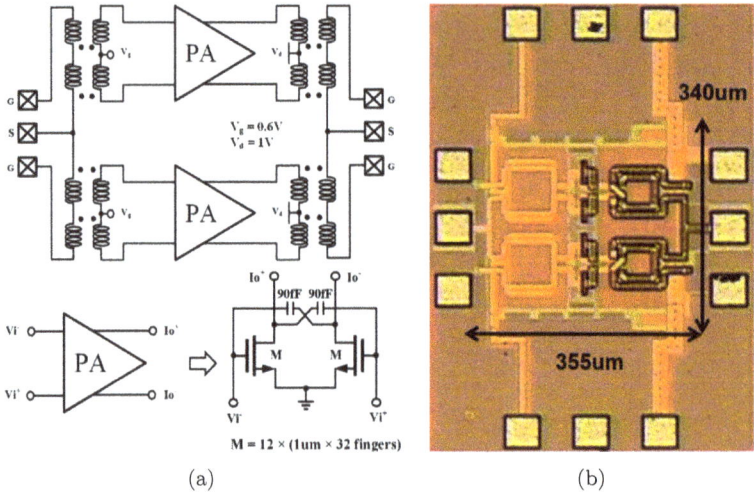

(a)   (b)

Fig. 6.34.   The LC tank-based power amplifier proposed in [7]. (a) The schematic of the PA. (b) A microphotograph.

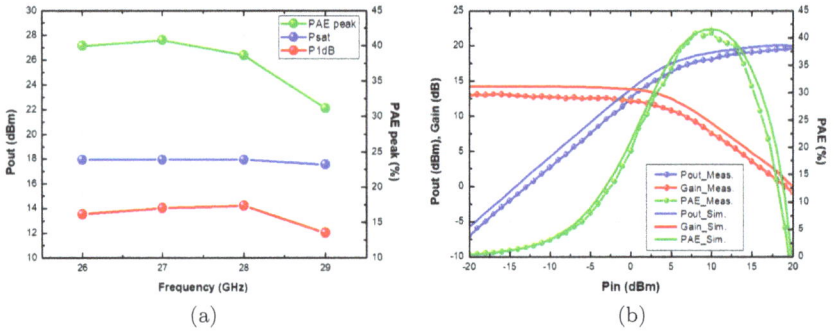

(a)   (b)

Fig. 6.35.   Measured results of the L-tank-based power amplifier proposed in [7].

standing-wave switches also demonstrate a filtering response and intrinsic electrostatic discharge (ESD) protection. At 60 GHz, the SPDT switch exhibits a 3.2 dB insertion loss and 20 dB isolation with a chip area of only 0.0675 mm². Both switches have good linearity with the input P_{1dB} of 21 dBm.

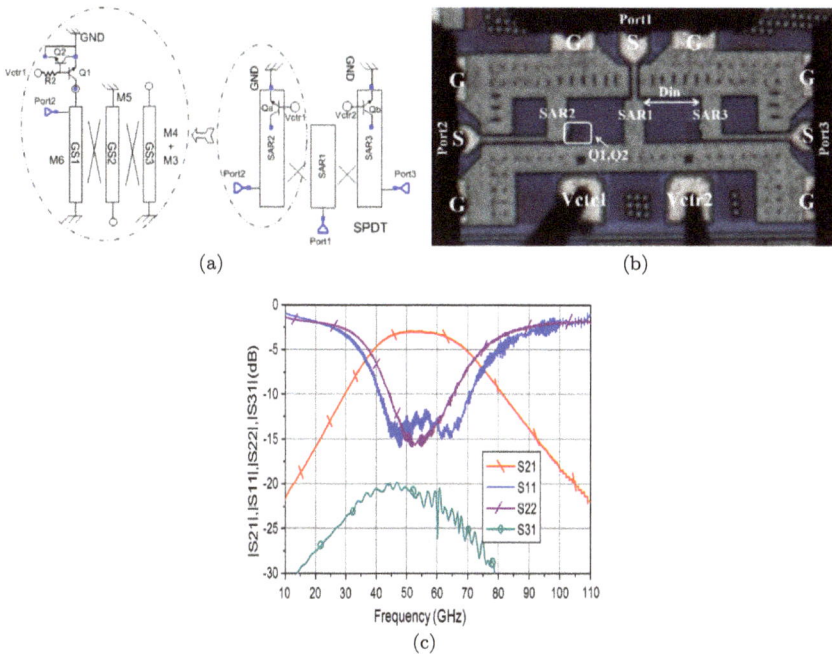

(a)

(b)

(c)

Fig. 6.36. A standing-wave SPDT filtering switch proposed in [13]: (a) Topology of the SPDT, (b) a die microphotograph, and (c) measurement results.

6.5.2. *SPDT switches artificial resonator based on multiple tanks with better isolation and small chip size in THz band*

Here, [14] introduces an SPDT switch using a magnetically switchable artificial resonator, which is illustrated in Fig. 6.37. The SPDT, which is modeled as a triple-coupled LC tank consisting of 3 main coupled lines and 2 secondary coupled lines. The artificial resonator mitigates the problematic $\lambda_g/4$ transmission line (T-line) and matching networks. The auxiliary switches have ON/OFF states. During the OFF state, strong magnetic coupling occurs between the main coupled lines. This can be used in the operation mode when the signal ON path is active. During the ON state, weak magnetic

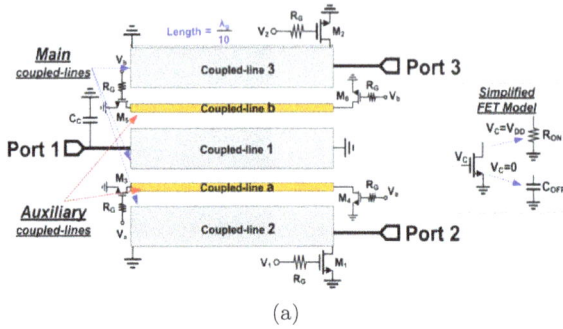

(a)

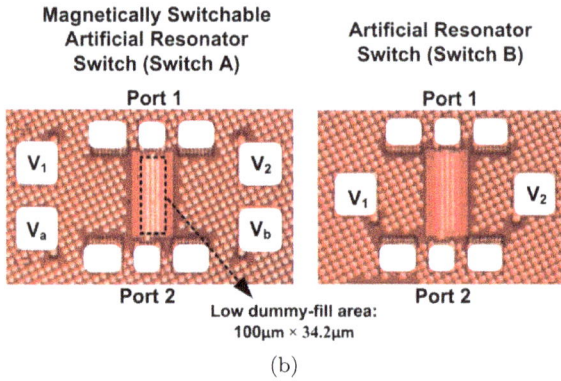

(b)

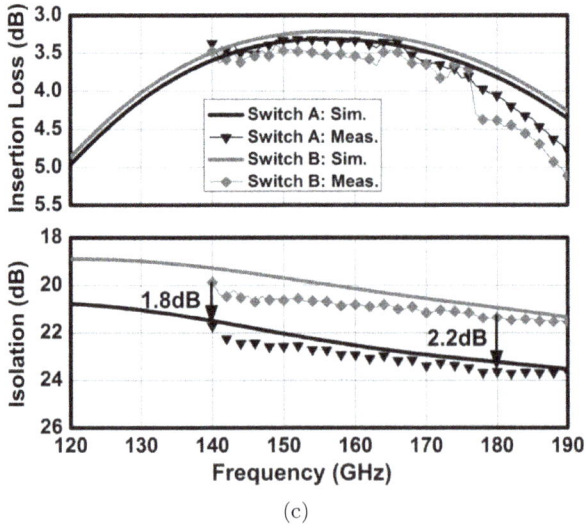

(c)

Fig. 6.37.　An SPDT switch proposed in [14]. (a) The configuration of the switch. (b) A die microphotograph of switch A and switch B. (c) Measuring insertion loss and isolation.

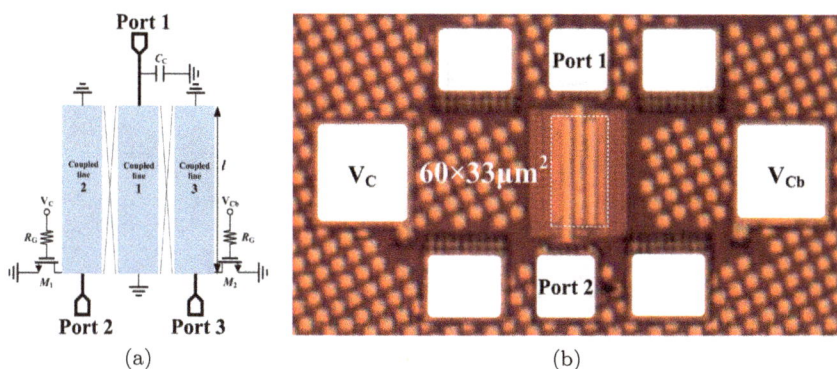

Fig. 6.38. (a) Configuration and (b) a die microphotograph of the SPDT switch using the switchable resonator concept proposed in [15].

coupling reduces the signal coupled through and hence it is selected for better isolation. The simulated switching ON and OFF times are 0.58 ns and 0.53 ns, respectively. [14] also compared the switch A with switch B (without auxiliary coupled lines), with the former achieving better measured isolation. The measured insertion losses of switch A and switch B are close, and switch A has a measured isolation of >21.8 dB from 140 to 180 GHz, which is a 2 dB improvement over switch B.

An SPDT switch using a switchable resonator concept with three coupled-lines topology is proposed in [15] to achieve high operating frequency and small chip size. The resonator utilizes normal RF nFET transistors and three coupled lines, which are closely spaced, as shown in Fig. 6.38. Figure 6.39 shows the measurement results of the proposed multi-tanks SPDT. Its insertion loss including RF pad losses and isolation are 4.2 dB and 19 dB, respectively, while the simulated switching ON and OFF times are 0.29 ns and 0.26 ns, respectively.

6.6. Multiple Tanks Technique in Transceiver SOC

6.6.1. *60-GHz low-power dual-chip wireless communication system*

The unlicensed band around 60 GHz is suitable for short-range multi-Gbps wireless communications, calling for standardization, such as

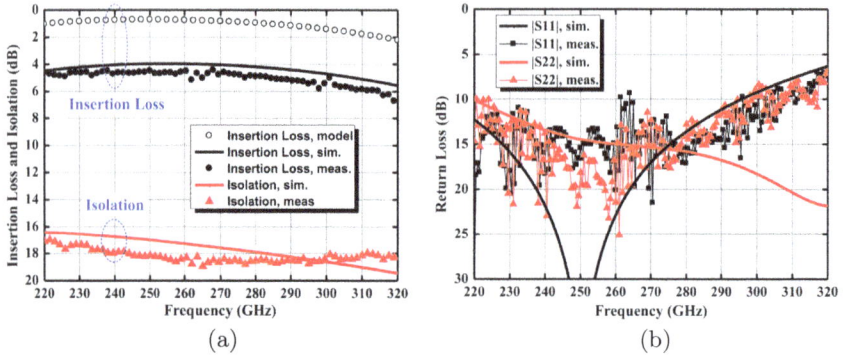

Fig. 6.39.   The SPDT switch using the switchable resonator concept proposed in [15]. (a) Insertion loss and isolation. (b) Return loss.

the IEEE 802.11ad standard. A two-chip solution for 60 GHz is available in [32]. The system consists of directional antennas and two communication systems, a baseband integrated circuit (BBIC) and a low-power 60 GHz RFIC, both following the IEEE 802.11ad standard. The overall system architecture is shown in Fig. 6.40. The BBIC uses an adaptive time-domain equalizer (TDE) to reduce power consumption and a generic 16-bit parallel host interface to support a maximum throughput of 4.6 Gbps. The 60-GHz RFIC includes a TX chain, synthesizer with LO network, SPI slave, and RX chain. The TX chain includes a cross-coupled LPF, drive digitally controlled variable gain amplifiers (DVGA), intermediate frequency variable gain amplifiers (IF VGA), a sub-harmonic mixer (SHM), and a 60-GHz power amplifier. The synthesizer includes an LC tank-based triple-coupled VCO, a transformer-based ILFD, and a phase-locked loop (PLL). The LO feed network includes SPDT switches, a broadband differential quadrature (DQ) converter, and gain controllable compensation amplifiers (CAMP). For a 1 GSymbol/s 16-QAM modulation signal, the error vector magnitude (EVM) of the 60-GHz dual-chip system is 9.8%.

6.6.2.   *29.5–33.4-GHz PLL synthesizer SOC*

Figure 6.41 shows a 30-GHz power-efficient PLL frequency synthesizer for 60-GHz applications proposed in [33]. The synthesizer

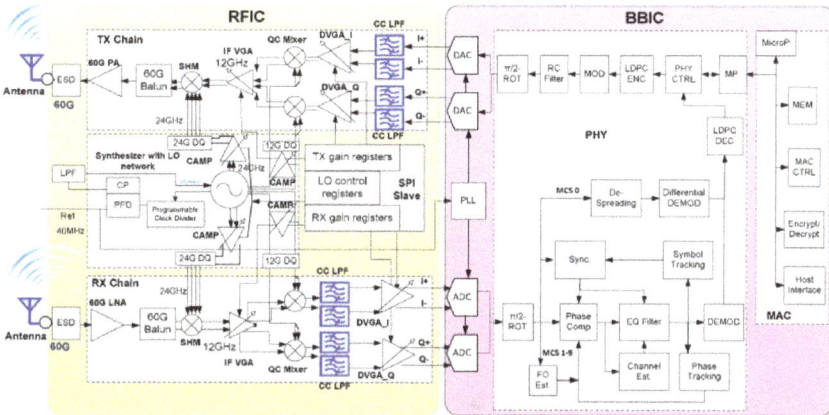

(a)

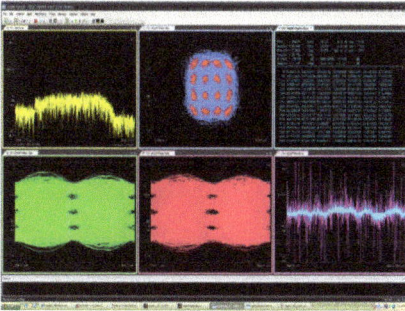

(b)

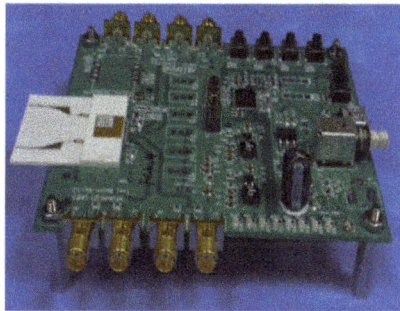

(c)

Fig. 6.40.   The 60-GHz low-power, dual-chip wireless communication system proposed in [32]. (a) Block diagram. (b) Measurements results. (c) A photograph.

includes a coupled LC tank-based VCO, a high PAE amplifier, a reconfigurable divider, an internal loop filter, a slave serial peripheral interface, and a programmable charge pump. The VCO with a coupled LC tank provides low-power consumption and low-phase noise. At the same time, a reconfigurable divider can realize fractional division ratios that have the characteristics of multiple-choice reference frequencies and low operating power. The PLL synthesizer consumes a low power of 63 mW at a single 1.8 V supply voltage, and it shows a phase noise of −97 dBc/Hz at 1 MHz offset with an output frequency from 29.5 GHz to 33.4 GHz.

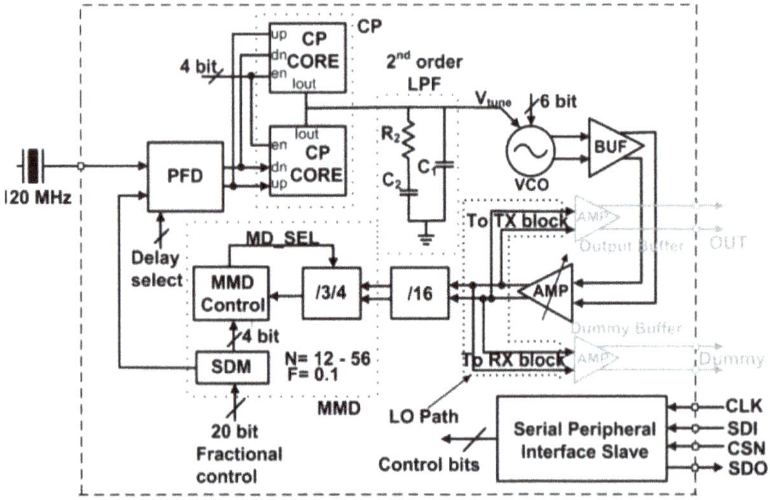

(a)

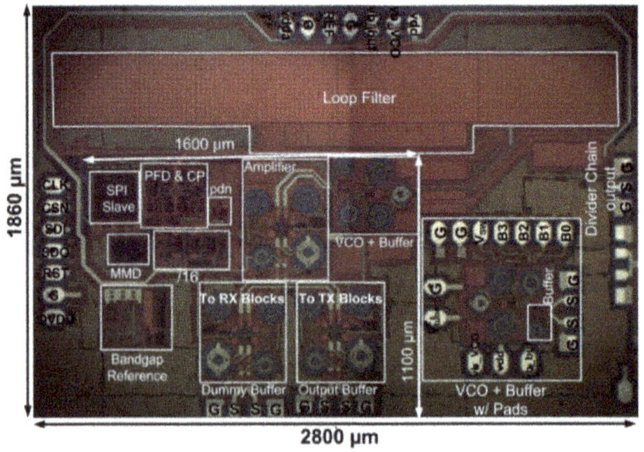

(b)

Fig. 6.41.   The 29.5–33.4-GHz PLL synthesizer SOC proposed in [33]. (a) Block diagram. (b) A die microphotograph.

References

[1] F. Gacim, "Modelling, characterization and optimization of substrate losses in RF switch IC design for WLAN applications," Normandie Université, 2017.

[2] J. N. Burghartz, "Progress in RF inductors on silicon — understanding substrate losses," *Proc. IEEE International Electron Devices Meeting (IEDM)*, San Francisco, CA, pp. 523–526, Dec. 1998.

[3] C. S. Kim, P. Park, J.-W. Park, N. Hwang and H. K. Yu, "Deep trench guard technology to suppress coupling between inductors in silicon RFIC," *IEEE MTT-S Int. Microw. Symp. Dig.*, vol. 3, pp. 1873–1876, Jun. 2001.

[4] K. Ma, M. Nagarajan, S. Mou and K. S. Yeo, "Integrated circuit architecture with strongly coupled LC tanks," US Patent Number 9331659, 3 May 2016.

[5] J.-A. Han, Z.-H. Kong, K. Ma and K. S. Yeo, "A 26.8 dB gain 1.7 dBm CMOS power amplifier using 4-way hybrid coupling combiner," *IEEE Microw. Wirel. Compon. Lett.*, vol. 25, no. 1, pp. 43–45, Jan. 2015.

[6] K. Ma, T. B. Kumar and K. S. Yeo, "A reconfigurable K-/Ka-band power amplifier with high PAE in 0.18-um SiGe BiCMOS for multi-band applications," *IEEE Trans. Microw. Theor. Tech.*, vol. 63, no. 12, pp. 4395–4405, Dec. 2015.

[7] K. Chiang, T. Tsai, I. Huang, J. Tsai and T. Huang, "A 27-GHz transformer based power amplifier with 513.8-mW/mm^2 output power density and 40.7% peak PAE in 1-V 28-nm CMOS," *2019 IEEE MTT-S Int. Microw. Symp. (IMS)*, Boston, MA, USA, pp. 1283–1286, 2019.

[8] Z. Zong, M. Babaie and R. B. Staszewski, "A 60 GHz frequency generator based on a 20 GHz oscillator and an implicit multiplier," *IEEE J. Solid-State Circuits*, vol. 51, no. 5, pp. 1261–1273, May 2016.

[9] M. Haghi Kashani, A. Tarkeshdouz, R. Molavi, A. Sheikholeslami, E. Afshari and S. Mirabbasi, "On the design of a high-performance mm-Wave VCO with switchable triple-coupled transformer," *IEEE Trans. Microw. Theor. Tech.*, vol. 67, no. 11, pp. 4450–4464, Nov. 2019.

[10] H. Jia, B. Chi, L. Kuang and Z. Wang, "A 47.6–71.0-GHz 65-nm CMOS VCO based on magnetically coupled π-type LC network," *IEEE Trans. Microw. Theor. Tech.*, vol. 63, no. 5, pp. 1645–1657, May 2015.

[11] X. Liu and H. C. Luong, "A 170-GHz 23.7% tuning-range CMOS injection-locked LO generator with third-harmonic enhancement," *IEEE Trans. Microw. Theor. Tech.*, vol. 68, no. 7, pp. 2668–2678, Jul. 2020.

[12] N. Mahalingam, K. Ma, K. S. Yeo and W. M. Lim, "Coupled dual LC tanks based ILFD with low injection power and compact size," *IEEE Microw. Wirel. Compon. Lett.*, vol. 24, no. 2, pp. 105–107, Feb. 2014.

[13] K. Ma, S. Mou and K. S. Yeo, "A miniaturized millimeter-wave standing-wave filtering switch with high P1dB," *IEEE Trans. Microw. Theor. Tech.*, vol. 61, no. 4, pp. 1505–1515, Apr. 2013.

[14] F. Meng, K. Ma and K. S. Yeo, "A 130-to-180GHz 0.0035mm 2SPDT switch with 3.3dB-loss and 23.7dB-isolation in 65nm bulk CMOS," *ISSCC Dig. Tech. Papers*, 2015.

[15] F. Meng, K. Ma, K. S. Yeo, C. C. Boon, W. M. Lim and S. Xu, "A 220–285 GHz SPDT switch in 65-nm CMOS using switchable resonator concept," *IEEE Trans. Terahertz Sci. Technol.*, vol. 5, no. 4, pp. 649–651, Jul. 2015.

[16] C. P. Yue and S. S. Wong, "Physical modeling of spiral inductors on silicon," *IEEE Trans. Electron. Dev.*, vol. 47, no. 3, pp. 560–568, Mar. 2000.

[17] B. Razavi, "A study of phase noise in CMOS oscillators," *IEEE J. Solid-State Circuits*, vol. 31, no. 3, pp. 331–343, Mar. 1996.

[18]   J. M. W. Rogers, J. W. M. Rogers and C. Plett, *Radio Frequency Integrated Circuit Design*. Artech House, Norwood, MA, 2010.

[19]   D. B. Leeson, "A simple model of feedback oscillator noise spectrum," *Proc. IEEE*, vol. 54, no. 2, pp. 329–330, Dec. 1966.

[20]   J. Yin and H. C. Luong, "A 57.5–90.1-GHz magnetically tuned multimode CMOS VCO," *IEEE J. Solid-State Circuits*, vol. 48, pp. 1851–1861, 2013.

[21]   Q. Zou, K. Ma, and K. S. Yeo, "A low phase noise and wide tuning range millimeter-wave VCO using switchable coupled VCO-cores," *IEEE Trans. Circuits Syst. I Regul. Pap.*, vol. 62, no. 2, pp. 554–563, 2015.

[22]   R. Han, J. Chen, A. Mostajeran, M. Emadi, H. Aghasi, H. Sherry, A. Cathelin and E. Afshari, "A SiGe terahertz heterodyne imaging transmitter with 3.3 mW radiated power and fully-integrated phase-locked loop," *IEEE J. Solid-State Circuits*, vol. 50, no. 12, pp. 2935–2947, Dec. 2015.

[23]   S. Mou, K. Ma, K. S. Yeo, N. Mahalingam and B. Kumar, "A low power wide tuning range VCO with coupled LC tanks," *IEEE Int. SOC Conf.*, Taipei, Taiwan, pp. 52–56, 2011.

[24]   A. Dec, H. Akima and K. Suyama, "A 5 GHz LC VCO with extended linear-range varactor in purely digital 0.15 μm CMOS process," *Proc. IEEE Radio Freq. Integr. Circuits Symp.*, Boston, MA, USA, Jun. 2009, pp. 567–570.

[25]   Z. Chen, M. Wang, J. Chen, W. Liang, P. Yan, J. Zhai and W. Hong, "Linear CMOS LC-VCO based on triple-coupled inductors and its application to 40-GHz phase-locked loop," *IEEE Trans. Microw. Theor. Tech.*, pp. 1–13, 2017.

[26]   M. H. Kashani, R. Molavi and S. Mirabbasi, "A 2.3-mW 26.3-GHz Gm-boosted differential colpitts VCO with 20% tuning range in 65-nm CMOS," *IEEE Trans. Microwav. Theor. Tech.*, vol. 67, no. 4, pp. 1556–1565, 2019.

[27]   M. H. Kashani, R. Molavi and S. Mirabbasi, "A wide-tuning-range low-phase-noise Colpitts oscillator with variable capacitive feedback," *Proc. IEEE Int. Symp. Circuits Syst. (ISCAS)*, Florence, Italy, pp. 1–5, May 2018.

[28]   N. Mahalingam, K. Ma, K. S. Yeo, S. X. Mou and T. B. Kumar, "A low power wide tuning range low phase noise VCO using coupled LC tanks," *Proc. Semicond. Conf. Dresden*, pp. 1–4, Sep. 2011.

[29]   N. Mahalingam, K. Ma, K. S. Yeo and W. M Lim, "K-band high-PAE wide-tuning-range VCO using triple-coupled LC tanks," *IEEE Trans. Circuits Syst. II Express Briefs*, vol. 60, no. 11, pp. 736–740, 2013.

[30]   N. Mahalingam, K. Ma and K. S. Yeo, "Coupled dual LC tanks based ILFD with low injection power and compact size," *IEEE Microw. Wirel. Compon. Lett.*, vol. 24, no. 2, pp. 105–107, 2014.

[31]   S. Lightbody, A. H. M. Shirazi, H. Djahanshahi, R. Zavari, S. Mirabbasi and S. Shekhar, "A 195 dBc/Hz FoMT 20.8-to-28-GHz LC VCO with transformer-enhanced 30% tuning range in 65-nm CMOS," *2018 IEEE Radio Freq. Integr. Circuits Symp. RFIC Dig. Tech. Pap.*, 2018.

[32]   K. Ma, S. Mou, N. Mahalingam, Y. Wang, B. K. Thangarasu, J. Yan, W. Ye, K. Wang, *et al.*, "An integrated 60GHz low power two-chip wireless system

based on IEEE802.11ad standard," *2014 IEEE MTT-S Int. Microw. Symp. (IMS2014)*, Tampa, FL, USA, pp. 1–4, 2014.

[33] N. Mahalingam, Y. Wang, B. K. Thangarasu, K. Ma and K. S. Yeo, "A 30-GHz power-efficient PLL frequency synthesizer for 60-GHz applications," *IEEE Trans. Microwav. Theor. Tech*, vol. 65, no. 11, pp. 4165–4175, Nov. 2017.

CHIP

The Future of Learning and Teaching is not just about finding the perfect solution, but rather seeing and understanding the problem perfectly.

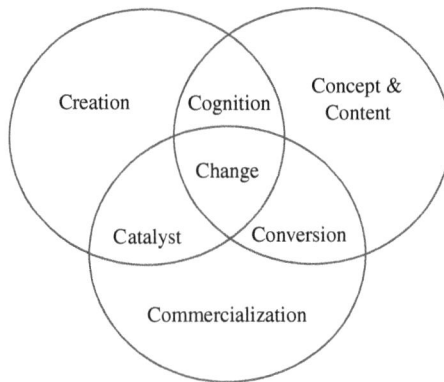

Creation — Cognition — Concept & Content — Change — Catalyst — Conversion — Commercialization

Kiat Seng Yeo

Chapter 7

Chip-to-Chip Communications Over a Galvanic Isolation Barrier

Richard Lum

7.1. Introduction

Isolation devices have been around for a few decades, and though there is not a lot of news and information related to this technology, its ubiquity in application is often overlooked as just another component in the electrification of human activities. With Industry 4.0, mass deployment of 5/6G communication technology, and the commercialization of Broadband Power devices, the role of an isolation device will increase exponentially as more human activities are electrified. So, what exactly is an isolation device, what does it do, and why is it important? In simple terms, an isolation device, also commonly known as an isolator, simply provides a communication channel across a galvanically isolated media so that both sides of the device can operate with very different power supply domains. Between the transmitter and receiver is a media, which can withstand a very high voltage potential difference, and yet, when it comes to communications, there is a robust link between them. The communication, therefore, must be *wireless* in nature for true galvanic isolation.

The ubiquity of isolation devices lies with the widespread use of switched-mode power conversion systems deployed today. Isolation serves two primary purposes in such applications — providing signaling capability between the low-voltage controller to the high-voltage switches and providing a safety barrier so that a man-machine interface is completely isolated from the dangers of high-voltage potential in the power system. Traditionally, it is required to isolate the 100–240 V AC mains from low-voltage panels where human interaction takes place. In modern power electronics systems, these voltages to be isolated are increasing in voltage levels, as using higher bus voltages will result in lower copper content needed for wiring and realization of magnetic systems such as inductors, transformers, and motors. New broadband power devices made of gallium nitride and silicon carbide technology are also capable of operating at higher voltage stress and power density levels, and their ability to switch at much higher frequencies will also impose a new criterion for robust communications between the controller and these power stages. In addition, the isolation barrier also breaks up the ground loop between the high-voltage power switching stages and the low-voltage sections. Without the breaking up of these ground loops, common mode currents generated when the power stages switches can easily affect the supplies and signals of the low voltage controllers used, resulting in system integrity and possible failure. Due to Moore's law, while the computation power of newer generations of CMOS technology has increased tremendously, it also means that their supply voltages are getting lower. On the other hand, as power systems get smaller and more efficient while delivering higher power, the voltages provided to such switches are increasing as well. This dichotomy is providing an impetus for isolation devices that will need to hold off higher voltages and communicate with higher data rates whilst being in an environment where significant common mode noises are ever present. This chapter, therefore, reviews the current state of the art in isolation components, discusses the methods, advantages, and disadvantages of various communication choices, and considers a future where highly integrated multi-functional isolation devices can be developed for the increased deployment of electrification of human activities.

7.2. Components for Isolation Technology

When AC power systems dominated the technology for power distribution, the power transformer operating at 50 Hz and 60 Hz was the first real isolation device between two domains of operating voltages. Its purpose was primarily to perform AC voltage scaling to reduce power loss when power is distributed across long distances via copper cables. With the invention of the transistor and semiconductor materials, one of the first isolation components to appear in the market is a combination of an incandescent lamp and a photo-sensitive resistor typically made up of materials such as cadmium sulfide. The brightness of the lamp shining onto the photoresistor provides a means to transfer a good linear voltage signal across a galvanic isolation barrier, which is typically air. Figure 7.1 shows a typical application circuit based on discrete optical isolation. The transmitter (consisting of a lamp, a variable resistor R, and a

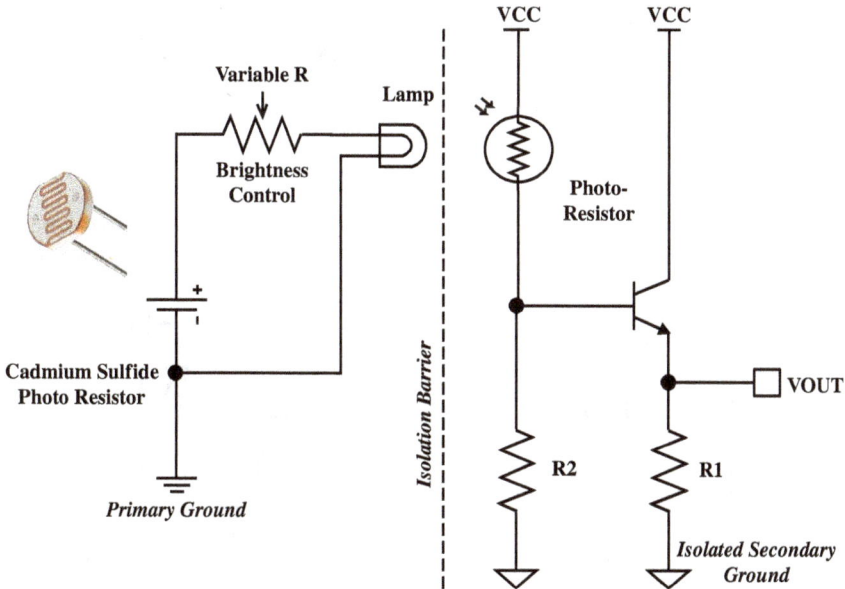

Fig. 7.1. Cadmium sulfide photoresistor and its application in isolation of linear signals.

battery) and the receiver (consisting of a photoresistor, two biasing resistors, R1 and R2, and a NPN transistor) are discrete devices operating as separate components.

7.2.1. *Optical-based isolation component*

With the invention of the LED (light-emitting diode) in the 1960s, a GaAs (gallium arsenide) LED emitting infrared light energy onto a silicon PN junction was first demonstrated, heralding the beginning of the first isolation component that embeds a signaling device, the LED, and a detector device, the light-sensitive PN junction diode in reverse bias, which was later extended to a silicon NPN phototransistor.

In Fig. 7.2, two different but commonly used arrangements of LEDs and photodiodes allow configuration of isolation for both

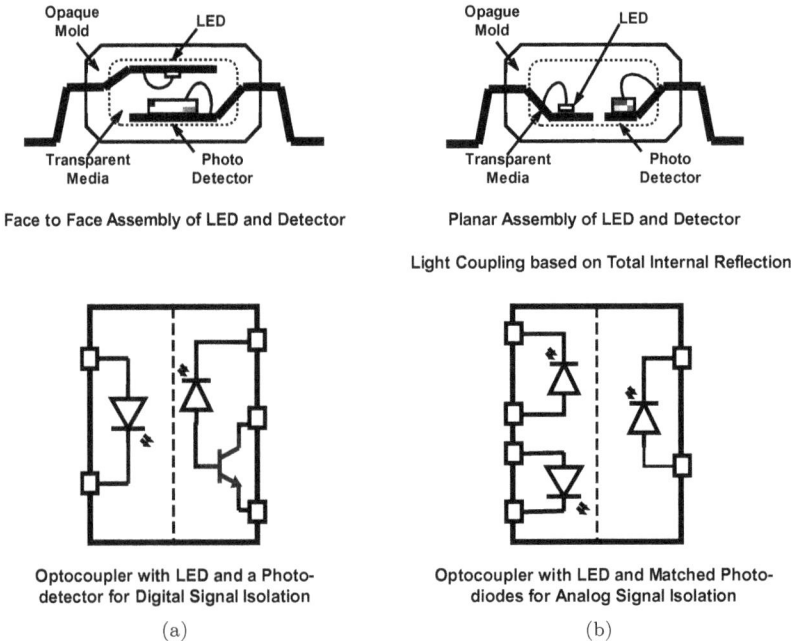

Face to Face Assembly of LED and Detector   Planar Assembly of LED and Detector

Light Coupling based on Total Internal Reflection

Optocoupler with LED and a Photo-   Optocoupler with LED and Matched Photodetector for Digital Signal Isolation   diodes for Analog Signal Isolation

(a)   (b)

Fig. 7.2.   (a) Optocoupler with LED and a photodetector for digital signal isolation. (b) Optocoupler with LED and matched photodiodes for analog signal isolation.

analog and digital signal transmission. With these two configurations, it is possible to transfer both analog as well as digital signals across the isolation barrier depending on how the LED light is controlled and how the signal from the diode or phototransistor is processed [1]. It is noteworthy that while these technologies were invented and commercialized in the 1960s, these same basic devices are still very much in use today and often in their original form and construction.

Figure 7.3 shows a typical schematic circuit of how analog signals are isolated. Two matched photodiodes are used — one in the input stage (PD1) and one in the output stage (PD2) — to greatly reduce nonlinearities and time-temperature instabilities. Isolation of analog signals is important for industrial automation and medical applications.

It is evident that signaling across a galvanic isolation remains a very critical and irreplaceable technology that will continue to exist as long as electrification of human activities remains at the forefront of driving civilizational progress.

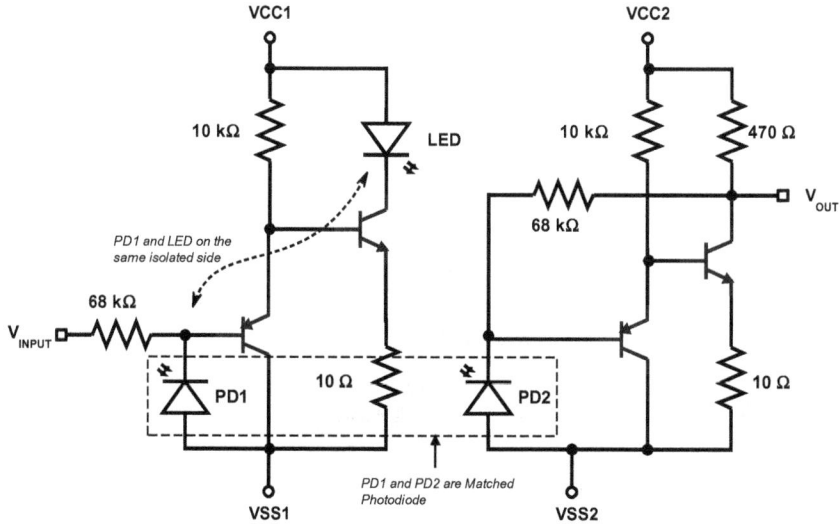

Fig. 7.3.   Isolating analog signals using an optocoupler with LED and matched photodiodes.

7.2.2. *Non-optical-based isolation component*

Using light to perform signaling is not the only possible way within an isolation component. In fact, the complexity of having to assemble a LED and a silicon detector and to provide the necessary high voltage separation requires special proprietary material systems and construction for managing the high voltage breakdown, the light transmission channel, light intensity, and detection sensitivity, all within a single package. Often, it is this complexity of construction that ultimately limits the performance and functionality of using optical-based technology.

7.2.3. *Isolation using magnetic and capacitive coupling*

Figure 7.4 shows two typical arrangements of component parts of a non-optical isolation device. Instead of light, an electromagnetic field becomes the obvious choice for getting signals across the galvanic barrier. While there were earlier demonstrations of magnetic-based transformer signaling, they do involve the packaging of a discrete toroidal signal transformer in between the transmitter chip and a receiver chip. Such components did not see much commercial success due to costs and performance limitations. It was not until analog devices showcased their $iCoupler^{TM}$ technology, where a set of isolated transformer coils was successfully constructed on top of a silicon chip, that the first true transformer-based solution became a commercially viable alternative to LED-based optoisolation devices [2]. In this case, the insulating material is a thin 20-um polyimide material on top of which the secondary coil is created, allowing magnetic coupling to perform communications across the insulation material.

(*iCoupler is a trademark of Analog Devices, Inc. for their magnetic coil-based isolation technology.*)

Figure 7.5 shows the cross-section on how galvanic isolation is achieved as an integration to regular CMOS technology. The technology requires additional processing after the silicon was fabricated. More recently, another class of non-optical isolation technology has emerged where the insulating material is based

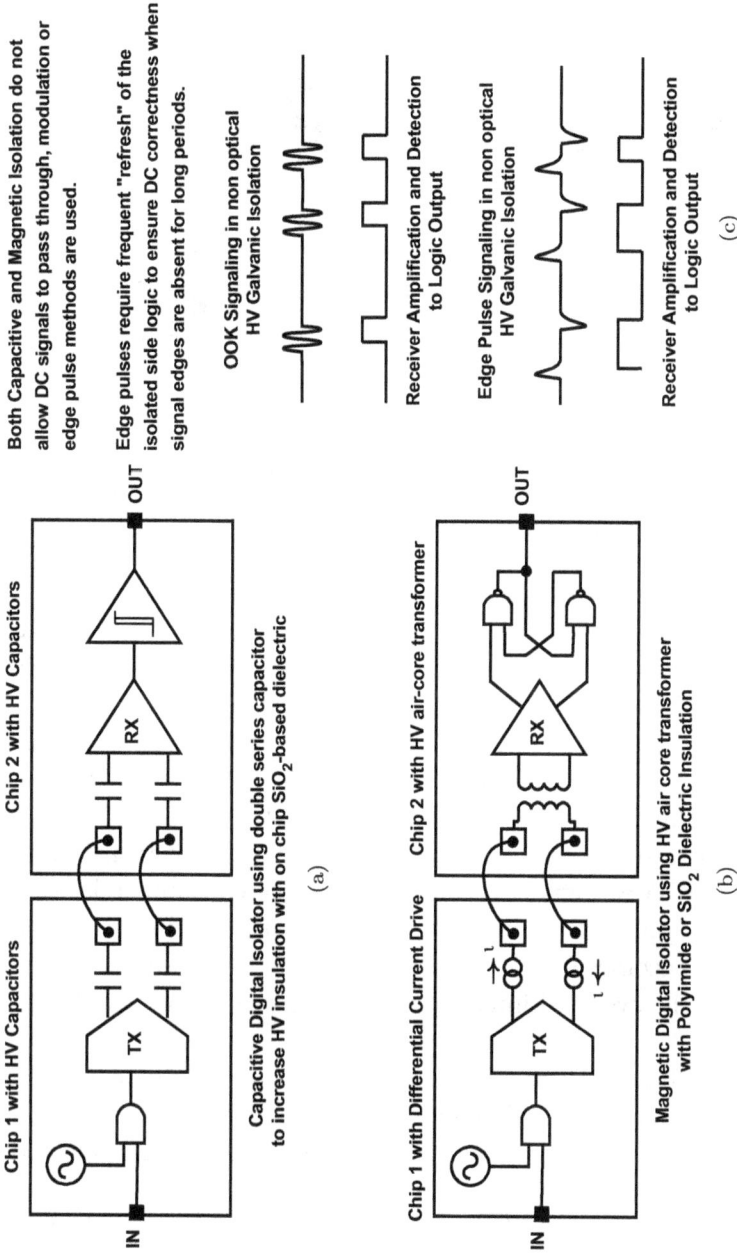

Fig. 7.4. (a) Capacitive digital isolator using a double series capacitor to increase HV insulation with on-chip SiO_2-based dielectric. (b) Magnetic digital isolator using an HV air core transformer with polyimide or SiO_2 dielectric insulation. (c) Signaling scheme used with non-optical isolation technology.

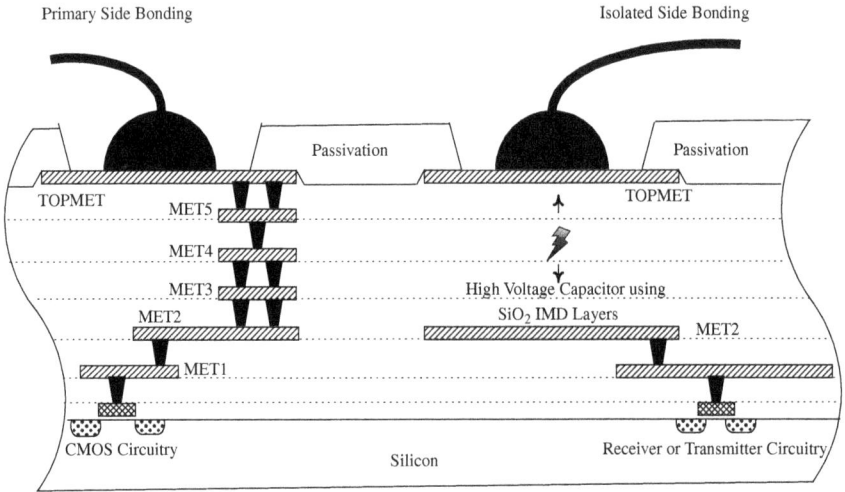

Fig. 7.5.   Capacitor isolation using SiO_2 IMD (inter-metal dielectric) layers in a multi-level metalization process.

on a thick deposition of silicon dioxide completed as part of the whole silicon fabrication process, making them truly monolithic in construction. The choice of isolation and their thickness is a whole science of high voltage breakdown capability, long-term reliability, and safe and unsafe failure modes that are beyond the scope of this chapter, which focuses primarily on the communication methods of isolated signaling.

With silicon dioxide as an insulating material, both transformer and capacitor-based signaling methods have shown commercial success in many applications. It is also evident that signaling at a much higher speed becomes feasible, and invariably, only digital signals are used, even if the eventual intent of the information to be transferred is analog.

With capability to integrate more functionality into a CMOS IC (integrated circuit), it is common that analog signals are first digitized, transferred across as digital signals, and re-constructed as analog signals after the isolation. This improves overall signal robustness and signal performance. Such an isolated product is illustrated in Fig. 7.6.

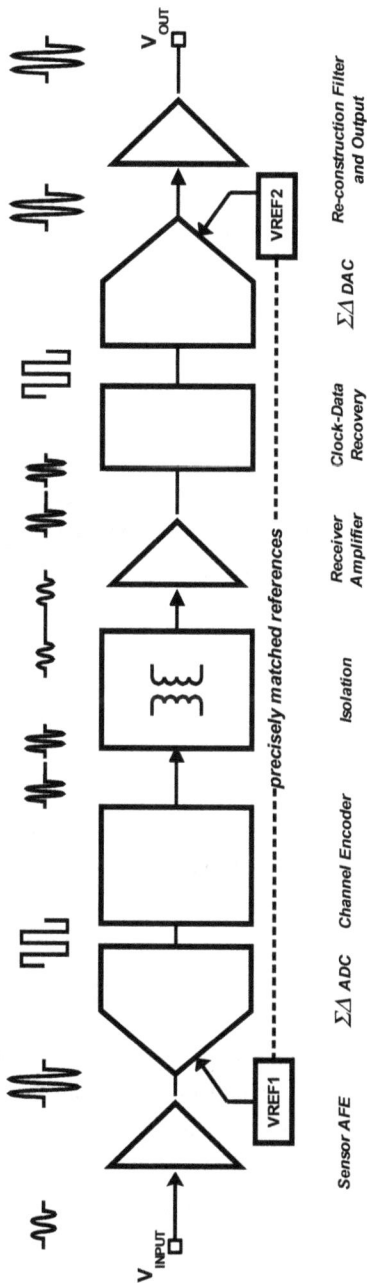

Fig. 7.6.   Digitalization of isolated communications — isolating analog signals.

7.3. Chip-to-Chip Communications over the Galvanic Isolation

7.3.1. *Optocouplers*

Most of the isolation components in existence today use the NRZ signaling scheme to transmit data from the primary side to the secondary side of the isolation barrier. In the case of the optocoupler device, the signaling method is level sensitive to the amount of transmitter LED irradiance that falls onto the detector photodiode, generating a proportional current for amplification and threshold detection back to a logic level. Typically, the LED used inside these components is in the near-infrared range of between 710 nm and 940 nm. Depending on the optical channel construction, the actual light level received at the detector is low, with a typical coupling efficiency of between 0.01% and 0.05% of the photo current to the applied LED current. This requires the addition of a good transimpedance amplifier to achieve a sufficient gain order to have a good noise margin for the quantizer to operate with a low bit error rate.

The optocoupler solution at the present state of the art is typically able to provide signaling rates from DC to around 25 Megabaud (MBd). Going beyond 25 MBd, the communication performance suffers from poorer data latency and timing skews between different communication channels. In addition, as the LED is essentially a current mode device, the power required to operate the LED at high speed would mean that the transmitter current into the LED will be significantly higher to overcome the junction capacitance of the LED, which can be in the range of between 11 pF and 30 pF or more.

As Moore's law dictates an ever-decreasing supply voltage for the integrated circuits used in the system board in components such as micro-controllers and DSPs, the forward voltage of the LED also presents a fundamental limit as to how low a signal can be applied to drive the LED for signal transfer. For 720 nm to 940 nm, the forward voltage of the LEDs is in the range of 1.2 V to 1.6 V, and this requires the input signal to be at least 2.5 V to drive the input of the optocoupler with a simple resistor configuration. As the supply voltages of the MCU gets lower due to smaller geometry silicon processes, the minimum voltage required to drive the LED becomes a major obstacle. At present, some system boards preserve a 3.3 V supply and use level shifters to provide a workable interface between the lower supply voltage MCU and the isolation component.

7.3.2. *Magnetic and capacitive isolator*

Magnetic and capacitive isolators, on the other hand, do not have fundamental physical limits such as LED forward voltage and can scale with better geometry process technology. However, both methods cannot operate at DC; therefore, some form of signal modulation is needed for signal transfer between the IC chips.

In addition, since signal coupling efficiency is dependent on the distance between the isolated electrodes, there is a trade-off between coupling efficiency and the amount of high voltage that can be held off with the material system of the components.

In certain applications where a prescribed distance through the insulation (DTI) is needed, such a magnetic or capacitive structure will be unable to meet similar requirements as optical-based solutions. Applications where insulation fails safe requirements in medical and high explosive environments are examples of where the DTI may be specified beyond 0.5 mm, which is much higher than the 20–70 um dielectric gap found in monolithic magnetic or capacitive isolators.

The advantage of these non-optical solutions is excellent speed, power, and their ability to operate at lower supply voltages. Current data rates are at 150 MBd, and even an isolated LVDS transceiver has been reported with a signaling rate of 10 Gbits/s [3, 4]. Table 7.1 shows the performance comparison of various technologies for signaling across a galvanic isolation barrier.

Table 7.1. Comparison of current available isolation component technology and possible future requirements.

	Optical	Magnetic	Capacitive	Future (Wish List)
Baud rate	50 MBd	100 MBd	150 MBd	Gigabit
HVCMR	35 kV/us	100 kV/us	150 kV/us	>150 kV/us
Latency	20 ns	10 ns	10 ns	10 ns
Power/channel	10 mA	1.5 mA	1 mA	1 mA
Bi-directional	Possible	Yes	Yes	Yes
Multi channels	2 Ch	4 to 6 Ch	4 to 6 Ch	High Speed SERDES
Signaling scheme	NRZ	Edge Pulse / OOK	Edge Pulse / OOK	Various
Operating temperature	−40/125°C	−40/125°C	−40/150°C	−40/150°C
EMC/EMI	Excellent	Poor	Moderate	To be Managed
Functional integration	Poor	Moderate	High	High
Cost/channel	High	Moderate	Low	Low
Process technology	IRLED + CMOS	CMOS with HV Coils	CMOS with HV Capacitors	Generic CMOS

7.4. Common Mode Noise

Unlike over-the-air wireless communications, the type of noise that can affect the transmitter and receiver of the isolator system is unique to its purpose. As both the transmitter and receiver are assembled within the same package, there are much fewer effects like channel fading and interference from non-related signals in the vicinity of the band.

The communication channel must handle what is commonly termed as high voltage common mode noise, which results from the fact that since both the transmitter and the receiver IC are galvanically isolated, it also means that they do not share a common ground and are susceptible to common mode noise in between their respective grounds.

This is particularly true in isolated power boards where an isolated gate driver is used to allow high voltage pulse width modulated switching of a power device. Figure 7.7 shows an isolated ground which can cause common mode noise events to the signals being isolated. The isolated gate driver will turn on the high-side power device, such as an Insulated Gate Bipolar Transistor (IGBT). This device is turned on and quickly rises to the power bus supply rails, which can be several hundreds of volts. This effectively raises the potential of the receiver ground to a new voltage level relative to the transmitter ground. Since the pair are galvanically isolated, there is no issue in holding off this voltage if the dielectric in between can withstand and operate at these potential differences. However, the slew rate of this voltage change will inject common mode currents into both the transmitter as well as receiver circuitries.

This potential change in a finite period generates a common mode current that will flow through the parasitic capacitors between the primary and secondary chips through the transmitter and receiver electrodes as well as via the lead frames, which include the supply and ground. In typical applications, this common mode slew rate can be in the order of 10–50 kV/μs, and in the newer gallium nitride or

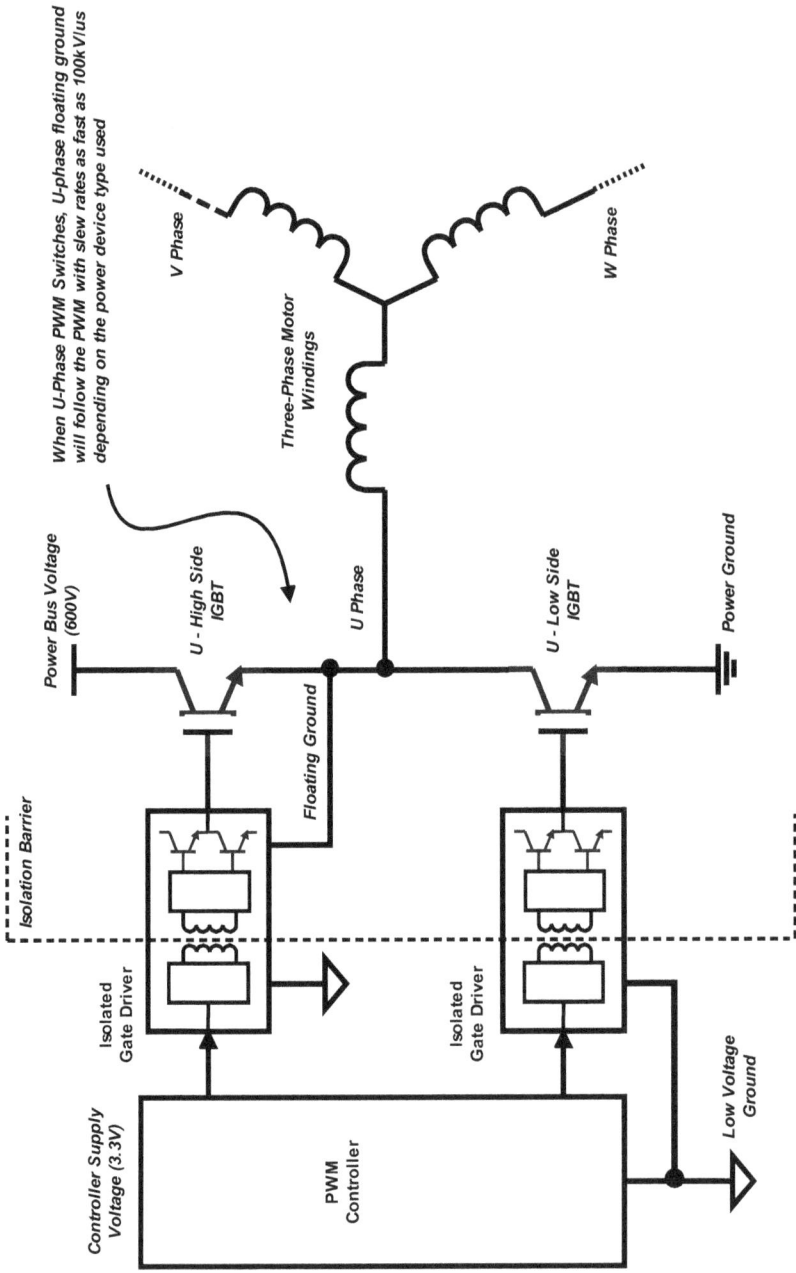

Fig. 7.7.   Common mode event during power switching of a typical 3-phase invertor motor Pulse Width Modulation (PWM) board.

silicon carbide power switches, the slew rate of the common mode noise can exceed 100 kV/μs.

For a primary to secondary coupling of 1 pF (which is typical between the isolated lead frame of the isolator device), the common mode current generated because of this transient can be in the order of 100 mA or more. This current injection persists for several tens of nanoseconds, and it can easily overload the transmitter and receiver electrodes to create sufficient disturbance to cause signaling errors in the channel.

7.5. Receiver Signal Strength versus Isolation Distance

In all three types of communications over the isolation barrier, the receiver signal strength is inversely proportional to the distance between the conductive electrodes, which are used to energize these elements on the transmitter side as well as the receiver side. Hence, there is a trade-off between good signal coupling and the amount of high voltage that can be held off.

Figure 7.8 shows the cross-section of a typical optocoupler device. Bond wire is used to supply the LED current from the transmitter chip and the photodiode on the receiver itself. The parasitic capacitance in between the LED and photodiode is the path in which the high voltage common mode transient can inject unwanted noise current and disturb both the LED current as well as the received photocurrent. In the photodiode, where detectable photocurrents are in the micro-amperes range, a transparent Faraday Shield covers the photodiode surface area, providing a shunt path to the milli-amperes of common mode current that can flow through using an HVCMR (high voltage common mode) event. While this can be done on the receiver stage, it is not possible to provide such a shunt path on the LED side, and it is often this reason that the LED needs to be driven at least to the same level or more than the common mode noise current. This severely limits the performance of the optocoupler in terms of speed and power considerations.

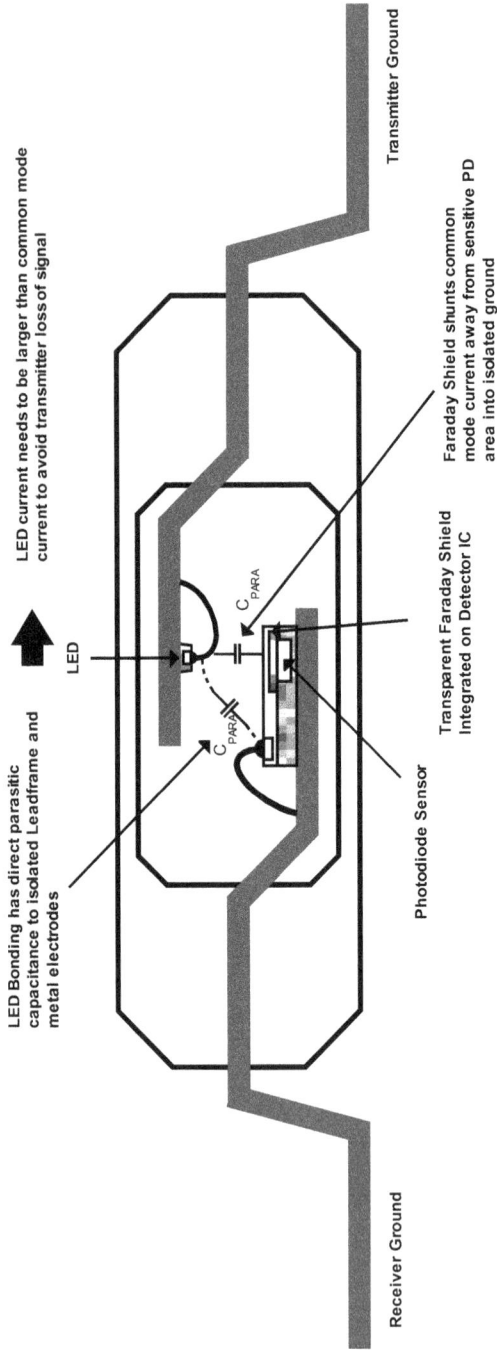

Fig. 7.8. Parastic capacitance and use of a transparent Faraday Shield to improve noise rejection in optocouplers.

The modern magnetic and capacitive structures also suffer from similar considerations. To get good signal coupling, the isolation distance is minimized to meet the minimum high voltage requirements. It is noted that since these devices' transfer characteristics fall off at DC, high frequency modulation is needed to allow signal transmission across the galvanic isolation.

In the application, since these are mostly single-function devices, the preferred mode of data communications is NRZ signals, where the isolated side tries to faithfully reproduce the exact same NRZ signals with a fix and time-invariant latency. Two common methods are used, OOK (On-OFF Keying) with a carrier frequency and one where the transmitter generates pulse edge information for the receiver detection and decoding of these pulses to re-generate the data. In NRZ transmission, there can be periods of long duration where there is no signal transition, and the receiver needs to have the memory to hold the present state until a valid transition occurs. This presents two issues, where upon the start up of the device, the initial state is unknown or forced by receiver design to power up on a known default state or in operations where noise corrupts the memory of the state and the state error persists until the next valid transition. Commercially available solutions address this by adding a watch dog time out, whereby if no valid signal transition occurs within a prescribed time, a "refresh" pulse is generated to ensure the state on the isolated side is correct if, indeed, a state error had occurred. Such signaling mechanisms are illustrated in Fig. 7.9.

The conventional method used is OOK modulation. In this case, where no signal is being generated, the output state can be interpreted as logic zero, and when there is a high-frequency signal presented to the isolator transmitter, a logic one is detected. With this implementation, the state consistency is preserved, and there is no need for a memory element to hold the state in the isolated side.

For edge triggered pulse signal across the isolation, there is lack of DC information in the isolated side when there is no activity for long periods of time. A memory latch will hold the last state which can be corrupted by noise. To avoid state errors and ensure DC correctness, a refresh pulse is generated to ensure that the DC state on both ends are always the same. This added pulse will have to be masked out and not presented to the output NRZ as a valid pulse.

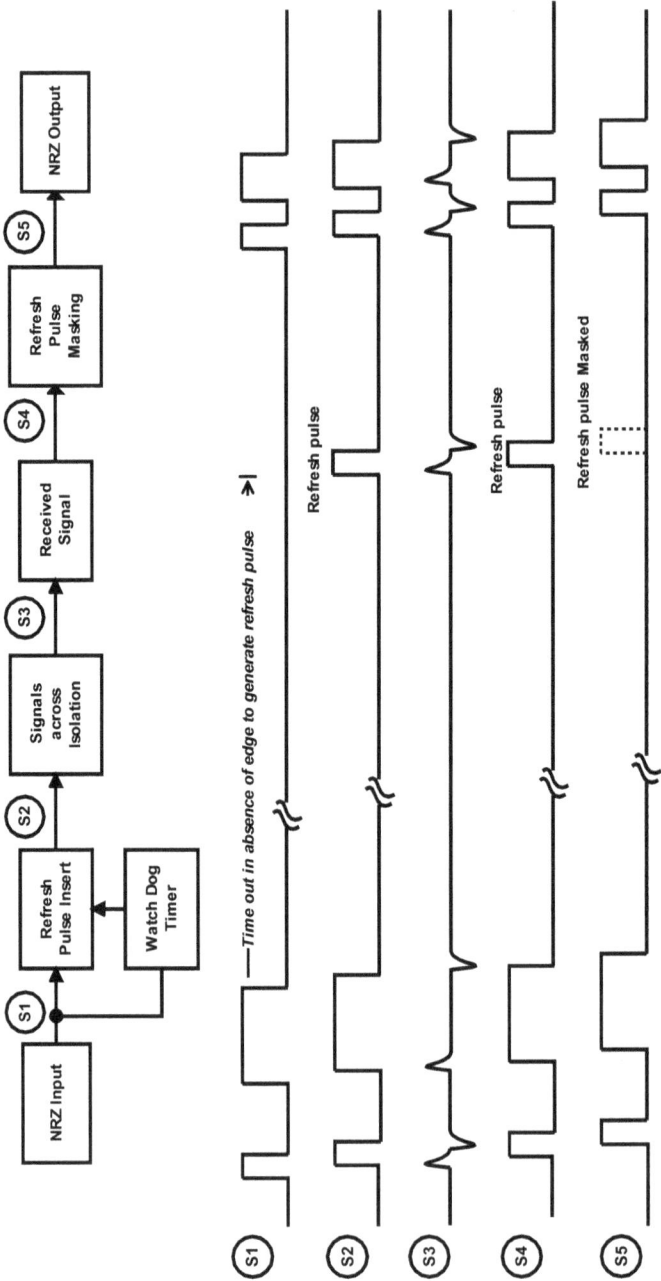

Fig. 7.9.   Edge trigger signaling requiring refresh pulse to assure DC correctness.

7.6. Effect of Common Mode Noise on Transmitter and Receiver

During a common mode event, both the transmitter and receiver will be affected by the common mode current injected into each electrode, respectively. Figure 7.10 shows a typical silicon dioxide based capacitive isolation receiver and transmitter structure using OOK as the transition modulation method. When the common mode event occurs, a large noise current is injected from the transmitter and flows into the receiver as shown. Depending on the polarity of the common mode transient, the reverse can occur as well.

Figure 7.11 shows an active clamping circuit to suppress effects of a common mode current injection. At high common mode slew rates, this current can be in the range of tens of milli-amperes, and both the transmitter and receiver are required to be terminated to keep the transmitter and receiver signals within the common mode range between the ground and supply of each chip on each side of the isolation. The simulation in Fig. 7.12 shows how the common

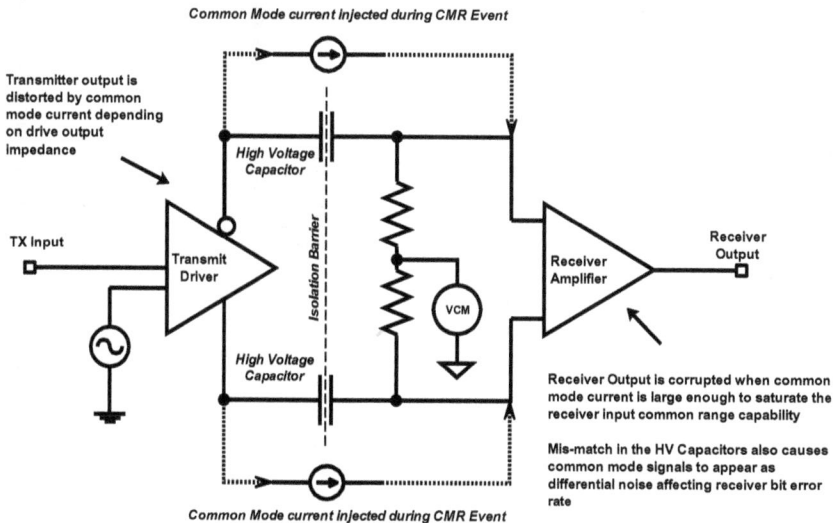

Fig. 7.10. Common mode current affecting both the transmitter and receiver in a capacitive isolator.

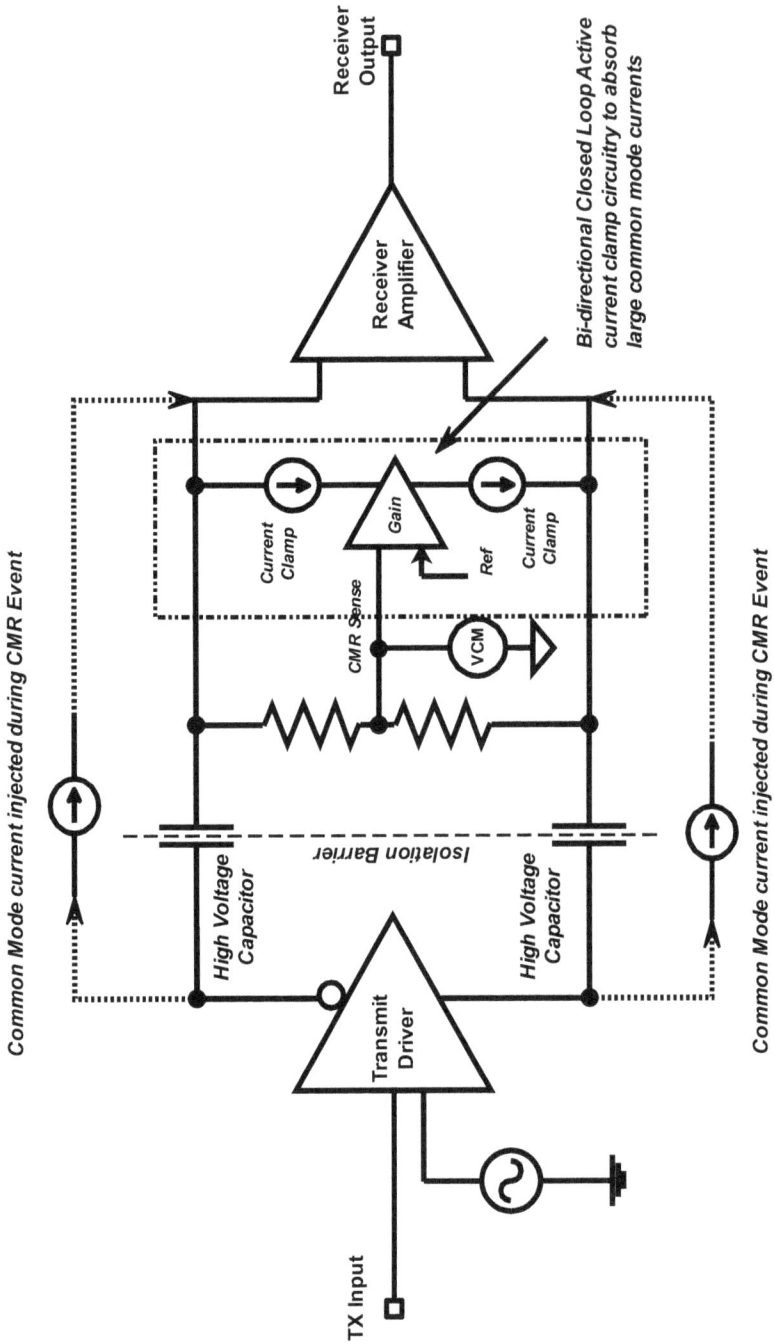

Fig. 7.11.   Active clamping circuit to suppress effects of a common mode current injection.

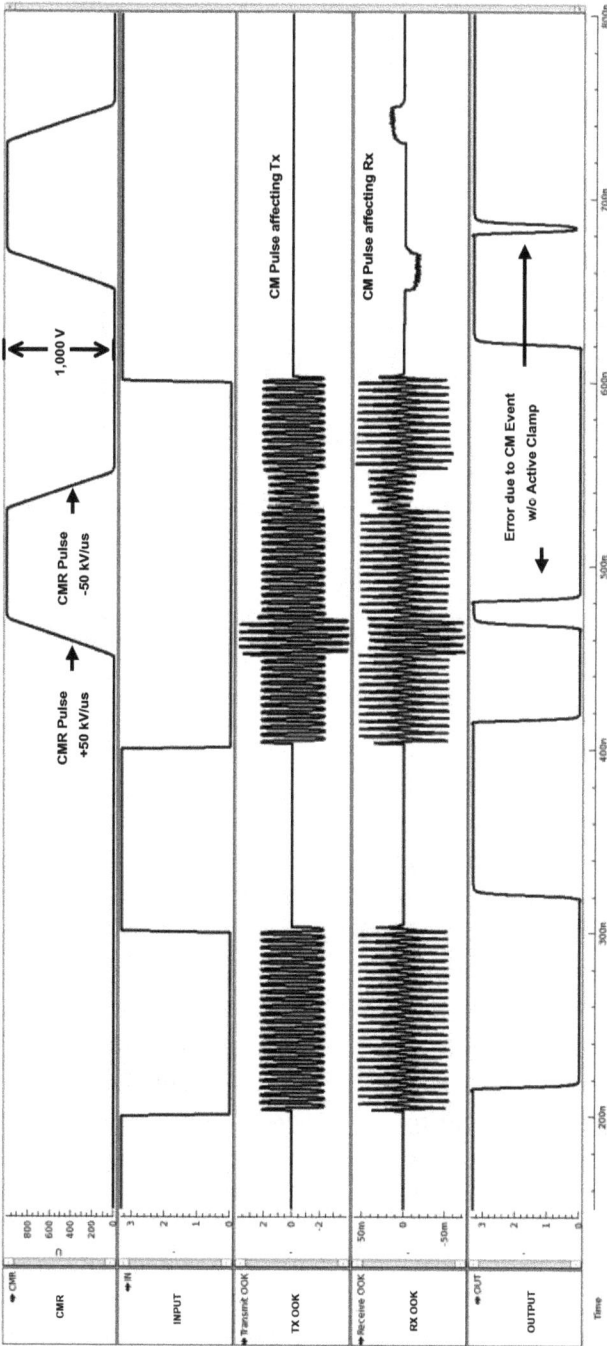

Fig. 7.12.   Common mode event affecting transmitter and receiver circuitry.

mode current generated can affect the transmitter and receiver, respectively.

In this simulation, a 3D Touchstone model of the differential high voltage isolation capacitors is obtained by 3D extraction of the structure using Keysight's PathWave software. The extracted models also include the parasitic capacitance and inductance of the bond wires and the lead frame on which the chips are mounted.

While a simple transmitter and terminated resistor can easily send and receive the OOK signal for detection and decoding into logic levels, they are usually unable to handle the large common mode current, as shown in Fig. 7.12. This results in disruption of the signal, and an error occurs in decoding the data.

To overcome these issues, the transmitter and receiver structures have additional common mode current absorber circuits built into the coupling device so that when the common mode event occurs, the electrodes are not overloaded, preventing the loss of signal. Figure 7.13 shows the simulation results of the technque.

7.7. Applications of Isolation Communications

The 3-phase isolated power inverter motor as shown in Fig. 7.14 is an example of how effective isolation can provide connectivity between the noisy power stage and the control unit, which contains the algorithm and signal processing capability that is typically operated at low supply voltages. This allows the usage of components with high computational power. The DSP or micro-controller controls the states of all the power switches, and these are done through isolated gate drivers. Isolated current sensors also provide current feedback so that the controller can process and optimize the pulse width modulated signals into each of the three phases of the inverter motor to control both speed and torque needed to drive a load. Therefore, in such systems, a bi-direction communication through the isolation is needed.

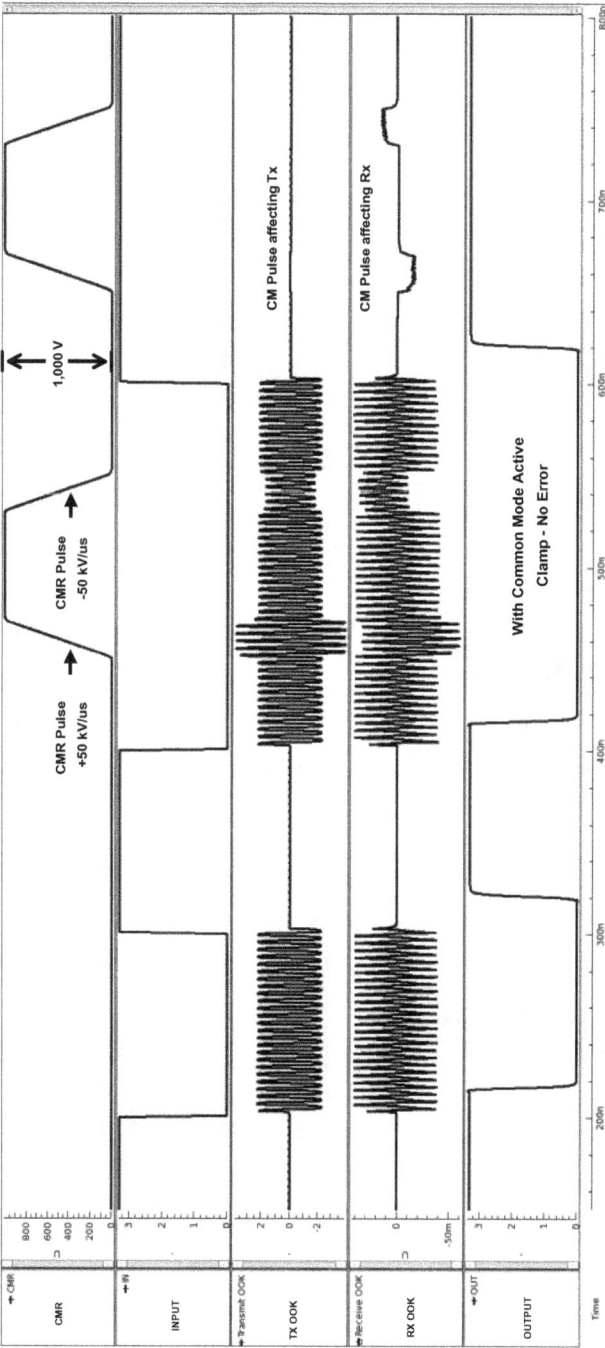

Fig. 7.13. Active common mode clamp circuit to prevent transmitter and receiver saturation during a Common Mode Rejection (CMR) event.

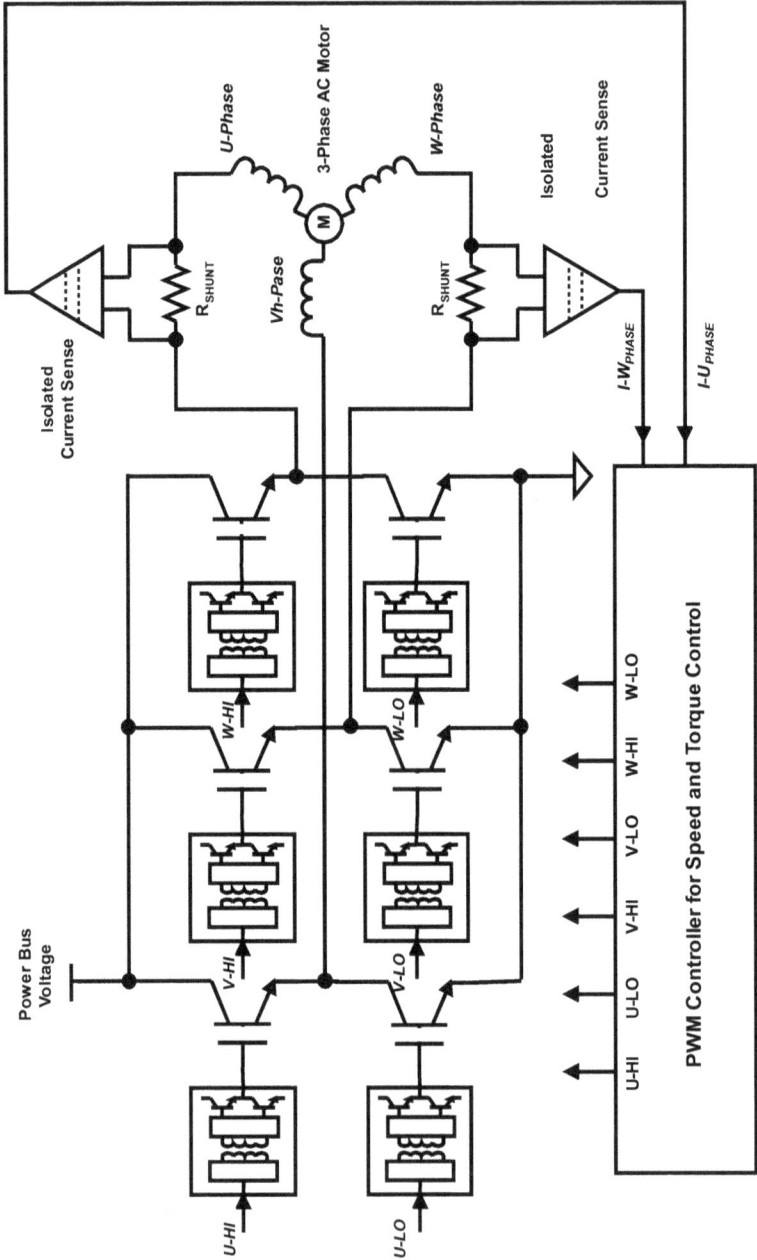

Fig. 7.14.   A 3-phase inverter motor application diagram with isolated gate drivers and current sensors.

If the communication through the isolation can reject the common mode noise generated by high-power switching of the power device, the system operates in a full closed loop configuration despite not sharing a common ground between the controller and the power stages. Such systems are commonly found in many sub-systems, such as power converters and variable speed inverter motors used in electric mobility and traction and renewable energy conversion.

The PWM signals are generated with dead time control such that the high side switch and the low side switch of each phase are never turned on at the same time. If it does, the power switches will have a large shoot-through current that can destroy the devices, leading to catastrophic system failure. Therefore, state correctness is not only necessary, but the timing between states of different isolated gate drivers needs to be well controlled for the dead time to be processed correctly. The trend for power devices is such that the switching frequency is increasing, and this gives the system advantage in increasing power with smaller-sized magnetics. Low timing skews between isolated devices become more critical, requiring faster isolation data transfer. The PWM is also being switched at higher voltages as new fast power devices based on silicon carbide and gallium nitride have been shown to be robust at these voltages. This exacerbates the common mode noise issue as well as raises demands on the propagation delays and timing skews of the isolators. These challenges will require new methods to provide robust and fast communications across the isolation channels [3].

The majority of current switching power stages are based on two-level pulse width modulated switching. There are trends where newer multi-level PWM topologies are used as in Fig. 7.15, leading to better harmonics control and overall better efficiency. These multi-level PWM converters also increase the total number of isolated channels, increasing system costs and adding demands to the performance and robustness needed [5].

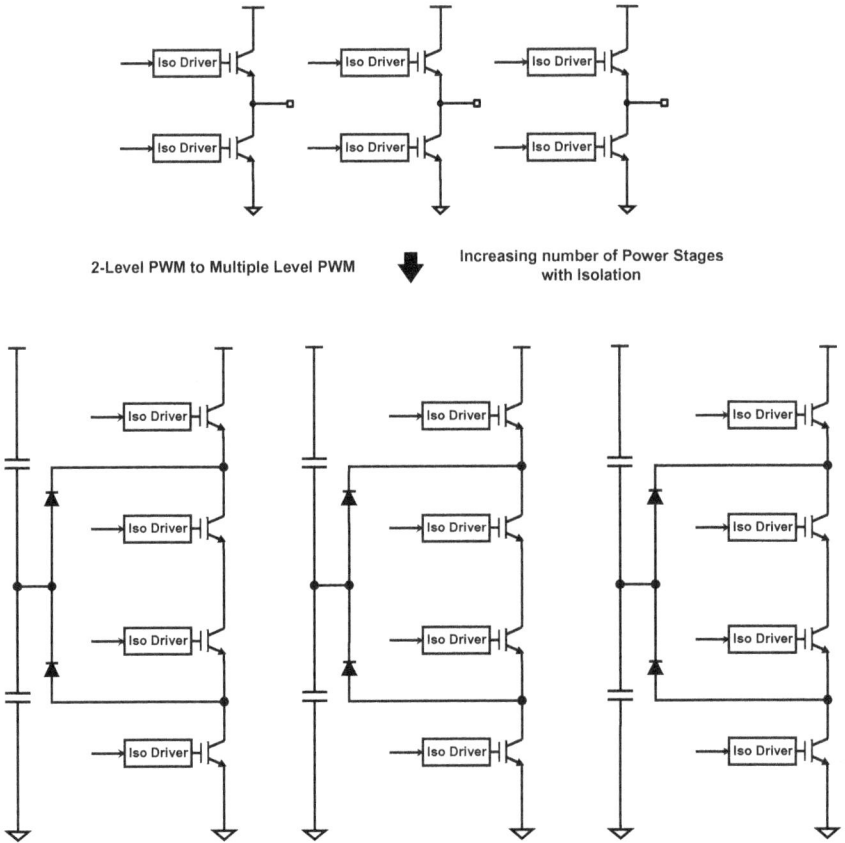

Fig. 7.15. Two-level PWM 3-phase inverter to tri-level diode clamped 3-phase multi-level PWM inverter.

7.8. Communications Topology

The present power system solution consists of the Digital Signal Processor (DSP) or microcontroller for controlling the switching of the power stages through the isolation and implementing the digital control loop for optimal feedback control. The interface between the DSP/MCU and the isolators are NRZ signals, and all timing control is handled by the MCU's digital algorithm. In this case, all the isolators are single-function devices that receive these NRZ

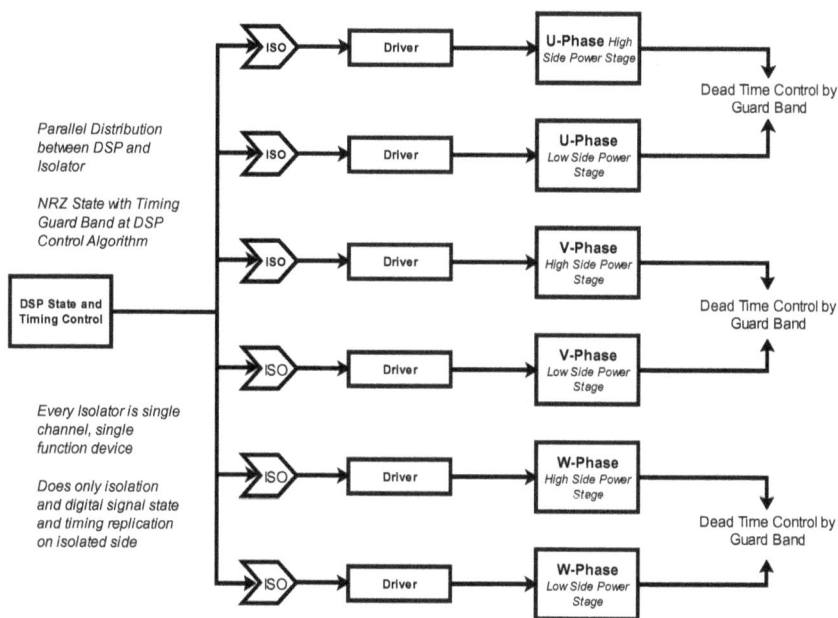

Fig. 7.16. Parallel centralized state and timing control.

signals from the controller and reproduce them correctly in the high-voltage isolated side where the power devices and magnetic circuits reside. Figure 7.14 shows a typical inverter motor driver system. The basic configuration isolates the gate drivers to the power stages as well as the phase current sensors. This forms a complete closed loop system where the control circuitries and user interface are completely isolated from the power stages. This is shown in Fig. 7.16.

As the processing power of the DSP/MCU increases with Moore's Law, it is possible to have a single MCU control several power systems control loops to reduce system complexity and cost. With the present configuration, the number of NRZ interface I/Os will have to increase.

This centralized topology with a large fan-out also presents physical limitations, as power devices and magnetic circuits tend to be large while computing electronics require only a small PCB footprint, and this leads to challenges in managing large fan-out of NRZ signals from a central location to power blocks. In addition, it is desirable for the gate driver to be placed near the power stages to reduce parasitic inductances, which can severely degrade the switching performance of the power devices. Large fan-out also presents challenges in meeting electromagnetic compatibility (EMC) and electromagnetic interference (EMI) requirements at the system level.

With new advances in communications technology, a serial topology offers much better system configuration and better physical planning and layout. This will require the isolators to be able to process beyond NRZ functions and increase their throughput in the isolated communications channels to a multi-functional component with bi-directional capability.

In such a configuration, there are two sets of communications between the DSP/MCU to isolation components and to the power stages. The first is high-speed differential wireline communication, which allows high throughput serial communications between the isolation device and the controller. Since these isolation devices are connected in a serial daisy chain topology, each of these components needs to be addressable to allow the MCU to manage these devices in real-time through the serial link. The new topology is illustrated in Fig. 7.17.

The second will be robust communications between the isolated side and the controller side. Bi-directional information will be needed to perform the state control on the isolated side, as well as receiving sensor feedback, status, and diagnostics from the power stages for feedback and fault handling.

7.8.1. *Primary side serial packetized data*

Having isolation devices on a serial link would require each of these devices to have a unique address for the MCU to control the state on the power side. The use of packetized information in this primary serial link also allows each node of the isolation to be aware of the state condition of one another, and this will allow better management of faults in the event of a power device failure.

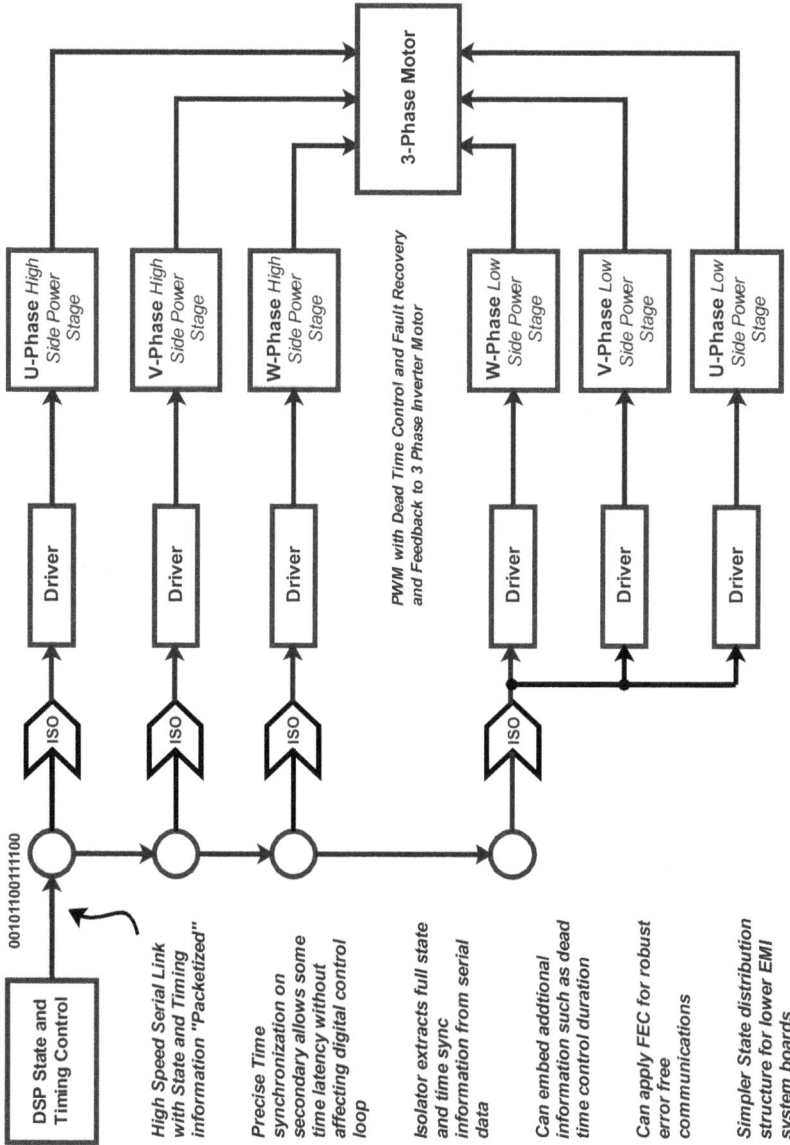

Fig. 7.17. Serial distribution with state and timing aware protocol.

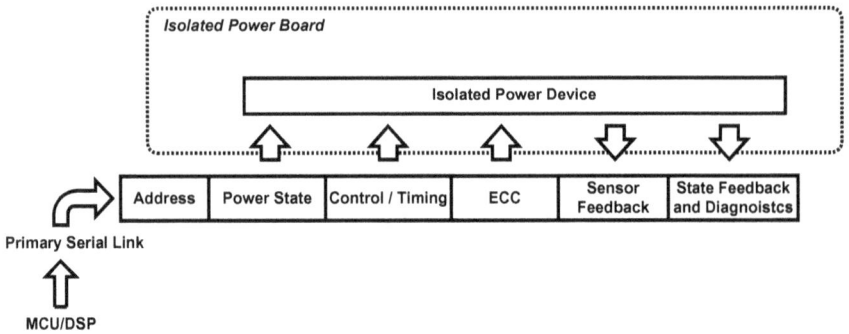

Fig. 7.18.   Serial link data communications between the MCU/DSP and isolated power board.

In Fig. 7.18, the packet could consist of the unique address of the isolator device, state information to be transferred across the isolation, sensors, and diagnostic data from the isolated side to be relayed back to the MCU. In this case, each power device can have awareness of the state and conditions of all other power devices, and this can lead to better control and fault management of the inverter.

On the isolated side, the states are used to drive the power switches. Dead time control and PWM edge timing control is still needed. For all the devices in the power side to have good timing control, there is a need to a common timing reference at the isolated side. This is another critical information to be communicated across the isolation to achieve better timing control than those with the traditional parallel NRZ approach.

When a hardware fault situation occurs on a power device, fast local shutdown circuitry will be needed to ensure that the power device does not fail catastrophically. This safe shutdown state and its fault diagnostics information is then transferred back across the isolation so that the MCU/DSP can handle the shutdown of the rest of the power devices in a safe, controlled manner.

7.9. Conclusion

It is evident that current isolation devices use rather basic methods of wireless communications and most of these are basic single-function components. As the technology evolves towards more complex systems where galvanic isolation is needed, newer proven communications technology can be deployed. The cost of deploying sophisticated signaling methods will be justified as isolation devices become more multi-functional and require higher data rates and volume to be transferred across the galvanic barrier. A complex integration of various technologies within an isolation component package will deliver better system-level optimization in terms of performance, reliability, and costs.

References

[1] Avago Technologies, "HCNR200/201 high-linearity analog optocouplers data sheet," Publication No. 5989-2137EN, p. 1, Feb. 2006.

[2] "Signal isolators using micro-transformers," US Patent US7719305 B2, Analog Devices, Jun. 2008.

[3] R. Ren, "iCoupler technology benefits gallium nitride (GaN) transistors in AC/DC designs," Analog Devices Inc Technical Article, May 2020.

[4] Skyworks Inc., "Isolator versus optocoupler technology," Technical White Paper, 2022.

[5] J. Rodriguez, J.-S. Lai and F. Z. Peng, "Multilevel inverters: A survey of topologies, controls, and applications," *IEEE Trans. Ind. Electron.*, vol. 49, no. 4, pp. 724–738, Aug. 2002.

When you ask questions, you learn better and faster.
When you question the answers, you have a deeper understanding of the problems.
When you question the questions, you will appreciate the problems better.

Kiat Seng Yeo

Index